AIDS Pathogenesis

Immunology and Medicine Series

VOLUME 28

Series Editors:

Dr. Graham Bird, *Churchill Hospital, Oxford, UK*
Professor Keith Whaley, *Kuwait University, Safat, Kuwait*

A list of titles in the series can be found at the end of this volume.

AIDS Pathogenesis

Edited by

Hanneke Schuitemaker

and

Frank Miedema

CLB and Laboratory for Experimental and Clinical Immunology,
University of Amsterdam,
Amsterdam, The Netherlands

SPRINGER-SCIENCE+BUSINESS MEDIA, B.V.

Library of Congress Cataloging-in-Publication Data is available.

ISBN 978-90-481-5407-4 ISBN 978-94-017-0685-8 (eBook)
DOI 10.1007/978-94-017-0685-8

Originally published by Kluwer Academic Publishers in 2000

AIDS PATHOGENESIS: TABLE OF CONTENTS

LIST OF CONTRIBUTORS

JAN ALBERT
Department of Clinical Virology
Swedish Institute for Infectious Disease Control
Karolinska Institute
S-10521 Stockholm, Sweden

BRIGITTE AUTRAN
Hôpital Pitié-Salpétrière
Lab. D'Immunol. Cell. Tissulaire URA CNRS 625
CERVI / 83 Boulevard de l'Hôpital
75651 Paris, France

THOMAS L. BENFIELD
Dept. of Infectious Diseases (144)
Hvidovre Hospital, University of Copenhagen
DK-2650 Hvidovre, DENMARK

SUSAN BUCHBINDER
AIDS Office
City and County of San Francisco
25 van Ness Ave., Suite 500
San Francisco, CA 94102, USA

SJOERD H. VAN DER BURG
Department of Immunohaematology and Blood Bank
Leiden University Hospital
Rijnsburgerweg 10
2333 AA Leiden, The Netherlands

DENNIS R. BURTON
Departments of Immunology and Molecular Biology
The Scripps Research Institute
10550 North Torrey Pines Road
La Jolla, CA 92037, USA

ANDREW CARR
University of New South Wales
HIV Medicine Unit and Centre for Immunology
St. Vincent's Hospital
376 Victoria Street
Sydney NSW 2010, Australia

DAWN R. CLARK
CLB and Lab. Exp. Clin. Immunology, University of Amsterdam
Plesmanlaan 125
1066 CX Amsterdam, The Netherlands

DAVID A. COOPER
University of New South Wales
HIV Medicine Unit and Centre for Immunology
St. Vincent's Hospital
376 Victoria Street
Sydney NSW 2010, Australia

ROLIEN H. ENTING
Dept. of Neurology
Academic Medical Centre, University of Amsterdam
Meibergdreef 9, 1105 AZ Amsterdam, The Netherlands

JANNIK HELWEG-LARSEN
Dept. of Infectious Diseases (144)
Hvidovre Hospital, University of Copenhagen
DK-2650 Hvidovre, DENMARK

HIDETO KANESHIMA
SyStemix
Director, HIV Gene Therapy
3155 Porter Drive
Palo Alto, CA 94304, USA

MARIE-JOSÉ KERSTEN
Dept. of Immunology – H7
Netherlands Cancer Institute
Plesmanlaan 121
1066 CX Amsterdam, The Netherlands

MICHÈL KLEIN
CLB and Lab. Exp. Clin. Immunology, University of Amsterdam
Plesmanlaan 125
1066 CX Amsterdam, The Netherlands

MAARTEN KOOT
CLB and Lab. Exp. Clin. Immunology, University of Amsterdam
Plesmanlaan 125
1066 CX Amsterdam, The Netherlands

RICHARD A. KOUP
Division of Infectious Diseases
University of Texas Southwestern Medical Center
5323 Harry Hines Boulevard
Dallas, TX 75235-9113, U.S.A.

JOEP M.A. LANGE
National AIDS Therapy Evaluation Centre (NATEC)
Academic Medical Centre, University of Amsterdam
Meibergdreef 9, 1105 AZ Amsterdam, The Netherlands

NORMAN L. LETVIN
Harvard Medical School
Beth Israel Hospital
Division of Viral Pathogenesis
330 Brookline Ave.
Boston, MA 02215, USA

JENS D. LUNDGREN
Dept. of Infectious Diseases (144)
Hvidovre Hospital, University of Copenhagen
DK-2650 Hvidovre, DENMARK

PAOLO LUSSO
Unit of Human Virology
Department of Biological and Technological Research
San Raffaele Scientific Institute
Via Olgettina n. 58
20132 Milano, Italy

FRANK MIEDEMA
CLB and Lab. Exp. Clin. Immunology, University of Amsterdam
Plesmanlaan 125
1066 CX Amsterdam, The Netherlands

JULIO S.G. MONTANER
Canadian HIV Trials Network
Centre for Excellence in HIV/AIDS St Paul's Hospital
University of British Columbia,
Vancouver, Canada

MARINUS H.J. VAN OERS
Department of Hematology
Academic Medical Center
Meibergdreef 9
1105 AZ Amsterdam, The Netherlands

NADINE G. PAKKER
CLB and Lab. Exp. Clin. Immunology, University of Amsterdam
Plesmanlaan 125
1066 CX Amsterdam, The Netherlands

PAUL W.H.I. PARREN
Departments of Immunology and Molecular Biology
The Scripps Research Institute
10550 North Torrey Pines Road
La Jolla, CA 92037, USA

WILLIAM A. PAXTON
Department of Human Retrovirology
Academic Medical Center, University of Amsterdam
Meibergdreef 15, 1105 AZ Amsterdam, The Netherlands

PETER PORTEGIES
Dept. of Neurology
Academic Medical Centre, University of Amsterdam
Meibergdreef 9, 1105 AZ Amsterdam, The Netherlands

MARIJKE TH.L. ROOS
CLB and Lab. Exp. Clin. Immunology, University of Amsterdam
Plesmanlaan 125
1066 CX Amsterdam, The Netherlands

QUENTIN SATTENTAU
Centre d'Immunologie de Marseille-Luminy
Case 906, 13288 Marseille Cedex 9, France

PETER TH.A. SCHELLEKENS
CLB and Lab. Exp. Clin. Immunology, University of Amsterdam
Plesmanlaan 125
1066 CX Amsterdam, The Netherlands

HANNEKE SCHUITEMAKER
CLB and Lab. Exp. Clin. Immunology, University of Amsterdam
Plesmanlaan 125
1066 CX Amsterdam, The Netherlands

THOMAS SCHULZ
Dept. of Med. Microb. and Genot. Urinaray Med.
Duncan Bld. Daulby Street
Liverpool, L693 GA, United Kingdom

INGRID SPIJKERMAN
Dept. of Public Health
Municipal Health Service
P.O. Box 20244
1000 HE Amsterdam, The Netherlands

ERIC VITTINGHOFF
AIDS Office
City and County of San Francisco
25 van Ness Ave., Suite 500
San Francisco, CA 94102, USA

FRANK DE WOLF
Lab. of Human Retrovirology
Academic Medical Centre, University of Amsterdam
Meibergdreef 9, 1105 AZ Amsterdam, The Netherlands

KATJA C. WOLTHERS
CLB and Lab. Exp. Clin. Immunology, University of Amsterdam
Plesmanlaan 125
1066 CX Amsterdam, The Netherlands

ANGÉLIQUE VAN 'T WOUT
CLB and Lab. Exp. Clin. Immunology, University of Amsterdam
Plesmanlaan 125
1066 CX Amsterdam, The Netherlands

CHAPTER 1

HIV-1 TRANSMISSION

ANGÉLIQUE VAN 'T WOUT[1], RICHARD A. KOUP[2] and WILLIAM A. PAXTON[3]

[1]*Department of Clinical Viro-Immunology of the CLB and Laboratory for Experimental and Clinical Immunology, Academic Medical Centre, University of Amsterdam, Amsterdam, The Netherlands*
[2]*Division of Infectious Diseases, University of Texas Southwestern Medical Center, Dallas, TX, USA*
[3]*Department of Human Retrovirology, Academic Medical Center, Amsterdam, The Netherlands*

1. Introduction

The increasing transmission rates of HIV-1 infection throughout many parts of the world poses one of the major health threats for the coming century. If the incidence of AIDS is to be reduced dramatically then preventing viral transmission is a necessary prerequisite. The administration of potent anti-HIV-1 drug cocktails will undoubtedly have an influence on HIV-1 transmission in the western world but is unlikely to be of great benefit in the developing world where their expense will restrict their widespread usage. It is in Africa, India, South America and South East Asia where the pandemic is accelerating rapidly and it is in these regions that a block to HIV-1 transmission is drastically needed. The development of a vaccine providing protection against the transmission of HIV-1 and HIV-2 is therefore essential for the ultimate control of HIV-1 infection and AIDS throughout the world.

Common modes of transmission of HIV-1 are either from mother to child (vertical) or via sexual contact. Heterosexual intercourse accounts for the largest number of new infections world-wide. The sharing of HIV-1-contaminated needles by intravenous recreational drug users (IDUs) still carries a major risk and has been predicted to account for approximately 35% of new HIV-1 cases in the USA [1]. Transmission through contaminated blood or blood products has all but been eradicated by the successful treatment or HIV-1 screening of all blood products. The medical practice of re-using needles is still a considerable threat in some countries of the world for viral transmission.

H. Schuitemaker and F. Miedema (ed.), Aids Pathogenesis, 1–17.

For all routes of HIV-1 exposure it is evident that a number of factors influence whether the virus is transmitted to the exposed individual or not, as summarized in Table 1.

TABLE 1. Factors influencing transmission of HIV-1	
1.	Environmental factors
-	social
-	cultural
-	political
2.	Host-related factors
-	immune response (donor and recipient)
	Th1 vs Th2, neutralizing Abs vs CTLs
-	genetic make up (donor and recipient)
	HLA, CCR5/Δ32, chemokine production
3.	Agent factors
-	size of inoculum
	blood transfusions vs needle sticks
-	viral genotype
	subtype E vs B
-	viral phenotype
	NSI/M-tropic vs SI/T-tropic

Th1 and Th2, T-helper lymphocytes type-1 and -2: Ab, antibodies: CTL, cytotoxic T lymphocytes: HLA, human leukocyte antigen: NSI, non-syncytium-inducing: SI, syncytium-inducing: M-tropic, macrophage-tropic: T-tropic, T-cell-tropic

These range from environmental factors (the social, cultural and political milieu), agent factors (viral phenotype, size of viral inoculum), to host-related factors (site of exposure, induction of an active immune response and the genetic make-up of the donor or recipient). In some extreme cases one of these factors may predominate to either enhance or protect against the acquisition of the virus, but in the majority of cases it will be a multifactorial effect. Therefore, there are many areas where viral transmission can be successfully blocked, and here we wish to address some of these areas and discuss where future attempts at preventing HIV-1 transmission could be directed.

2. Cofactors Influencing HIV-1 Transmission

There are many sociological, behavioural and clinical factors (i.e. co-factors) associated with the transmission of HIV-1. In America the largest growing HIV-1-positive AIDS populations are amongst the ethnic groups in society and especially those residing within the inner cities and the poorer neighbourhoods [1]. Alcohol and recreational drug use have been widely reported to play a role with the transmission of HIV-1 infection by altering behavioural patterns and safe sex practices. More recently -within the USA- crack cocaine use linked with prostitution has been shown to be associated with a high incidence of HIV-1 [1]. The introduction and/or maintenance of programs aimed at providing comprehensive health education and drug abuse rehabilitation are therefore essential within the inner cities as a preventative measure against further viral transmission. The introduction in some countries of needle exchange programs has been associated with a reduction in the incidence of HIV-1 transmission amongst IDUs, but their effectiveness is still hotly contested amongst advocates who believe that these programs themselves promote drug usage thereby exacerbating the problem [1].

Vertical transmission may occur *in utero*, *intrapartum* or *postpartum* via breast milk (reviewed in [2]). Children can be infected *in utero*, possibly after ingestion of amniotic fluid or through virus produced by placental macrophages or lymphocytes. However, several studies suggest that the majority of vertical transmissions occur during delivery, most likely through exposure to contaminated maternal blood and cervico-vaginal secretions. Women in the western world have the possibility to abstain from breast feeding, however, this is usually not an option elsewhere.

The incidence of sexually transmitted diseases (STDs) has been implicated in the spread of HIV-1. A trial conducted in Tanzania successfully demonstrated that the treatment of STDs reduced the incidence of HIV-1 by approximately 40% [1]. The presence of genital sores, often associated with STDs, can partly explain this result. But new findings also show that the treatment of STDs, such as urethritis, can reduce the HIV-1 viral load present in semen and cervico-vaginal secretions, thereby posing a block to viral transmission at the level of the donor. These results indicate that the successful treatment of STDs in areas where HIV-1 is prevalent will have a sizable effect in reducing the number of transmissions.

3. Inoculum Size

The quantity of HIV-1 that an individual is exposed to appears to correlate with whether or not HIV-1 will be transmitted (reviewed in [2]). This is very apparent in blood exposure incidents. Transfusion of a unit of HIV-1-contaminated blood will inevitably result in the infection of the recipient, whereas needle-stick injuries resulting in the transfer of a small volume of blood will only rarely result in a seroconversion [3,4].

Individuals in clinical stages that are associated with a high viral load, such as primary HIV-1 infection or AIDS, are most likely to transmit their virus (reviewed in [2,5]). Moreover, as mentioned above, other infections which cause lesions at the sites of transmission, such as vaginal/rectal ulcers or syphilis-induced ruptures of the placental barrier, are associated with increased risk of transmission [6,7].

HIV-1 vertical transmission risk is increased by many factors in the mothers, including decreased CD4 lymphocyte counts, lack of HIV-1-neutralizing antibody, advanced clinical disease and also increased viral load (reviewed in [2,8]). The significance of viral load became apparent when it was reported that the administration of zidovudine to mothers during gestation and delivery reduced the level of HIV-1 transmission by almost 70% [9]. It should be noted, however, that the children were treated for a period of 6 months post-delivery and this may have provided a greater protective effect than the decreased viral load. More recent studies on large cohorts of HIV-1-positive mothers suggest that there may not be a strong correlation between the viral load in the peripheral blood of the mother and the potential for that mother to transmit HIV-1 [10-12]. This does not rule out differences in the tissue, whereby levels of virus present in the placenta may correlate with viral transmission. Additional support of the importance of viral load comes from the observation that children born by natural birth are more likely to become infected than those delivered by caesarean section, a less traumatic procedure [13, 14]. The same holds true for second born twins, for whom the first twin has already "cleansed" the birth canal [13].

4. Viral Genotype and Phenotype

The properties of HIV-1 itself may also influence transmission. HIV-1 variants can differ in replication rate, cell tropism as determined by co-receptor usage and cytopathicity (reviewed in [15]). Non-syncytium-inducing (NSI) variants, which are generally macrophage-tropic and predominantly utilize the CC-chemokine receptor CCR5 as a viral co-receptor, are found early in HIV-1 disease and may be better adapted to spreading than the more T-cell-tropic syncytium-inducing (SI) variants which utilize the CXC chemokine receptor CXCR4 (reviewed in [15]). The identification of specific amino acid residues in the third variable domain (V3) of the HIV-1 envelope gp120 molecule as major determinants for SI capacity and cell tropism [16-18] provided the possibility for a biologically significant interpretation of available sequence data. Comparison of the V3-genotype and phenotype of HIV-1 variants present in donors and their recipient, revealed a homogeneous population of NSI/M-tropic viruses in recipients (and in most individuals experiencing primary HIV-1 infection), irrespective of the presence of SI/T-tropic viruses in their respective donors [19-22]. This seems to hold true for sexual, vertical and parenteral transmission [22]. During vertical transmission, however, in some cases not one but several of the mothers' highly M-tropic variants are detected in their children [22,23] possibly due to the long and large exposure *in utero* or during delivery. The selective effect of zidovudine on NSI variants may also in part explain the efficacy of this drug in reducing vertical transmission [24-26].

In an attempt to explain the selection for M-tropic variants, HIV-1 populations present in genital secretions were compared with those in the peripheral blood. It seemed that only a selection of the variants present in blood were also detected in genital secretions [27,28]. It has also been postulated that macrophages, present in mucosa and placental tissues, may be the first target cells that HIV-1 encounters [29]. This was supported by the observation that HIV-1 infection of uterine cervical explants is only achieved by using the HIV-1 M-tropic strain Ba-L, and is restricted to macrophages present in the explants [30]. However, this selective replication in genital secretions cannot be the only explanation since the same selection for M-tropic variants is also observed in individuals that are directly exposed to all virus populations in the blood of their donor [22]. The increasing immune response in the newly infected individual apparently also selects for M-tropic variants during the early rounds of viral replication. Further evidence that NSI/M-tropic variants are responsible for establishing infection in most individuals came from studies showing the relative resistance to infection after sexual or parenteral exposure of individuals with a homozygous mutation in their CCR5 gene [31-33], as will be discussed later.

The different subtypes of HIV-1 have distinct geographic distributions, with A, C, D and E predominant in sub-Saharan Africa and Asia, and B predominant in the United States, the Caribbean, South America and Western Europe [34]. This distribution has led to speculation that biologic differences between the variant subtypes may establish differences in transmission efficiency. Subtype E, the most common subtype in Thailand, was reported to have a greater tropism for Langerhans' cells than subtype B [35], which was thought to contribute to the rapid epidemic spread of HIV-1 through Thailand and the high per contact transmission rate observed there [36]. However, several other groups have not been able to confirm this finding nor were any other biological differences detected between the other different subtypes of HIV-1 [37,38].

5. Protective Immunity

It is still unclear whether or not superinfection with multiple strains of HIV-1 occurs in humans. Co-infection during the period of primary infection before the development of a full immune response with different types (HIV-1/HIV-2), subtypes (clades A-H), or subtype strains (clade B strains) has been documented on numerous occasions (reviewed in [39]). In an extensive analysis of published HIV-1 sequences, almost 10% apparently were recombinants between different HIV-1 clades [40]. Once an infection with one variant has been established, it is unclear whether new variants can still establish superinfection in such an infected individual. Clearly, this is an issue of importance as vaccinees should be protected against superinfection with not just their own vaccinating strain but also against other unrelated strains that they may be exposed to at a later time. Individuals infected with HIV-2 appear to be somewhat protected from a subsequent infection with HIV-1 [41], supporting the concept of a block in superinfection.

For the development of an effective vaccine against HIV-1, it is also desirable to identify the correlates of protection occurring naturally in HIV-1 infection. The two obvious groups to study are those long-term survivors (LTSs) who have kept low viral loads and constant CD4 counts whilst remaining disease-free for many years and those individuals who have been multiply exposed to HIV-1 but who have resisted infection and are referred to as exposed-uninfected (EU) individuals. Since an HIV-1 vaccine that prevents the establishment of HIV-1 infection is desired, then the EUs are an important group to focus attention on.

5.1. EXPOSED-UNINFECTED COHORTS

Several different cohorts of EU individuals have been described where multiple and varied immune responses towards HIV-1 have been identified. Included in these groups are those haemophilia patients who were multiply exposed to known HIV-1-contaminated batches of factor VIII before the introduction of heat treatment around the mid-1980's. Another group of individuals extensively studied are those HIV-1-negative infants born to HIV-1-positive mothers. Undoubtedly, the most extensive group studied to date are those individuals who have been exposed to HIV-1 through high-risk sexual practices and it is from this group of individuals that the most has been learned. In this section we wish to review and discuss what is known of the many immunological factors that may be associated with the lack of HIV-1 transmission.

There is strong epidemiological evidence indicating that factors other than chance alone are involved in the non-acquisition of HIV-1. The most convincing evidence was obtained in a study on a cohort of commercial sex workers from Nairobi, Kenya, where it was shown that a small group of women were significantly protected against HIV-1 seroconversion [42]. After one year of work, 50% of women seroconverted but this rate did not remain constant with approximately 5% of the cohort remaining negative over the subsequent years of study. All the women were shown to have similar exposure histories in terms of the number of partners, sexual practices and condom usage. However, there will always be individuals who will remain negative as a result of chance alone. This seems more likely when the estimates for the number of high-risk sexual exposures required to establish an infection are taken into consideration, with some as high as 1 in every 100 [43,44]. Given this figure, it would be an unwise practice for any individual to regard themselves as immune to this virus and the practice of safe sex should therefore be advocated at all times for all individuals.

5.2. CELLULAR-MEDIATED IMMUNE RESPONSES

The absence of any HIV-1-specific antibody responses in these cohorts of exposed individuals suggests the absence of an induced systemic B-cell response. It cannot be ruled out, however, that a localised or low-frequency protective response has been stimulated in some individuals through exposure to low levels of viral antigen [45]. High antibody neutralisation responses have been previously identified in LTSs where viral levels of replication are extremely low [46]. If such neutralising responses could exist at low levels at the site of exposure, then a systemic B-cell response may go unidentified.

It is the $CD4^+$ lymphocytes or T-helper (Th) cells which control the immune response. Th lymphocytes recognise through their T-cell receptor foreign antigen in the context of the major histocompatibility complex (MHC) class-II molecule and provide their function through the differential production of cytokines (reviewed in [2]). Th responses are typically associated with the stimulation of the MHC class-I-restricted cytotoxic T-lymphocyte (CTL) responses via interleukin-2 (IL-2) production, whereas Th2 responses, through the secretion of IL-4, IL-5 and IL-10, are predominantly associated with the enhancement of the humoral immune response. Both MHC class-I and class-II responses have been identified in HIV-1 EU individuals.

MHC class-I presentation is restricted to endogenously expressed antigen that has been transported through the endoplasmic reticulum, therefore the presence of an HIV-1-specific CTL response would be indicative of exposure to replicating virus. Since CTL responses have been shown to protect against a number of other viral infections, it would not be surprising to find that they have a protective role to play in HIV-1 infection. HIV-1-specific CTL responses have been associated with the acute seroconversion stage of HIV-1 infection and are believed to play a role in controlling viral replication and disease progression. Indeed, a number of studies have now reported on the identification of HIV-1-specific CTLs in EU individuals, indicating that they may be involved in preventing either the establishment of infection or viral dissemination from the site of exposure (reviewed in [2]).

CTL activity has been described in a small group of commercial sex workers from the Gambia who are multiply HIV-1-exposed but seronegative [47]. These responses were shown to be HLA-B35 class-I-restricted and were present in three of six individuals in the study but were absent from HLA-matched control individuals [47]. The HIV-1-specific responses were only detected in the peripheral blood mononuclear cell fractions (PBMC) after repeated *in vitro* peptide stimulations, possibly reflecting the low levels of virus replication. HIV-1-specific CTL responses have also been identified in a group of HIV-1 highly exposed gay men [48]. A study by Langlade-Demoyen *et al.* reported nef-specific CTL responses in a group heterosexual individuals who were partners of HIV-1-positive individuals and who were considered as being exposed [49]. Env-specific CTL responses have also been identified in a group of health care workers who were exposed to the virus through a single occupational accident [50]. CTL responses have also been correlated with vertical transmission of the virus after Rowland-Jones *et al.* demonstrated transient gag-specific HIV-1 responses in a negative child born to an HIV-1-positive mother [51]. Whether the described CTL responses in such individuals reflects their true ability to clear active infection or whether they are only the consequence of an abortive viral infection, brought about by some other unrelated reason, remains to be elucidated, but the induction of such a response can only be beneficial to the exposed individual. It will be of relevance to determine to what extent these responses are present at the active site of infection opposed to the periphery in these individuals and how such responses could be induced in naive non-exposed individuals and how they could be specifically targeted to those sites.

Within the Th response, Th1 and Th2 lymphocytes have been shown to down-regulate each other and it would not be wholly surprising to find individuals with a strong Th1 response lacking a detectable HIV-1 antibody response. This scenario could explain the lack of seroconversion in some individuals possessing strong protective CTL activity. While this is not the case in infected individuals who demonstrate both antibody and CTL responses against HIV-1 antigen, it could reflect a limited infection at the site of exposure with no resultant viral dissemination in the EUs. The presence of a predominant Th1 response has been described previously in a number of reports (reviewed in [2]). PBMC from six gay men who were at risk of HIV-1 infection showed increased IL-2 production when stimulated *in vitro* with env antigen and peptides opposed to the control non-exposed individuals [52]. This same observation was made when health care workers who had been exposed to HIV-1 through needle-stick injuries were studied [53]. Similar results were also found when studying infants born to HIV-1-infected mothers suggesting a protective role for the Th lymphocytes against the vertical transmission of HIV-1 [54]. Unlike the CTL responses, which require actively replicating virus for correct antigen processing and presentation, the Th responses can be mounted through exposure to either defective virus or viral antigen. Obviously, an acquired protective Th response would be beneficial for the development of an HIV-1 vaccine since replicating virus would not be needed as the source of antigen.

Specific MHC alleles have been associated with increasing or decreasing the risk of HIV-1 seroconversion, lending support to the concept of a cellular immune response providing protection in those EU individuals [55-59]. Exactly how this is mediated is not known, but presumably reflects the ability of an individual with a certain MHC type to mount an effective antigen-specific immune response against HIV-1. The implications for the development of an HIV-1 vaccine are not known but, obviously, any successful vaccine will have to function for all MHC types as well as against all the variant clades of virus.

6. Resistance to Infection

The last few years have seen some remarkable advances in understanding host-viral interactions with relation to viral transmission and disease pathogenesis. The identification of some CC and CXC chemokine receptors functioning, in conjunction with the CD4 molecule, as co-receptors for HIV-1 entry into $CD4^+$ cells has answered some long standing questions and at the same time posed many more.

It was known for many years that soluble factors secreted by $CD8^+$ lymphocytes were capable of inhibiting HIV-1 replication in a non-MHC-restricted and non-cytolytic manner [46,60-62]. Not only was the presence of such factors associated with the asymptomatic stage of infection, it was also observed in EU individuals [46,60-62]. One report correlated such activity with non-infected individuals who were multiply exposed to a known HIV-1-positive partner through high-risk sexual behaviour [62]. In another study it was determined that viral inhibitory factors were present in infants born to HIV-1-infected mother [63].

Cocchi *et al.* determined that the β-chemokines, RANTES, MIP-1α and MIP-1β, were capable of suppressing HIV-1 replication in an M-tropic viral-specific manner, the viruses predominantly associated with HIV-1 transmission (reviewed in [15]). Closely following this discovery came the observation that the CXC chemokine receptor, CXCR4, functioned as a co-receptor for T-tropic viral isolates into $CD4^+$ lymphocytes. These two studies led to five reports showing that the CC chemokine receptor, CCR5, functioned as the main co-receptor for the HIV-1 M-tropic viruses (reviewed in [15]). Within these reports was the identification that other CC chemokine receptors, namely CCR2b and CCR3, could also function as HIV-1 co-receptors, albeit in a more restrictive manner. It is now evident that a multitude of both CC and CXC chemokine receptors, such as CCR8, TYMSTER, BONZO and BOB, can function as co-receptors for HIV-1 entry into $CD4^+$ lymphocytes (reviewed in [15]).

$CD4^+$ lymphocytes isolated from a small cohort of HIV-1 EU individuals were relatively resistant to infection with M-tropic viral isolates in comparison to non-exposed control persons [64]. Variations in susceptibilities of PBMCs to HIV-1 infection had previously been reported [65,66]. Two individuals in the above study, designated EU2 and EU3, had CD4$^+$ lymphocytes completely refractory for the replication of M-tropic viruses but not T-tropic isolates. The cells were shown to secrete 5-10-fold higher levels of the CC chemokines RANTES, MIP-1α and MIP-1β over what was observed for the control cells [64]. When the CCR5 genes from these individuals were PCR-amplified and utilised in an *in vitro* expression and infection assay they were shown to lack HIV-1 co-receptor function. DNA sequencing revealed a deletion of 32 bp (Δ32) within the region corresponding to the second extracellular loop domain of the CCR5 protein, providing for an amino acid frame shift [32]. The resultant Δ32 CCR5 protein was found to be lacking normal physiological activity as demonstrated by its inability to support MIP-1β cell signalling. Further analysis demonstrated that both EU2 and EU3 were homozygous for the deletion (Δ32/Δ32) with them each carrying two copies of the defective gene [32].

A number of epidemiological studies have shown that individuals with the Δ32/Δ32 genotype are highly protected against the transmission of HIV-1 [31, 33,67-70]. The Δ32 allele has a relatively high frequency amongst Caucasian populations of European decent (allelic frequency of 0.08-0.1), with 1% of the population being Δ32/Δ32 and 15-20% being Δ32/CCR5 [31-33,67-70]. The allele was found at a much lower frequency amongst African Americans and was absent from individuals of African or Asian ancestry. No Δ32/Δ32 individuals were identified within the HIV-1-positive cohorts screened and there was a higher percentage of individuals amongst the EU cohorts. In one study where EU individuals were segregated into different groups based on their sexual risk behaviour, a higher frequency of the Δ32/32 genotype was identified amongst the group with the highest exposure risk [68]. Collectively, these studies indicate that the Δ32/Δ32 genotype provides protection against the transmission of HIV-1.

Recently, there have been reports describing HIV-1-infected individuals that are homozygous for the deleted genotype [71-73]. Since the CD4$^+$ lymphocytes from Δ32/Δ32 individuals are infectable with T-tropic viral isolates, it may be that these individuals were infected with a virus that utilises an HIV-1 co-receptor other than CCR5. One epidemiological study described a slight protective effect against HIV-1 transmission and the Δ32/ CCR5 genotype but this was not observed in other studies [31]. A possible explanation for this phenomenon comes from a detailed study of twenty monogamous homosexual couples of whom ten had concordant and ten had discordant sero-status, despite high-risk sexual behaviour [74]. Compared with PBMCs from healthy blood donors, eight of ten non-recipients (three with Δ32/CCR5 genotype) and three of eight recipients had PBMC with reduced susceptibility to *in vitro* infection with NSI HIV-1 variants isolated from their partner. No difference in susceptibility was observed for infection with an SI variant. All three recipients with reduced susceptible PBMC had partners with high cellular virus load and, conversely, both non-recipients with normally susceptible PBMC had partners with a very low load. This suggests that a combination of susceptibility of target cells and inoculum size largely determines whether HIV-1 infection is established upon homosexual exposure. Promiscuous sexual behaviour increases the chance of exposure to a high viral inoculum, which may override relative insusceptibility. Whether the relative insusceptibility of PBMC from most non-recipients is due to reduced expression of CCR5, higher expression of β-chemokines, or combination of both, will be discussed below.

These results support the earlier discussion that HIV-1 primary isolates which utilise the CCR5 co-receptor are preferentially transmitted. Accordingly, individuals heterozygous or homozygous for a mutation in CCR2b, a co-receptor used by some but not all HIV-1 strains, are not protected from infection, although they do have reduced rates of disease progression once infected in most cohorts studied thus far [75-80]. The cell surface expression pattern of the co-receptors on the cells at the site of infection may influence viral transmission patterns as well. The initial cell type, monocyte or T cell, required to establish an ongoing productive infection may be important for viral dissemination, or the controlling immune response may suppress the replication of T-tropic but not M-tropic viral isolates.

Some HIV-1-infected Δ32/CCR5 individuals have been shown to progress slower to disease in comparison to homozygous CCR5 wild-type individuals [31,33,67-70]. These Δ32/CCR5 individuals are more likely to harbour lower viral loads, have slower rates of CD4 decline and live disease-free longer than persons with the CCR5/CCR5 genotype. Individuals heterozygous for Δ32/CCR5 have been shown to have $CD4^+$ lymphocytes that are less infectable with M-tropic viral isolates of HIV-1 and this was correlated to CCR5 cell-surface expression levels [81]. Therefore, the protective effect appears to be at the phenotype level of CCR5 expression rather than the genotype level. CCR5/CCR5 individuals had variable PBMC cell-surface expression levels of CCR5 which showed some degree of overlap with the expression levels of the Δ32/CCR5 individuals [81]. There is evidence that $CD4^+$ lymphocytes from both Δ32/CCR5 and CCR5/CCR5 individuals can have reduced surface expression levels of the CCR5 protein, be more sensitive to the HIV-1-blocking effects of recombinant chemokines whilst secreting higher levels of endogenous RANTES [82].

These *in vitro* results taken together promote the concept that a complex network of interactions between chemokines, their receptors and HIV-1 can influence the extent of viral infection. There is *in-vivo* evidence to support the above; SCID-hu mice reconstituted with PBMCs from Δ32/CCR5 individuals develop lower end-point viral loads with slower rates of viral replication than those mice reconstituted with CCR5/CCR5 PBMCs [83].

In conclusion, the down-regulation of the CCR5 chemokine receptor or its blockage by either natural chemokines or by antagonists could have a desirable effect on HIV-1 transmission rates or in delaying disease progression. This seems more feasible when it is taken into consideration that individuals with the Δ32/Δ32 genotype appear to be perfectly healthy, therefore blocking of the CCR5 receptor in healthy individuals may have no detrimental side effects. However, it still remains to be seen whether the blocking or down-regulation of CCR5 expression has a similar null effect on CCR5/CCR5 individuals.

7. Animal Model Systems

The prospect of developing a successful HIV-1 vaccine for humans has been heightened by recent advances describing successes with the SIV macaque model system. When making comparisons between HIV-1 and SIV, it should be noted that these two viruses have some differences in their co-receptor usage, HIV-1 utilises CCR5 and CXCR4 as its main receptors, whereas SIV utilizes CCR5, BONZO and BOB, which may limit the relevance to HIV-1 (reviewed in [15]). In one study rhesus macaques challenged with a high infectious dose of SIV seroconverted whilst those given a low dose did not seroconvert over a two-year period [84]. These low-dose-challenged monkeys were protected against seroconversion when re-challenged with the higher viral dose. These animals were shown to have greater $CD8^+$ cell-mediated SIV suppressor activity than the non-infected control animals [84]. Another study demonstrated that low-dose mucosal inoculation resulted in protection from subsequent higher dose SIV challenges which could be correlated to anti-SIV-specific proliferative responses [85]. There is also evidence from macaque studies that SIV-specific CTL can be detected at the vaginal mucosa of monkeys infected with SIV [86]. Although these animals are infected, it does demonstrate that a cell-mediated immune response can exist at the vaginal epithelium, a site where a localised response would be expected to be induced.

An important study describes an induced protective immune response in a group of macaques immunised with a SIV env and p27-gag subunit vaccine [87]. These monkeys were vaccinated via the iliac lymph node and the protection was only observed in those monkeys exposed to a rectal challenge of SIV and not vaginal. This result suggests that a protective immune response was stimulated in the regional draining lymph node near the site of vaccination. This immune response consisted of both cell-mediated responses and an IgG and IgA antibody response [87]. Although a detectable antibody response was observed in these monkeys, the results may have significance with relation to the HIV-1-exposed individuals. An individual mounting a localised immune response at the site of exposure may be protected from subsequent infection at that site but not others, therefore when considering vaccination strategies the likely route of viral transmission should be taken into consideration.

8. Concluding Remarks

No two exposures to HIV-1 can be regarded as identical and we have discussed here some of the many factors that may contribute to HIV-1 transmission. These factors can range from obvious differences, such as route of exposure, size of the viral inoculum to the viral phenotype present. MHC class-I and -II cellular immune responses have been correlated with groups of EU individuals and whether it is these responses that are providing protection is still not known. The identification that individuals homozygous for a defective CCR5 gene are significantly protected against the acquisition of HIV-1 is a clear indication of the importance that host genetic factors can have on HIV-1 transmission. Multiple combinations of the above undoubtedly exist to influence whether the virus is successfully transmitted or not. An individual with relatively resistant $CD4^+$ lymphocytes for M-tropic viral isolates may acquire an active infection at the site of exposure, this virus is slow to disseminate allowing for the development of an active anti-HIV-1 cellular immune response which can control viral replication in a site-specific manner. This individual would then be protected from subsequent exposures at the same site. An alternative would be when the $CD4^+$ lymphocytes are more susceptible to infection, this individual would encounter rapid viral dissemination, lack an effective controlling immune response and would seroconvert. T-tropic isolates, because of the cellular milieu at the site of infection, may not disseminate thereby establishing a protective cellular response that is capable of preventing the subsequent infection with an M-tropic viral isolate.

We have discussed here a variety of social and medical interventions by which HIV-1 transmission can be prevented (summarized in Table 2). Although social and cultural policies may decrease HIV-1 transmission rates, it is unrealistic to believe that this alone will dramatically alter the epidemic. What is needed is the development of an effective vaccine against HIV-1. This vaccine should stimulate a protective immune response or alter the levels of β-chemokine receptor expression or the synthesis of β-chemokines at the site of exposure. Only effective vaccines will overcome the social and economic obstacles and interrupt the chain of viral transmission which has devastated populations in both the developed and developing worlds.

TABLE 2. Opportunities to block HIV-1 transmission	
1.	
-	Health education programs
-	Drug rehabilitation programs
-	Needle exchange programs
2.	
-	Induction of HIV-1-specific, localized or systemic, cell-mediated immune responses by vaccination
-	Up-regulation of β-chemokine production
-	Down-regulation of HIV-1 co-receptor expression
-	Antagonists to block HIV-1 co-receptors
3.	
-	Treatment of HIV-1-infected pregnant women
-	Caesarean birth
-	Treatment of STDs
-	Anti-retroviral drug treatment of infected individuals

STDs: sexually transmitted diseases

9. Acknowledgements

The authors thank Dr. Charles R. Makay, Dr. Lijun Wu and Stan Kang for contributions to this work and Charla A. Andrews for helpful comments on the manuscript. This work was supported by grants and contracts from the National Institutes of Health, RO1 A135522, RO1 A142397 and R21 A142630. R.A.K. is an Elizabeth Glaser Scientist of the Pediatric AIDS Foundation.

10. References

1. Coutinho, R.A., Prins, M., Spijkerman, I.J.B., Geskus, R.B., Keet, I.P.M., Fennema, H.S.A. and Strathdee, S.A.: Summary of track C: epidemiology and public health, *AIDS* **10** (suppl. 3), (1996) S115-S121.
2. Paxton, W.A. and Koup, R.A.: Mechanisms of resistance to HIV infection, *Springer Semin. Immunopathol.* **18**, (1997) 323-340.
3. Marcus, R.: Surveillance of health care workers exposed to blood from patients infected with HIV, *New Engl. J. Med.* **319**, (1988) 1118-1123.
4. Ward, J.W., Bush, T.J., Perkins, H.A., Lieb, L.E., Allen, J.R., Goldfinger, D., Samson, S.M., Pepkowitz, S.H., Fernando, L.P., Holland, P.V., Kleinman, S.H., Grindon, A.J., Garner, J.L., Rutherford, G.W. and Holmberg, S.D. : The natural history of transfusion-associated infection with human immunodeficiency virus: factors influencing the rate of progression to disease, *New Engl. J. Med.* **321** (1989), 947-952.
5. DeGruttola, V., Seage, G.R., Mayer, K.H. and Horsburgh, C.R.: Infectiousness of HIV between male homosexual partners, *J. Clin. Epidemiol.* **42** (1989), 849-856.
6. Laga, M., Manoka, A., Kivuvu, M., Malele, B., Tuliza, M., Goeman, J., Behets, F., Batter, V., Alary, M., Heyward, W.L., Ryder, R.W. and Piot, P.: Non-ulcerative sexually transmitted diseases as risk factors for HIV-1 transmission in women: results from a cohort study, *AIDS* **7** (1992), 95-102.
7. Nair, P., Alger, L., Hines, S., Seiden, S., Hebel, R. and Johnson, J.P.: Maternal and neonatal characteristics associated with HIV infection in infants of seropositive women, *J. Acq. Immune Def. Syndr. Human Retrovir.* **6** (1993), 298-302.
8. Borkowsky, W., Krasinski, K., Cao, Y., Ho, D.D., Pollack, H., Moore, T., Chen, S.H., Allen, M. and Tao, P.T.: Correlation of perinatal transmission of human immunodeficiency virus type 1 with maternal viremia and lymphocyte phenotypes, *J. Pediatr.* **125** (1994), 345-351.
9. Connor, E.M., Sperling, R.S., Gelber, R., Kiselev, P., Scott, G., O'Sullivan, M.J., VanDyke, R., Bey, M., Shearer, W., Jacobson, R.L., Jimenez, E., O'Neill, E., Bazin, B., Delfraissy, J.-F., Culnane, M., Coombs, R.W., Elkins, M., Moye, J., Stratton, P., Balsley, J. and for the Pediatric AIDS Clinical Trials Group Protocol 076 Study Group: Reduction of maternal-infant transmission of human immunodeficiency virus type 1 with zidovudine treatment, *New Engl. J. Med.* **331** (1994), 1173-1180.
10. Burchett, S.K., Kornegay, J., Pitt, J., Landesman, S., Rosenblatt, H., Hillyer, G., Garcia, P., Kalish, L., Burns, D., Davenney, K. and Lew, J.: Assessment of maternal plasma HIV viral load as a correlate of vertical transmission, 3rd *Conf. Retrovir. Opport. Infect.* Abstract (1996).
11. Sperling, R.S., Shapiro, D.E., Coombs, R.W., Todd, J.A., Herman, S.A., McSherry, G.D., O'Sullivan, M.J., VanDyke, R., Jimenez, E., Rouzioux, C., Flynn, P.M., Sullivan, J.L. and for the Pediatric AIDS Clinical Trials Group Protocol 076 Study Group: Maternal viral load, zidovudine treatment, and the risk of transmission of human immunodeficiency virus type 1 from mother to infant, *New Engl. J. Med.* **335** (1996), 1621-1629.
12. Cao, Y., Krogstad, P., Korber, B.T.M., Koup, R.A., Muldoon, R., Macken, C., Song, J., Jin, Z., Zhao, J.-Q., Clapp, S., Chen, I.S.Y., Ho, D.D., Ammann, A.J. and the French Federation of AIDS Reference Centers: Maternal HIV-1 load and vertical transmission of infection: the Ariel Project for the prevention of HIV transmission from mother to infant, *Nature Medicine* **3** (1997), 549-552.
13. Goedert, J.J., Duliege, A.-M., Amos, C.I., Felton, S. and Biggar, R.J.: High risk of HIV-1 infection for first-born twins, *Lancet* **338** (1991), 1471.
14. European Collaborative Study: Risk factors for mother-to-child transmission of HIV-1, *Lancet* **339** (1991), 1007-1012.
15. Unutmaz, D., KewalRamani, V.N. and Littman, D.R.: G protein-coupled receptors in HIV and SIV entry: new perspectives on lentivirus-host interactions and on the utility of animal models, *Seminars in Immunology* **10** (1998), 225-236.
16. De Jong, J.J., Goudsmit, J., Keulen, W., Klaver, B., Krone, W.J.A., Tersmette, M. and De Ronde, A.: Human immunodeficiency viruses type-1 chimeric for the envelope V3 domain are distinct in syncytium formation and replication capacity, *J. Virol.* **66** (1992), 757-765.
17. Fouchier, R.A.M., Groenink, M., Kootstra, N.A., Tersmette, M., Huisman, J.G., Miedema, F. and Schuitemaker, H.: Phenotype-associated sequence variation in the third variable domain of the human immunodeficiency virus type 1 gp120 molecule, *J. Virol.* **66** (1992), 3183-3187.

18. Chesebro, B., Wehrly, K., Nishio, J. and Perryman, S.: Macrophage-tropic human immunodeficiency virus isolates from different patients exhibit unusual V3 envelope sequence homogeneity in comparison with T-cell-tropic isolates: definition of critical amino acids involved in cell tropism, *J. Virol.* **66** (1992), 6547-6554
19. McNearney, T., Hornickova, Z., Markham, R., Birdwell, A., Arens, M., Saah, A.J. and Ratner, L.: Relationship of human immunodeficiency virus type 1 sequence heterogeneity to stage of disease, *Proc. Natl. Acad. Sci. USA* **89** (1992), 10247-10251.
20. Zhang, L.Q., MacKenzie, P., Cleland, A., Holmes, E.C., Leigh-Brown, A.J. and Simmonds, P.: Selection for specific sequences in the external envelope protein of HIV-1 upon primary infection, *J. Virol.* **67** (1993), 3345-3356.
21. Wolfs, T.F.W., Zwart, G., Bakker, M. and Goudsmit, J.: HIV-1 genomic RNA diversification following sexual and parenteral virus transmission, *Virology* **189** (1992), 103-110.
22. Van 't Wout, A.B., Kootstra, N.A., Mulder-Kampinga, G.A., Albrecht-van Lent, N., Scherpbier, H.J., Veenstra, J., Boer, K., Coutinho, R.A., Miedema, F. and Schuitemaker, H.: Macrophage-tropic variants initiate human immunodeficiency virus type 1 infection after sexual, parenteral and vertical transmission, *J. Clin. Invest.* **94** (1994), 2060-2067.
23. Lamers, S.L., Sleasman, J.W., She, J.X., Barrie, K.A., Pomeroy, S.M., Barrett, D.J. and Goodenow, M.M.: Persistence of multiple maternal genotypes of human immunodeficiency virus type 1 in infants by vertical transmission, *J. Clin. Invest.* **93** (1994), 380-390.
24. Tudor-Williams, G., St. Clair, M.H., McKinney, R.E., Maha, M., Walter, E., Santacroce, S., Mintz, M., O'Donnell, K., Rudoll, T., Vavro, C.L., Connor, E.M. and Wilfert, C.M.: HIV-1 sensitivity to zidovudine and clinical outcome in children, *Lancet* **339** (1992), 15-19.
25. Koot, M., Schellekens, P.Th.A., Mulder, J.W., Lange, J.M.A., Roos, M.Th.L., Coutinho, R.A., Tersmette, M. and Miedema, F.: Viral phenotype and T-cell reactivity in human immunodeficiency virus type 1-infected asymptomatic men treated with zidovudine, *J. Infect. Dis.* **168** (1993), 733-736.
26. Van 't Wout, A.B., De Jong, M.D., Kootstra, N.A., Veenstra, J., Lange, J.M.A., Boucher, C.A.B. and Schuitemaker, H.: Changes in cellular virus load and zidovudine resistance of syncytium-inducing and non-syncytium-inducing human immunodeficiency virus populations under zidovudine pressure: a clonal analysis, *J. Infect. Dis.* **174** (1996), 845-849.
27. Zhu, T., Wang, N., Carr, A., Nam, D.S., Moor-Jankowski, R., Cooper, D.A. and Ho, D.D.: Genetic characterization of human immunodeficiency virus type 1 in blood and genital secretions: evidence for viral compartimentalization and selection during sexual transmission, *J. Virol.* **70** (1996), 3098-3107.
28. Overbaugh, J., Anderson, R.J., Ndinya-Achola, J.O. and Kreiss, J.K.: Distinct but related human immunodeficiency virus type 1 variant populations in genital secretions and blood, *AIDS Res. Human Retrovir.* **12** (1996), 107-115.
29. Braathen, L.R., Ramirez, G., Kunze, R.O.F. and Gelderblom, H.R.: Langerhans cells as primary target cells for HIV infection, *Lancet* **ii** (1987), 1094.
30. Palacio, J., Souberbielle, B.E., Shattock, R.J., Robinson, G., Manyonda, I. and Griffin, G.E.: *In-vitro* HIV-1 infection of human cervical tissue, *Res. Virol.* **145** (1994), 155-161.
31. Samson, M., Libert, F., Doranz, B.J., Rucker, J., Liesnard, C., Farber, C., Saragosti, S., Lapouméroulie, C., Cognaux, J., Forceille, C., Muyldermans, G., Verhofstede, C., Burtonboy, G., Georges, M., Imai, T., Rana, S., Yi, Y., Smyth, R.J., Collman, R.G., Doms, R.W., Vassart, G. and Parmentier, M.: Resistance to HIV-1 infection in causacian individuals bearing mutant alleles of the CCR-5 chemokine receptor gene, *Nature* **382** (1996), 722-725.
32. Liu, R., Paxton, W.A., Choe, S., Ceradini, D., Martin, S.R., Horuk, R., MacDonald, M.E., Stuhlmann, H., Koup, R.A. and Landau, N.R.: Homozygous defect in HIV-1 coreceptor accounts for resistance of some multiply-exposed individuals to HIV-1 infection, *Cell* **86** (1996), 1-20.
33. Zimmerman, P.A., Buckler-White, A.J., Alkhatib, G., Spalding, T., Kubofcik, J., Combadiere, C., Weissman, D., Cohen, O., Rubbert, A., Lam, G., Vaccarezza, M., Kennedy, P.E., Kumaraswami, V., Giorgi, J.V., Detels, R., Hunter, J., Chopek, M., Berger, E.A., Fauci, A.S., Nutman, T.B. and Murphy, P.M.: Inherited resistance to HIV-1 conferred by an inactivating mutation in CC chemokine receptor 5: studies in populations with contrasting clinical phenotypes, defined racial background, and quantified risk, *Molecular Medicine* **3** (1997), 23-36.
34. Hu, D.J., Dondero, T.J. and Rayfield, M.A.: The emerging genetic diversity of HIV: the importance of global surveillance for diagnostics, research and prevention, *JAMA* **275** (1996), 210-216.

35. Soto-Ramirez, L.E., Renjifo, B., McLane, M.F., Marlink, R., O'Hara, C., Sutthent, R., Wasi, C., Vithayasai, P., Vithayasai, V., Apichartpiyakul, C., Auewarakul, P., Cruz, V.P., Chui, D.S., Osathanondh, R., Mayer, K., Lee, T.H. and Essex, M.: HIV-1 Langerhans' cell tropism associated with heterosexual transmission of HIV, *Science* **271** (1996), 1291-1293.
36. Kunanusont, C., Foy, H.M., Kreiss, J.K., Rerks-Ngarm, S., Phanuphak, P., Raktham, S., Pau, C.-P. and Young, N.L.: HIV-1 subtypes and male-to-female transmission in Thailand, *Lancet* **345** (1995), 1078-1083.
37. Dittmar, M.T., Simmons, G., Hibbitts, S., O'Hare, M., Louisirirotchanakul, S., Beddows, S., Weber, J., Clapham, P.R. and Weiss, R.A.: Langerhans cell tropism of human immunodeficiency virus type 1 subtype A through F isolates derived from different transmission groups, *J. Virol.* **71** (1997), 8008-8013.
38. Pope, M., Frankel, S.S., Mascola, J.R., Trkola, A., Isdell, F., Birx, D.L., Burke, D.S., Ho, D.D. and Moore, J.P.: Human immunodeficiency virus type 1 strains of subtypes B and E replicate in cutaneous dendritic cell-T-cell mixtures without displaying subtype-specific tropism, *J. Virol.* **71** (1997), 8001-8007.
39. Burke, D.S.: Recombination in HIV: an important viral evolutionary strategy, *Emerging Infectious Diseases* **3** (1997), 253-259.
40. Robertson, D.L., Sharp, P.M., McCutchan, F.E. and Hahn, B.H.: Recombination in HIV-1, *Nature* **374** (1995), 124-126.
41. Travers, K., Mboup, S., Marlink, R., Guèye-Ndiaye, A., Siby, T., Thior, I., Traore, I., Dieng-Sarr, A., Sankalé, J.L., Mullins, C., Ndoye, I., Hsieh, C.C., Essex, M. and Kanki, P.: Natural protection against HIV-1 infection provided by HIV-2, *Science* **268** (1995), 1612-1615.
42. Taylor, R.: Quiet clues to HIV-1 immunity: do some people resist infection?, *J. NIH Res.* **6** (1994), 29-31.
43. O'Brien, T.R., Busch, M.P., Donegan, E., Ward, J.W., Wong, L., Samson, S.M., Perkins, H.A., Altman, R., Stoneburner, R.L. and Holmberg, S.D.: Heterosexual transmission of human immunodeficiency virus type 1 from transfusion recipients to their sex partners, *J. Acq. Immune Def. Syndr. Human Retrovir.* **7** (1994), 705-710.
44. Prevots, D.R., Ancelle-Park, R.A., Neal, J.J. and Remis, R.S.: The epidemiology of heterosexually acquired HIV infection and AIDS in western industrialized countries, *AIDS* **8** (1994), S109-S117.
45. Jehuda-Cohen, T., Slade, B.A., Powell, J.D., Villinger, F., De, B., Folks, T.M., McClure, H.M., Sell, K.W. and Ahmed-Ansari, A.: Polyclonal B-cell activation reveals antibodies against human immunodeficiency virus type 1 (HIV-1) in HIV-1-seronegative individuals, *Proc. Natl, Acad. Sci. USA* **87** (1990), 3972-3976.
46. Cao, Y., Qin, I., Zhang, L., Safrit, J.T. and Ho, D.D.: Virologic and immunologic characterization of long-term survivors of human immunodeficiency virus type-1 infection, *New Engl. J. Med.* **332** (1995), 201-208.
47. Rowland-Jones, S.L., Nixon, D.F., Aldhous, M.C., Gotch, F.M., Ariyoshi, K., Hallam, N., Kroll, J.S., Froebel, K.S. and McMichael, A.J.: HIV-specific cytotoxic T-cell activity in an HIV-exposed but uninfected infant, *Lancet* **341** (1993), 860-861.
48. Detels, R., Liu, Z., Hennessey, K., Kan, J., Visscher, B.R., Taylor, J.M.G., Hoover, D.R., Rinaldo, C.R., Phair, J.P., Saah, A.J. and Giorgi, J.V.: Resistance to HIV-1 infection, *J. Acq. Immune Def. Syndr. Human Retrovir.* **7** (1994), 1263-1269.
49. Langlade-Demoyen, P., Ngo-Giang-Huong, N., Ferchal, F. and Oksenhendler, E.: HIV-1 nef-specific cytotoxic T lymphocytes in noninfected heterosexual contact of HIV-infected patients, *J. Clin. Invest.* **93** (1994) 1293-1297.
50. Pinto, L.A., Sullivan, J.L., Berzofsky, J.A., Clerici, M., Kessler, H.A., Landay, A.L. and Shearer, G.M.: Env-specific cytotoxic T-lymphocyte responses in HIV-contaminated body fluids, *J. Clin. Invest.* **96** (1995) 867-876.
51. Rowland-Jones, S.L., Sutton, J., Ariyoshi, K., Dong, T., Gotch, F.M., McAdam, S., Whitby, D., Sabally, S., Gallimore, A., Corrah, T., Takiguchi, M., Schulz, T., McMichael, A.J. and Whittle, H.C.: HIV-specific cytotoxic T cells in HIV-exposed but uninfected Gambian women, *Nature Medicine* **1** (1995), 59-64.
52. Clerici, M., Giorgi, J.V., Chou, C.-C., Gudeman, V.K., Zack, J.A., Gupta, P., Ho, H.N., Nishanian, P.G., Berzofsky, J.A. and Shearer, G.M.: Cell-mediated immune response to human immunodeficiency virus (HIV) type 1 in seronegative homosexual men with recent sexual exposure to HIV-1, *J. Infect. Dis.* **165** (1992), 1012-1019.

53. Clerici, M., Levin, J.M., Kessler, H.A., Harris, A., Berzofsky, J.A., Landay, A.L. and Shearer, G.M.: HIV-specific T-helper activity in seronegative health care workers exposed to contaminated blood, *JAMA* **271** (1994), 42-46.
54. Clerici, M., Sison, A.V., Berzofsky, J.A., Rakusan, T.A., Brandt, C.D., Ellaurie, M., Villa, M.L., Colie, C., Venzon, D.J., Sewver, J.L. and Shearer, G.M.: Cellular immune factors associated with mother-to-infant transmission of HIV, *AIDS* **7** (1993), 1427-1433.
55. Just, J.J., Louie, L.G., Abrams, E., Nicholas, S.W., Wara, D.W., Stein, Z. and King, M.C.: Genetic risk factors for perinatally acquired HIV-1 infection, *Paedriatr. Perinat. Epidemiol.* **6** (1997), 215-224.
56. Kaslow, R.A., Duquesnoy, R., Van Raden, M., Kingley, L., Marrari, M., Friedman, H., Su, S., Saah, A.J., Detels, R., Phair, J.P. and Rinaldo, C.R.: A1, Cw7, B8, DR3 HLA antigen combination associated with rapid decline of T-helper lymphocytes in HIV-1 infection, *Lancet* **335** (1990), 927-930.
57. Kilpatrick, D.C., Hague, R.A., Yap, P.L. and Mok, J.Y.: HLA antigen frequencies in children born to HIV-infected mothers, *Dis. Markers* **9** (1991), 21-26.
58. Klein, M.R., Keet, I.P.M., D'Amaro, J., Bende, R.J., Hekman, A., Mesman, B., Koot, M., De Waal, L.P., Coutinho, R.A. and Miedema, F.: Associations between HLA frequencies and pathogenic features of human immunodeficiency virus type-1 infection in seroconverters from the Amsterdam Cohort of homosexual men, *J. Infect. Dis.* **169** (1994), 1244-1249.
59. Steel, C.M., Ludlam, C.A., Beatson, D., Peutherer, J.F., Cuthbert, R.J.G., Simmonds, P., Morrison, H. and Jones, M.: HLA haplotype A1 B8 DR3 as a risk factor for HIV-related disease, *Lancet* **1** (8596) (1988), 1185-1188.
60. Levy, J.A., Mackewicz, C.E. and Barker, E.: Controlling HIV-1 pathogenesis: the role of the non-cytotoxic anti-HIV-1 response of $CD8^+$ T cells, *Immunol. Today* **17** (1996), 217-224.
61. Walker, C.M., Moody, D.J., Stites, D.P. and Levy, J.A.: $CD8^+$ lymphocytes can control HIV infection *in vitro* by suppressing virus replication, *Science* **234** (1986), 1563-1566.
62. Mackewicz, C. and Levy, J.A.: $CD8^+$ cell anti-HIV activity: nonlytic suppression of virus replication, *AIDS Res. Human Retrovir.* **8** (1992), 1039-1050.
63. Levy, J.A.: HIV pathogenesis and long-term survival, *AIDS* **7** (1993), 1401-1410.
64. Paxton, W.A., Martin, S.R., Tse, D., O'Brien, T.R., Skurnick, J., Vandevanter, N.L., Padian, N., Braun, J.F., Kotler, D.P., Wolinsky, S.M. and Koup, R.A.: Relative resistance to HIV-1 infection of CD4 lymphocytes from persons who remain uninfected despite multiple high-risk sexual exposures, *Nature Medicine* **2** (1996), 412-417.
65. Ometto, L., Zanotto, C., Maccabruni, A. et al.: Viral phenotype and host-cell susceptibility to HIV-1 infection as risk factors for mother-to-child HIV-1 transmission, *AIDS* **9** (1995), 427-434.
66. Lederman, M.M., Jackson, B.J., Kroner, B.L., White, G.C., Eyster, M.E., Aledort, L.M., Hilgartner, M.W., Kessler, C.M., Cohen, A.R., Kiger, K.P. and Goedert, J.J.: Human immunodeficiency virus (HIV) type-1 infection status and *in-vitro* susceptibility to HIV infection among high-risk HIV-1-seronegative hemophiliacs, *J. Infect. Dis.* **172** (1995), 228-231.
67. Dean, M., Carrington, M.N., Winkler, C., Huttley, G.A., Smith, M.W., Allikmets, R., Goedert, J.J., Buchbinder, S.P., Vittinghoff, E., Gomperts, E., Donfield, S., Vlahov, D., Kaslow, R.A., Saah, A.J., Rinaldo, C.R., Detels, R., Hemophilia Growth and Development Study, Multicenter AIDS Cohort Study, Multicenter Hemophilia Cohort Study, San Francisco City Cohort, ALIVE Study and O'Brien, S.J.: Genetic restriction of HIV-1 infection and progression to AIDS by a deletion allele of the CKR5 structural gene, *Science* **273** (1996), 1856-1862.
68. Huang, Y., Paxton, W.A., Wolinsky, S.M., Neumann, A.U., Zhang, L., He, T., Kang, S., Ceradini, D., Jin, Z., Yazdanbakhsh, K., Kunstman, K., Erickson, D., Dragon, E., Landau, N.R., Phair, J., Ho, D.D. and Koup, R.A.: The role of a mutant CCR5 allele in HIV-1 transmission and disease progression, *Nature Medicine* **2** (1996), 1240-1243.
69. De Roda Husman, A.-M., Koot, M., Cornelissen, M.T.E., Brouwer, M., Broersen, S.M., Bakker, M., Roos, M.Th.L., Prins, M., De Wolf, F., Coutinho, R.A., Miedema, F., Goudsmit, J. and Schuitemaker, H.: Association between CCR5 genotype and the clinical course of HIV-1 infection, *Ann. Intern. Med.* **127** (1997), 882-890.
70. Michael, N.L., Chang, G., Louie, L.G., Mascola, J.R., Dondero, D., Birx, D.L. and Sheppard, H.W.: The role of viral phenotype and CCR-5 gene defects in HIV-1 transmission and disease progression, *Nature Medicine* **3** (1997), 338-340.
71. Biti, R., French, R., Young, J., Bennetts, B., Stewart, G. and Liang, T.: HIV-1 infection in an individual homozygous for the CCR5 deletion allele, *Nature Medicine* **3** (1997), 252-253.
72. O'Brien, T.R., Winkler, C., Dean, M., Nelson, J.A., Carrington, M., Michael, N.L. and White, G.C.: HIV-1 infection in a man homozygous for CCR5 D32, *Lancet* **349** (1997), 1219.

73. Theodorou, I., Meyer, L., Magierowska, M., Katlana, C. and Rouzioux, C.: HIV-1 infection in an individual homozygous for CCR5 delta-32, *Lancet* **349** (1997), 1219.
74. Blaak, H., Van 't Wout, A.B., Brouwer, M., Cornelissen, M.T.E., Kootstra, N.A., Albrecht-van Lent, N., Keet, I.P.M., Goudsmit, J., Coutinho, R.A. and Schuitemaker, H.: Infectious cellular load in HIV-1-infected individuals and susceptibility of PBMC from their exposed partners to NSI HIV-1 as major determinants for HIV-1 transmission in homosexual couples, *J. Virol.* **72** (1998), 218-224.
75. Smith, M.W., Dean, M., Carrington, M.N., Winkler, C., Huttley, G.A., Lomb, D.A., Goedert, J.J., O'Brien, T.R., Jacobson, L.P., Kaslow, R.A., Buchbinder, S.P., Vittinghoff, E., Vlahov, D., Hoots, K., Hilgartner, M.W., Hemophilia Growth and Development Study, Multicenter AIDS Cohort Study, Multicenter Hemophilia Cohort Study, San Francisco City Cohort, ALIVE and O'Brien, S.J.: Contrasting genetic influence of CCR2 and CCR5 variants on HIV-1 infection and disease progression, *Science* **277** (1997), 959-965.
76. Michael, N.L., Louie, L.G., Rohrbaugh, A.L., Schulz, K.A., Dayhoff, D.E., Wang, C.E. and Sheppard, H.W.: The role of CCR5 and CCR2 polymorphisms in HIV-1 transmission and disease progression, *Nature Medicine* **3** (1997), 1160-1162.
77. Smith, M.W., Carrington, M.N., Winkler, C., Lomb, D.A., Dean, M., Huttley, G.A. and O'Brien, S.J.: CCR2 chemokine receptor and AIDS progression, *Nature Medicine* **3** (1997), 1052-1053.
78. Van Rij, R.P., De Roda Husman, A.-M., Brouwer, M., Goudsmit, J., Coutinho, R.A. and Schuitemaker, H.: Role of CCR2 genotype in the clinical course of syncytium-inducing (SI) or non-SI human immunodeficiency virus type-1 infection and in the time to conversion to SI virus variants, *J. Infect. Dis.* **178** (1998), 1806-1811.
79. Hendel, H., Henon, N., Lebuanee, H., Lachgar, A., Poncelet, H., Caillat-Zucman, S., Winkler, C.A., Smith, M.W., Kenefic, L., O'Brien, S.J., Lu, W., Andrieu, J.-M., Zagury, D., Schachter, F., Rappaport, J. and Zagury, J.F.: Distinctive effects of CCR5, CCR2 and SDF1 genetic polymorphisms in AIDS progression, *J. Acq. Immune Def. Syndr. Human Retrovir.* **19** (1998), 381-386.
80. Eugen-Olsen, J., Iversen, A.K.N., Benfield, T.L., Koppelhus, U. and Garred, P.: Chemokine receptor CCR2b 641 polymorphism and its relation to CD4 T-cell counts and disease progression in a Danish cohort of HIV-infected individuals. Copenhagen AIDS Cohort, *J. Acq. Immune Def. Syndr. Human Retrovir.* **18** (1998), 110-116.
81. Wu, L., Paxton, W.A., Kassam, N., Ruffing, N., Rottman, J.B., Sullivan, N., Choe, H., Sodroski, J., Newman, W., Koup, R.A. and Mackay, C.R.: CCR5 levels and expression pattern correlate with infectability by macrophage-tropic HIV-1, *in vitro*, *J. Exp. Med.* **185** (1997), 1681-1691.
82. Paxton, W.A., Liu, R., Kang, S., Gingeras, T.R., Landau, N.R., Mackay, C.R. and Koup, R.A.: Reduced HIV-1 infectability of $CD4^+$ lymphocytes from exposed-uninfected individuals: association with low expression of CCR5 and high production of β-chemokines, *Virology* **244** (1998), 66-73.
83. Picchio, G.R., Gulizia, R.J. and Mosier, D.E.: Chemokine receptor CCR5 genotype influences the kinetics of human immunodeficiency type-1 infection in human PBL-SCID mice, *J. Virol.* **71** (1997), 7124-7127.
84. Salvato, M.S., Emau, P., Malkovsky, M., Schultz, K.T., Johnson, E. and Pauza, C.D.: Cellular immune responses in rhesus macaques infected rectally with low-dose simian immunodeficiency virus, *J. Med. Primatol.* **23** (1994), 125-130.
85. Clerici, M., Clark, E.A., Polacino, P., Axberg, I., Kuller, L., Casey, N.I., Morton, W.R., Shearer, G.M. and Benveniste, R.E.: T-cell proliferation to subinfectious SIV correlates with lack of infection after challenge of macaques, *AIDS* **8** (1994), 1391-1395.
86. Lohman, B.L., Miller, C.J. and McChesney, M.B.: Antiviral cytotoxic T lymphocytes in vaginal mucosa of simian immunodeficiency virus-infected macaques, *J. Immunol.* **155** (1995), 5855-5860.
87. Lehner, T., Wang, Y., Cranage, M., Bergmeier, A., Mitchell, E., Tao, L., Hall, G., Dennis, M., Cook, N., Brookes, R., Klavinskis, L., Jones, I., Doyle, C. and Ward, R.: Protective mucosal immunity elicited by targeted iliac lymph-node immunization with a subunit SIV envelope and core vaccine in macaques, *Nature Medicine* **2** (1996), 767-775.

CHAPTER 2

PRIMARY HIV INFECTION

ANDREW CARR, M.D[1]. and DAVID A. COOPER, D.SC., M.D.[1,2]

[1]From the HIV Medicine Unit and Centre for Immunology, St. Vincent's Hospital (A.C., D.A.C.), and [2]National Centre in HIV Epidemiology and Clinical Research, University of New South Wales (D.A.C.), Sydney, Australia.

1. Introduction

Primary human immunodeficiency virus (HIV) infection defines a brief period soon after inoculation with HIV that is characterised by intense viraemia, a subsequent immune response and, in the majority of patients, a brief, febrile illness [17]. Understanding the immune response during primary HIV infection has improved our insight in the two-way interaction between HIV and the immune system. Evolution in molecular diagnostics has transformed the speed and accuracy of diagnosing HIV infection, although the most appropriate diagnostic algorithm has not been defined. Zidovudine monotherapy initiated during primary HIV infection confers modest clinical benefit [55]. Preliminary reports now suggest that combination antiretroviral therapy of primary HIV infection is safe, tolerated and far more effective in suppressing viral replication [47,65,73] and results in greater preservation of immune function [83]. The clinical impact of early combination therapy, however, is unknown. Whether such therapy can eradicate HIV or is more effective than delayed therapy is currently under investigation. Because there are minimal data on primary HIV-2 infection, only primary HIV-1 infection is considered in this review; nevertheless, the same biological, diagnostic and therapeutic principles would apply.

2. Clinical Findings

2.1. GENERAL FEATURES

An acute clinical illness associated with primary HIV infection occurs in 53% to 93% of individuals [17,71,97,98,105]. In a prospectively studied group of homosexual men, primary HIV-1 infection was the second most common cause (after

H. Schuitemaker and F. Miedema (ed.), Aids Pathogenesis, 19–40.

influenza) of an acute febrile illness lasting more than three days [34]. Although asymptomatic primary HIV infection does occur, a high index of clinical suspicion and prior experience with primary HIV infection greatly increase the recognition rate. Indeed, in our experience many patients present because they believe that they may have primary HIV infection reflecting a high awareness in both primary care physicians and high risk groups.

The vast majority of individuals are infected sexually, by sharing infected needles or peripartum. Nevertheless, infection in a health care setting [45], including by patient to patient transmission [21], and from breast milk [116] can occur; there is also potential for infection across undamaged oral mucosa [5]. The time from exposure to HIV-1 until the onset of the acute clinical illness is typically 2 to 4 weeks although incubation periods of six days to six weeks are not rare [24,31, 98,104]. There are isolated reports of primary HIV illness and seroconversion up to 12 months after presumed inoculation [81]. The clinical illness is acute in onset and lasts from 1 to 4 weeks [12,17,27,40,46,54,71,89,97,98,105]. In the largest cohort analysis to date, Vanhems *et al.* reported a median illness duration of 20 days that was not influenced by age, gender nor risk factor [105].

2.2. CLINICAL MANIFESTATIONS

The main clinical features of primary HIV infection reflect the broad cellular tropism of HIV-1 during this period. In a review of 139 cases of primary HIV-1 infection [24], the most common physical signs and symptoms were fever (97%), adenopathy (77%), pharyngitis (73%), rash (70%), and myalgia or arthralgia (58%). In another series of 46 patients [87], the most common features were fever (94%), fatigue (90%), pharyngitis (72%), weight loss (70%), myalgias (60%) and headaches (55%). Nevertheless, most patients in these reports were not diagnosed at the time of their illness but at subsequent routine testing, stressing the need for awareness on the part of the diagnosing doctor particularly the primary care physician.

In the largest and unselected cohort of patients with primary HIV infection [105] the most common features were different: fever (77%), fatigue (66%), rash (56%), myalgia (55%) and headache (51%) (Table 1). These findings emphasise that many patients with primary HIV infection do not have the classic "mononucleosis-like" illness as first described in 1985. [27]. In fact, only 16% of patients with primary HIV infection had what might be described as classic features, namely fever, pharyngitis and cervical lymphadenopathy. Furthermore, 10% of patients had none of these three features. The most common features in febrile patients without pharyngitis or lymphadenopathy were headaches (52%), mouth ulcers (25%), abdominal pain (25%) and diarrhoea (17%). These data suggest that increasing awareness by experienced diagnosing physicians increases the likelihood of diagnosis in those with few or atypical symptoms.

Other important features regarding the clinical features of primary HIV infection [17] are:

- # most clinical manifestations are self-limited, although some symptoms, such as fatigue may persist for months;
- # the illness is generally of rapid onset;
- # fever may or may not be associated with night sweats;

- # lymphadenopathy is most common in the second week of the illness, especially of the axillary, occipital, and cervical nodes. Splenomegaly is also reported;
- # the classic erythematous, nonpruritic, maculopapular rash (Figure 1) is generally symmetrical, has lesions 5 to 10 mm in diameter and affects the face or trunk, but can affect the extremities or be generalised;
- # other skin lesions noted include a roseola-like rash, diffuse urticaria, a vesicular, pustular exanthem, desquamation of the palms and soles, alopecia and erythema multiforme (desquamation and alopecia typically occur in the second month);
- # ulceration of the oropharynx is common (and may less commonly involve the oesophagus, anus or penis) is generally oval and well demarcated;
- # although headache is very common in many acute viral illnesses, retro-orbital pain exacerbated by eye movements appears fairly specific for primary HIV infection;
- # photophobia may reflect an underlying aseptic meningitis;
- # clinical and biochemical hepatitis may occur but generally resolves within 3 months
- # oral or oesophageal candidiasis, often self-limiting, probably reflects an appreciable decline in $CD4^+$ lymphocyte count;
- # the presence of candidiasis or neurological involvement during primary HIV infection has been found to have adverse prognostic significance (Table 2);
- # an illness lasting longer than 14 days is also associated with a worse prognosis.

TABLE 1. Clinical manifestations of primary HIV-1infection in 218 patients (modified from Vanhems *et al.*; ref.105)

	frequency (%)	duration (mean days)
>50%		
fever >38°C	77	17
fatigue	66	24
erythematous maculopapular rash	56	15
myalgia	55	18
headache	51	26
25-50%		
pharyngitis	44	12
cervical lymphadenopathy	39	15
arthralgia	31	23
oral ulcer	29	13
odynophagia	28	16

continued

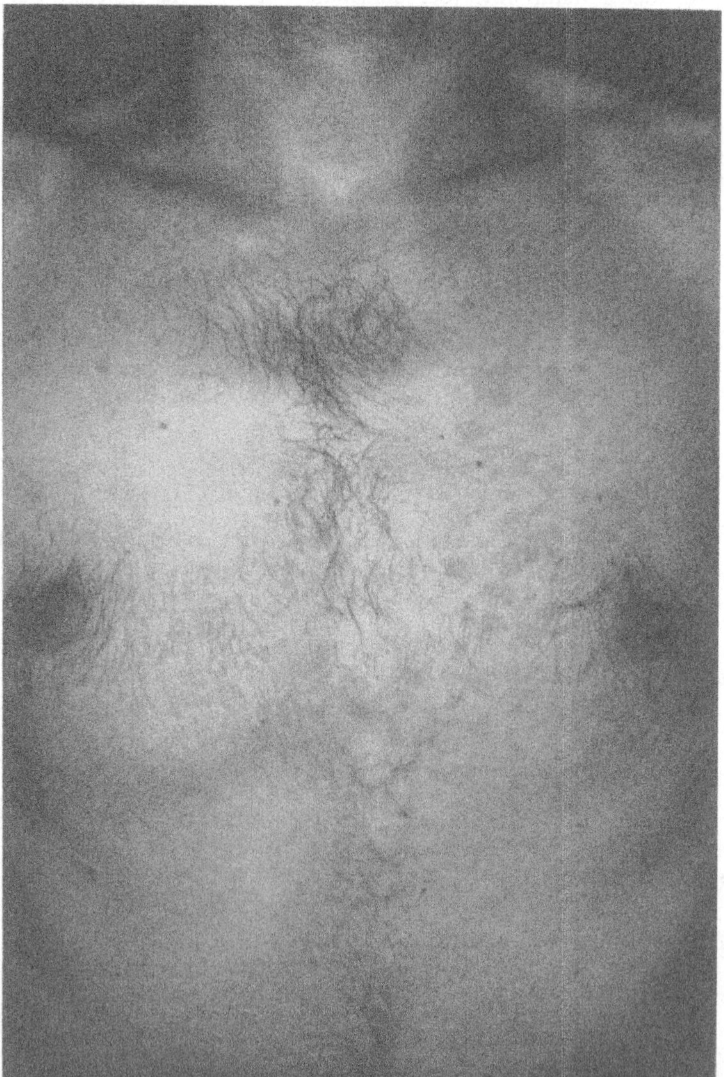

Figure 1. Rash of primary HIV infection

Table 1 (continued)

5-25%		
axillary lymphadenopathy	24	164
weight loss	24	29
nausea	24	18
diarrhoea	23	13
night sweats	22	15
cough	22	18
anorexia	21	15
inguinal lymphadenopathy	20	9
abdominal pain	19	15
oral candidiasis	17	10
vomiting	12	10
photophobia	12	11
sore eyes	12	13
genital ulcer	7	14
tonsillitis	7	13
depression	6	23
dizziness	6	11

TABLE 2. Useful prognostic variables at or soon after primary HIV infection

Clinical (during PHI)	Reference
presence of symptoms	60
duration of symptoms (>14 days)	60, 71
number of symptoms (>3 symptoms)	105
acquisition of HIV from an index with advanced HIV disease	53, 110
candidiasis	70, 105
neurological involvement	11
immunodeficiency at time of infection	71, 106
Laboratory (at or 3 months after* onset of PHI)	
$CD4^+$ lymphocyte count*	87
HIV RNA*	66
beta-2 microglobulin*	75
infection with SI viral strain	67, 82
persistent p24 antigenaemia	53
higher HIV RNA viraemis post seroconversion	66
Reported but not usually evaluated	
low p24 antibody titres	23
high gp120 antibody titres	23
high CD38 expression on $CD8^+$ lymphocytes	52
Infection with zidovudine-resistant HIV does affect prognosis	48

*Only 16% of patients had the "classic" triad of fever, pharyngitis and cervical lymphadenopathy and 10% of patients had none of these features.

2.3. PREDISPOSING FACTORS

Several factors have been associated with an increased relative risk for symptomatic primary HIV infection, including a low pre-existing $CD4^+$ cell count or $CD4^+$:$CD8^+$ ratio [61] anergy to the neoantigen dinitrochlorobenzene (DNCB) [63] (but not to recall antigens), and infection from an index case with advanced HIV-1 disease [110]. Acute infection with other viruses such as cytomegalovirus are often symptomatic in adults and so acute co-infection should increase the likelihood of symptoms [8] as well as accelerate the expression of HIV [76].

2.4. DIAGNOSIS

The most important step in the diagnosis of primary HIV infection is to suspect the possibility. In turn, the most important clinical data are the knowledge that an individual has had an HIV exposure, usually in the preceding 2 to 8 weeks, and the presence of one or more suggestive symptoms or signs. Once the possibility is entertained, the diagnosis can be confirmed in the laboratory usually within days of the initial presentation.

There are several advantages of early diagnosis of HIV infection. Firstly, it allows the patient the opportunity to access therapies at a time when the immune system has suffered the least amount of damage. Secondly, some clinical and laboratory variables at or soon after primary HIV infection have prognostic value and help define patients who might merit early or more aggressive antiretroviral therapy. Lastly, a positive diagnosis (combined with counselling and perhaps antiviral therapy) should help minimise further transmission of HIV.

Although originally described as "mononucleosis-like" [27] and described as such in the Centers for Disease and Prevention classification of HIV disease [19], symptomatic primary HIV infection is a distinct and recognisable clinical syndrome. The major symptoms and signs that strengthen the diagnosis include mucocutaneous ulceration, maculopapular rash, candidiasis, lymphadenopathy, and non-suppurative pharyngitis.

TABLE 3. Major differential diagnoses

Primary HIV-1 infection
Epstein-Barr virus mononucleosis
Cytomegalovirus mononucleosis
Toxoplasmosis
Rubella
Vital hepatitis
Secondary syphilis
Disseminated genococcal infection
Primary herpes simplex virus infection
Other viral infection
Drug reaction

The differential diagnoses of primary HIV infection are listed in Table 3. Skin eruptions are rare in patients with EBV or CMV infections (unless antibiotics such as amoxicillin have been given) and toxoplasmosis, do not affect the palms and soles in patients with rubella, and are typically scaly pruritic in pityriasis rosea. Mucocutaneous ulceration is a fairly distinctive finding because it is unusual in most of the other differential diagnoses. Clinical evidence of immune compromise such as candidiasis or Pneumocystis carinii pneumonia is highly suggestive of HIV infection rather than of other viral infections.

The major differences between primary HIV infection and EBV mononucleosis have been detailed by Gaines *et al.* [40] and are summarised in Table 4. Although serological testing for HIV and EBV usually provides the definitive diagnosis, clinicians should be aware that false-positive tests for heterophil antibodies may occur during primary HIV infection. Of course, patients can be acutely infected with HIV and another virus such as CMV or EBV [8,18].

TABLE 4. Clinical differences between primary HIV-1 infection and Epstein-Barr Virus (EBV) mononucleosis (modified from Gaines, H. *et al.*; ref. 40)

Primary HIV-1 Infection	EBV Mononucleosis
acute onset	insidious onset
little or no tonsillar hypertrophy	marked tonsillar hypertrophy
enanthema on hard palate	enanthema on border of hard and soft palates
exudative pharyngitis uncommon	exudative pharyngitis common
mucocutaneous ulcers common	no mucocutaneous ulcers
rash common	rash rare (in absence of ampicillin)
jaundice rare	jaundice (8%)
diarrhoea possible	no diarrhoea
opportunistic infections occasionally	no opportunistic infections

Because HIV remains predominantly a sexually transmitted disease (STD) and because the incubation period of primary HIV infection is comparable to that of most common STDs, primary care physicians managing other STDs should be particularly alert to the clinical manifestations of primary HIV infection and include this in their differential diagnosis. Patients suspected of sexually acquired HIV infection should also be screened for other STDs; for example, rash, mouth ulcers and lymphadenopathy can also be seen with syphilis and pharyngitis with orally acquired gonorrhoea).

In patients with meningo-encephalitis, other causes of aseptic meningo-encephalitis should be excluded rapidly.

3. Laboratory Features

A number of laboratory variables should be measured during primary HIV infection to assist in diagnosis. Some of these variables should be measured in the following months to assist in determining prognosis and so identify those individuals most likely to gain from intervention with antiretroviral therapy.

3.1. SEROLOGY

3.1.1. *Antibody Testing*

Primary HIV infection must be confirmed serologically by the detection of serum antibodies directed against specific internal and surface proteins of HIV [41]. Such antibodies are usually detectable within the first few weeks of onset of the acute illness by enzyme-linked immunosorbent assay (ELISA) [28,39-41]. IgM antibodies to HIV gag or env proteins appear within 2 weeks of infection, precede the IgG response, reach peak titers at 2 to 5 weeks, and then decline to undetectable levels within approximately 3 months. IgG antibody is then detected, usually 2 to 6 weeks after the onset of illness. Separate detection for HIV-specific IgM is not required as most antibody ELISAs detect both HIV-specific IgG and IgM [15,33,39,58]. Differences in the length of the window period according to the type of the ELISA screening tests mandates the consistent use of sensitive screening tests.

The 3 to 6 month "window period" of seroconversion commonly cited is by necessity conservative, particularly if other tests are used in conjunction. Rare reports of up to 12 months HIV infection without seroconversion have important implications for public health and our understanding of HIV immunopathogenesis [81].

HIV ELISAs are designed for maximum sensitivity. Nevertheless, negative results in early primary HIV infection are not uncommon. Clinical information on a request form that suggests possible primary HIV infection (e.g. "febrile illness" or "flu-like illness") should result in automatic testing as per Table 5, maximising diagnostic sensitivity and allowing for rapid turnaround time. IgG detection in saliva samples may have comparable sensitivity for HIV seroconversion as compared with licensed serum ELISAs [14]. Antibodies to regulatory proteins (including those produced from the rev, tat, nef, vpu and vpr genes) develop early in infection [3, 22,32,77-79], but are not measured routinely.

3.1.2. *Immunoblotting*

Figure 2 shows the typical evolution of antibody development demonstrated by serial immunoblotting in a patient with primary HIV infection. Immunoblotting usually first shows antibody to p24 or gp41 with virtually all sera obtained 2 weeks or more after onset of the acute illness being positive [40,41]. Patients can rarely have positive immunoblots prior to ELISA reactivity, perhaps because most immunoblots involve an overnight incubation of patient serum rather than a few hours in most ELISAs.

TABLE 5. Diagnosis of Primary HIV Infection

Setting(s)

At least 1 clinical feature of primary HIV infection
and/or
exposure to HIV within preceding 12 weeks
and/or
indeterminate HIV ELISA antibody

Algorithm

HIV-specific antibodies (at least one ELISA)[1]
and
HIV western blot[1]
and
HIV-1 p24 antigen[2]
and
T-cell subset enumeration[3]
and
HIV DNA PCR (especially if serology indeterminate)
and
HIV RNA quantitation (especially if antiviral therapy contemplated)
and
exclude other causes of viral illness (*e.g.* CMV, EBV)

For inconclusive samples, retest every 2 to 4 weeks until diagnosis confirmed or excluded (most sensitive and cost effective interval unknown)

Assays of uncertain utility

genotypic/phenotypic assays for antiretroviral drug resistance
virus isolation
SI/NSI phenotyping

[1]An overnight incubation method should be used to maximise assay sensitivity; either the ELISA(s) or western blot should be able to detect HIV-2.
[2]If antibody assays are inconclusive, sample should be neutralised to ensure p24 assay specificity.
[3]A small number (percentage unknown) p24-negative/antibody-nègative patients may have $CD4^+$ lymphocyte counts <500 or inverted $CD4^+$:$CD8^+$ ratios (<1.0) [112].

Immunoblotting is essential to confirm the diagnosis of HIV infection as biological fals-positive or inconclusive ELISA results are not rare. Immunoblots, however, can also yield indeterminate results (*e.g.* the presence of one or more non-glycoprotein bands) especially during primary HIV infection or advanced HIV disease, or in those with biological fals-positive ELISA results but the pattern of reactivity and the clinical history can usually distinguish between these possibilities. One possible cause for an indeterminate result may be abortive HIV infection, suggested from molecular and virological analyses of recipients of HIV-infected blood products [42].

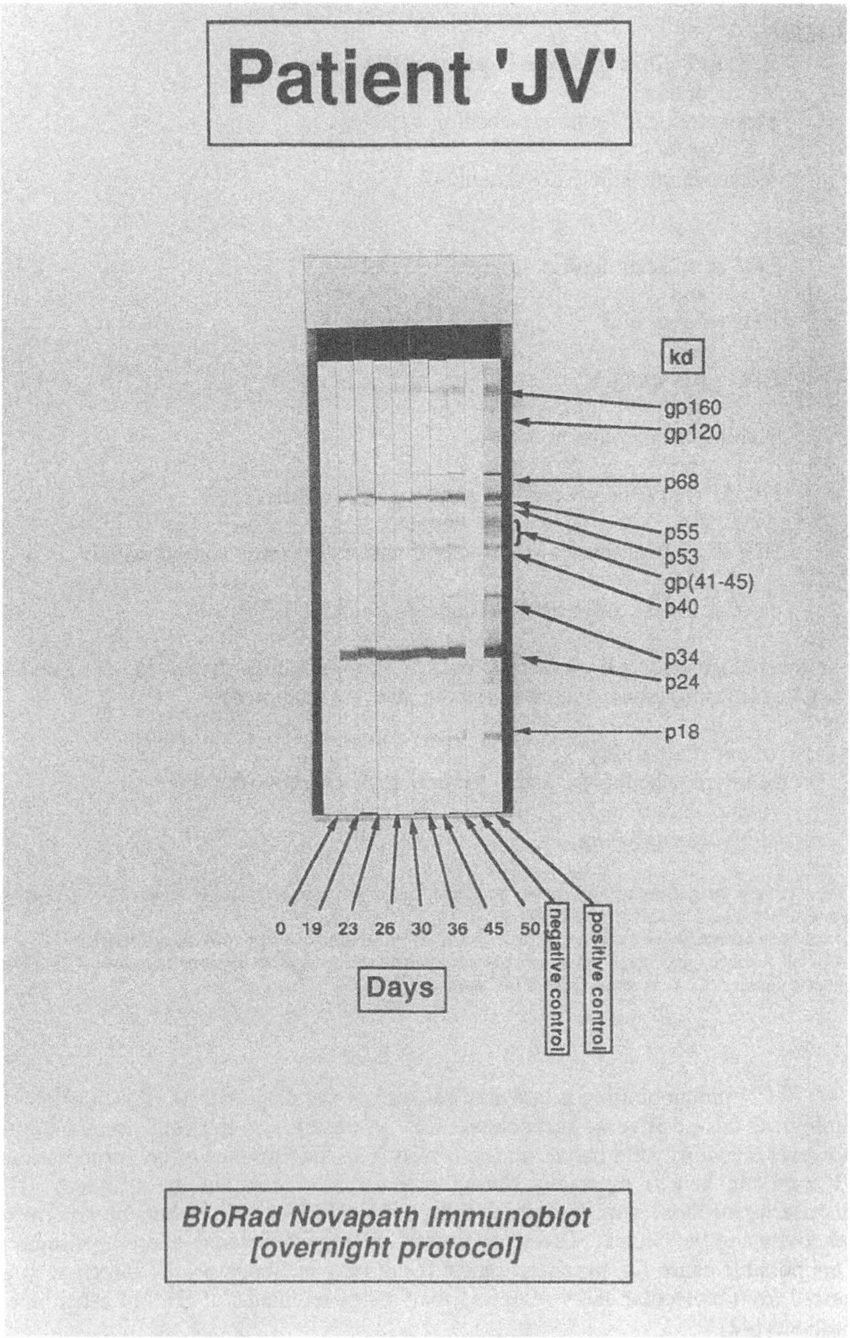

Figure 2. Serial immunoblots in a patient with primary HIV infection

For patients with a history suggestive of primary HIV infection, an equivocal ELISA and immunoblot should also suggest the possibility of primary HIV-2 infection, although some commercially available HIV immunoblot strips can detect both HIV-2 and HIV-1 infection.

3.2. VIROLOGY

3.2.1. *HIV-1 p24 Antigen*

HIV-1 p24 antigen can be detected in serum and cerebrospinal fluid in the period before gag and env seroconversion, in most patients during the first few weeks of symptoms, as early as 24 hours after the onset of acute illness and a mean 7 days prior to HIV antibodies [1,15,16,27,38,43,50,54,64,93,107]. Serum p24 antigen levels typically decreases as immune complexes develop [27,89] and the levels of serum HIV antibodies increase [27, 28,31,50,54]. Persistent antigenaemia or the reappearance of antigenaemia at a later time is associated with an increased risk for the development of severe HIV disease [53,72].

Assay for serum HIV-1 p24 antigen is essential if the differential diagnosis includes primary HIV infection. The assay is easy, cheap, widely available, sensitive and specific and can be performed in a few hours. False-positive p24 antigen tests rarely occur, however, and the specificity should be confirmed by neutralisation.

To minimise the risk of transfusion-associated HIV infection (estimated at 1 in 493000 in the United States), blood banks in the United States now screen for HIV antibodies and p24 antigen in the one assay [88]. In developing countries such as Thailand, where the risk of transfusion-associated HIV infection has been estimated to be much greater (1 in 2644) [56], such combined assays may be of greater value. One alternative practice in some countries is to not pay blood donors, eliminating any financial incentive to high risk individuals to donate blood.

3.2.2. *HIV RNA and Proviral DNA Detection*

During primary HIV infection, there are very high HIV RNA levels (about 10^6 copies/ml plasma) similar to those seen in patients with advanced HIV disease. These levels decrease spontaneously and precipitously concurrently with resolution of symptoms and the appearance of anti-HIV-1 antibodies. Nadir ("set point") levels are not reached for a mean 3 to 4 months, however, and levels measured prior to the nadir have been shown not to have prognostic value [66,87]. The viral clearance during primary HIV-1 infection is likely to be due to the emergence of an effective host immune response, but may also reflect reduced numbers of infectable target cells [74].

Assays to detect HIV-specific RNA or DNA have revolutionised HIV diagnosis and therapy. With respect of diagnosis, HIV DNA or RNA may be detected 2 to 3 days prior to p24 antigen [15]. DNA PCR detection is cheaper than HIV RNA detection and rapidly assists resolving sera with indeterminate immunoblot reactivity and diagnosis of perinatal transmission. Therefore, although the sensitivity and specificity of such an approach is currently unclear, we routinely perform HIV DNA detection in those with indeterminate or negative serology as part of our diagnostic algorithm of primary HIV infection. This has the potential to shorten the period of diagnostic uncertainty from several weeks to a few days.

HIV RNA should probably be measured only in those who are contemplating antiretroviral therapy at this time with the aim of determining the most appropriate combination of antivirals to use; the higher the viral load, the more potent therapy would be appropriate.

3.2.3. *Strain Selection during Transmission*

A single strain of HIV is transmitted in general [113], although co-infection has been described in individual cases of sexual and transfusional transmission [15,115]. Transmitted strains tend to have an non-syncytium-inducing (NSI) phenotype [48, 87] but may not be the most common strain in the blood or semen of the source subject [113,114], suggesting that one or more structural, viral or immunologic feature may influence strain transmission.

3.2.4. *Viral Phenotype*

The ability of HIV to be syncytium-inducing (SI) in MT-2 cell culture correlates with a worse prognosis [80]. However, the vast majority of HIV isolates at seroconversion are NSI and determining whether a virus is SI or NSI is slow and unlikely to alter therapy [48].

Transmission of zidovudine and nevirapine-resistant HIV has been documented [36,48,49]. A study of 61 individuals in Sydney found that 10% of seroconverters infected since 1990 had phenotypic and genotypic evidence of infection with high level zidovudine-resistant HIV [48]. A rapid, simple assay can now determine the presence of plasma HIV resistant to most reverse transcriptase inhibitors [94]. If antiretroviral therapy is to be used in those with primary HIV infection, it would be advantageous to determine the drug susceptibility prior to commencing therapy. Similar assays assessing protease inhibitor resistance are in development.

3.3. IMMUNOLOGY

3.3.1. *T-cell Subsets*

Primary HIV infection is characterised by rapid changes in peripheral T-lymphocyte subsets [29,69,82,112]. During the first 1 to 2 weeks there characteristically is lymphopenia affecting both $CD4^+$ and $CD8^+$ subsets. The nadir of total, $CD4^+$ and $CD8^+$ lymphocytes occurs a median of 9 days following onset of illness and may approach those seen with advanced HIV disease [17]. There does not appear to be selective depletion of specific V-beta subfamilies of either $CD4^+$ or $CD8^+$ lymphocytes [30]. This transient lymphopenia is followed at 3 to 4 weeks by lymphocytosis; in less than 50% of patients atypical lymphocytes are seen in the peripheral blood film [40,59,92]. The decrease in $CD4^+$ lymphocyte numbers mostly represents a decline in those expressing CD45RA (*i.e.* "naive" T lymphocytes) [112]. Although both $CD4^+$ and $CD8^+$ cells contribute to the lymphocytosis, the increase in $CD8^+$ cells is relatively greater, leading to a sustained inversion of the CD4+:$CD8^+$ ratio. Many $CD8^+$ lymphocytes also express the activation markers HLA-DR, CD38 and CD11a/CD18 markers that correlate with a cytotoxic phenotype.

The inverted CD4:CD8 ratio may have some diagnostic value for primary HIV infection, as an inverted ratio is not seen in most other acute viral illnesses (except acute EBV infection) and may develop prior to detectable HIV-specific antibodies or p24 antigenaemia [112]. Nevertheless, the changes in $CD8^+$ lymphocyte counts are similar to those seen in primary CMV and EBV infections [18].

3.3.2. *Beta-2 Microglobulin*

Elevated serum beta-2 microglobulin levels a mean of 4 months after primary HIV infection correlate with a worse prognosis [75]. Nevertheless, the values during primary infection, like $CD4^+$ counts and RNA levels at this time, have no known prognostic value.

3.4. HISTOPATHOLOGY

Lymph-node biopsies show a reduction of extrafollicular B cells, $CD8^+$ cell follicular infiltration, and minimal activation and proliferation of germinal center cells. The env proteins gp120 and gp160 have been found in interfollicular and follicular lymphocytes, endothelial cells, and interdigitating and dendritic reticulum cells [89]. The relative normality of the structure of the germinal centers during primary HIV-1 infection contrasts with the follicular hyperplasia associated with established HIV-1 infection.

4. Immune Response to HIV

4.1. CD8 CELL RESPONSES

The increase in the number of $CD8^+$ cells during primary HIV-1 infection occurs concomitant with resolution of clinical symptoms and a decrease in the detectable levels of serum p24 antigen [29], suggesting that the $CD8^+$ cell response to primary HIV-1 infection has a role in controlling viral replication *in vivo* as it has been shown to have *in vitro* [109].

Activated $CD8^+$ cells were found in all patients with primary HIV-1 infection [112] but were only transiently elevated in those patients infected with SI variants of HIV [82]. SI variants may be selectively cleared by the host immune response; the persistent isolation of SI variants may reflect failure of the $CD8^+$ response.

HIV-specific, $CD8^+$ lymphocyte-mediated, HLA-restricted cytotoxic activity appeared early in HIV seroconversion at a time when HIV-specific neutralizing antibody activity could not be detected [4,9,57]. These cells appear to present with relatively high precursor frequency (0.1 to 1%). Analysis of V-beta T-cell receptor gene usage in these $CD8^+$ lymphocytes has demonstrated different responses between individuals. The V-beta response may be monoclonal, oligoclonal or polyclonal, with patients generating a polyclonal response appearing to have the best $CD4^+$ count outcome over the medium term [35, 44].

$CD8^+$ lymphocytes may suppress HIV replication by secretion of one or more soluble non-cytolytic chemokines, such as RANTES, MIP-1α, MIP-1β, and IL-16 which inhibit HIV replication [6,25]. This soluble suppressor activity has been observed in a small number of HIV seroconverters [111]. However, this activity was also observed in some individuals months prior to seroconversion, suggesting that this activity is not sufficient to prevent infection [4].

4.2. PROLIFERATIVE RESPONSES

Severe lymphocytic proliferative hyporesponsiveness to both mitogens and antigens occurs during primary HIV-1 infection [31,69,95] and persists after resolution of the acute illness in the absence of therapy. Pokeweed-mitogen (PWM) responses were depressed in most patients at some point in the acute illness, and reached the lowest levels at 6 to 8 weeks after onset of illness [69]. Responses subsequently improved but remained sub-normal at one year of follow-up. Phytohaemagglutinin (PHA) responses showed a similar pattern, but were less affected. Rapid and persistent impairment of B-cell function also follows primary HIV-1 infection [95,96].

4.3. NEUTRALISING ANTIBODIES

Neutralising antibodies may contribute to viral clearance [2,4,10,103], although a direct correlation between decline in viral load and development of neutralising antibody has not been demonstrated. Limited data suggest that serum-neutralising antibodies are not present when viral load falls [57,103]. Antibodies that inhibit syncytium formation and antibodies that mediate antibody-dependent cellular cytotoxicity (ADCC) against virally-infected cells also develop soon after infection [7]. HIV-1 immune complexes appear in the blood during the period of declining concentrations of p24 antigen and increasing concentration of IgM and IgG antibodies [107] and may be detected before overt seroconversion.

4.4. CYTOKINES

Increased levels of interferon-alpha (IFN-α), tumor necrosis factor-alpha (TNF-α), neopterin, and beta-2 microglobulin have been detected in blood and CSF during primary HIV-1 infection [41,90,108], reflecting activation of the cellular immune system. High circulating levels of these cytokines are also presumably responsible for the pathogenesis of some of the major clinical manifestations of primary HIV-1 infection (*e.g.*, the symptoms of primary HIV-1 infection, such as fevers, chills, myalgia, headache, fatigue, leukopenia, and weight loss are very similar to those found in people receiving exogenous IFN-α).

p24 antigen levels rapidly declined following the IFN-α peak response that in turn occurred before the development of HIV-specific antibodies and before the $CD8^+$ cell lymphocytosis, suggesting that it is a first line of defence against HIV-1 infection, a finding which has some support in a murine model [86]. The order of appearance of these markers in relation to HIV-1 seroconversion in humans remains unclear.

5. Prognosis

After resolution of the acute clinical illness, most patients enter a stage of asymptomatic infection that lasts from many months to years. The development of severe HIV-1-related disease within the first 2 years after infection is unusual, although such cases have been reported [50]. Several studies have highlighted clinical, virological and immunological parameters at or soon after primary HIV infection that have prognostic significance (Table 2). In contrast, infection with zidovudine-resistant HIV-1 (*i.e.* HIV with an IC_{90} of greater than 1 μM) did not affect the clinical or $CD4^+$ lymphocyte outcome at 1 year after documented seroconversion [48].

6. Management

6.1. ANTIRETROVIRAL THERAPY

Antiretroviral therapy of primary HIV infection has progressed greatly in recent years. Although not yet proven, it is possible that antiretroviral intervention during primary HIV-1 infection might prevent persistent HIV-1 infection or lessen the initial viral load and subsequently improve long-term prognosis. Several observations suggest that HIV could be eradicated at this time: the reduction in perinatal transmission of HIV by 70% by the use of zidovudine monotherapy [26], the reduction in occupational transmission of HIV after needle-stick injury by 80% by zidovudine monotherapy [20], the occasional spontaneous clearance of HIV following perinatal transmission [13] and in some animal retroviral models, zidovudine has prevented persistent infection or significantly altered disease progression if it was given soon after inoculation [62,85,86]. By the time symptomatic primary HIV-1 infection develops, however, widespread viral dissemination has already occurred and so it is less likely that treatment at this stage would prevent persistent HIV infection. Nevertheless, studies evaluating the possibility of eradicating HIV at this time are in progress.

The first report of the effect of antiretroviral therapy during symptomatic primary HIV infection involved 11 patients who were treated with zidovudine 1 g daily for a median period of 8 weeks [101]. Compared with a group of historical controls, no clear evidence of any clinical benefit in terms of resolution of the clinical illness and no indication that the intervention would prevent persistent infection emerged. Subjects treated with zidovudine also had a significantly lower $CD8^+$ response in the second month than did a series of untreated, historical controls [99]. It is not clear whether this was due to decreased viral replication and therefore decreased stimulus for $CD8^+$ cell proliferation, or whether zidovudine dampened the normal, and presumably protective, $CD8^+$ cell response.

A subsequent double-blind, placebo-controlled trial of zidovudine for 6 months found that zidovudine was safe although it did not shorten the duration of seroconversion illness [55]. After 6 months therapy, patients receiving zidovudine had higher $CD4^+$ lymphocyte counts (by about 140 cells/mm3), lower HIV RNA load (by about 0.5 log), and a significant reduction in the number of HIV-related minor opportunistic infections during 18 months of follow-up.

More recent reports have shown that combination therapy yields more potent antiviral effects, similar to antiviral effects seen in those with established HIV infection [47,65,73,84]. Patients who received and tolerated zidovudine, lamivudine and one of the protease inhibitors ritonavir, indinavir or nelfinavir all had plasma RNA levels fall to undetectable levels within 6 months. Enthousiasm for such therapy has been dampened by anecdotal reports of viral replication becoming detectable within weeks in those who interrupted therapy.

The immunological consequences of such therapy are not clear. It is noteworthy, however, that the combination therapy appeared to result in less intense bands on immunoblots, suggesting substantial reduction in HIV antigen presentation. Combination therapy during acute HIV infection also appeared to result in greater preservation of HIV-specific CD4-proliferative responses and CD8-mediated cytotoxic activity than in treatment with equivalent combinations during established infection [83]. The clinical significance of these findings is unknown, but sug-gests that patients treated during primary HIV infection may have a greater chance of withdrawing therapy with the hope of an intact immune response suppressing any further viral replication.

Other unresolved treatment issues include the best time to initiate therapy, whether therapy can be safely reduced to maintenance doses, what combination is most potent in those with drug-resistant HIV at primary HIV infection and what potent combination is best tolerated (Table 6).

It will presumably not be long before transmission of protease inhibitor-resistant HIV is documented. The implications for therapy are obvious and need evaluation.

TABLE 6. Unresolved issues in primary HIV infection

Diagnosis
- most sensitive, rapid and cost-effective diagnostic algorithm

Treatment
- value of treatment
- clinical impact
- when to start
    - pre/post full blot
    - pre/post set point
- when to withdraw or reduce therapy
- possibility of eradication
- best treatment
    - most potent
    - fewest drugs
    - fewest side effects
- anti-retroviral drug resistance assays
- withdrawal of therapy (complete/maintenance)

6.2. COUNSELLING

A diagnosis of HIV infection is likely to be associated with profound psychosocial consequences. Specific factors that must be considered in counselling persons with primary HIV-1 infection include the acute physical distress that many persons experience, the tentative nature of the diagnosis before seroconversion occurs, self-reproach resulting from recent risk behaviour, and the potential development of mood disorders during primary HIV-1 infection.

The identification of a patient with primary HIV-1 infection necessarily implies the existence of a source case of infection [51]. When possible, contact tracing should be implemented to identify this source case. If he or she is identified, counselling for the source person is also indicated. Such persons may be unaware of their infection, may be in the process of seroconverting themselves, or may be unaware of what constitutes safe sex or safe-injecting practices. Counselling of such patients also must address their emotional reactions to having infected another individual.

7. References

1. Alain, J.-P., Laurian, Y., Paul, D.A., et al: Serological markers in early stages of human immunodeficiency virus infection in haemophiliacs, *Lancet* **ii** (1986), 1233-1236
2. Albert, J., Abrahamsson, B., Nagy, K., *et al*: Rapid development of isolate-specific neutralizing antibodies after primary HIV-1 infection and consequent emergence of virus variants which resist neutralization by autologous sera, *AIDS* **4** (1990), 107-112.
3. Ameisen, J.-C., Guy, B., Chamaret, S., *et al*: Antibodies to the nef protein and to nef peptides in HIV-1-infected seronegative individuals, *AIDS Res. Human Retrovir.* **5** (1989), 279-291.
4. Ariyoshi, K., Harwood, E., Chiengsong-Popov, R. and Weber, J.: Is clearance of HIV-1 at seroconversion mediated by neutralising antibodies?, *Lancet* **340** (1992), 1257-1258.
5. Baba, T.W., Trichel, A.M., An, L., *et al*: Infection and AIDS in adult macaques after nontraumatic oral exposure to cell-free SIV, *Science* **272** (1996), 1486-1489.
6. Baier, M., Werner, A., Bannert, N., Metzner, K. and Kurth, R.: HIV suppression by interleukin-16, *Nature* **378** (1995) 563.
7. Bolognesi, D.P.: Prospects for prevention of and early intervention against HIV, *JAMA* **261** (1989), 3007-3013.
8. Bonnetti, A., Weber, R., Vogt, M.W., *et al*: Co-infection with human immunodeficiency virus-type 1 (HIV-1) and cytomegalovirus in two intravenous drug users, *Ann. Intern. Med.* **111** (1989), 293-296.
9. Borrow, P., Lewicki, H., Hahn, B.H., Shaw, G.M. and Oldstone, M.B.A.: Virus-specific $CD8^+$ cytotoxic T-lymphocyte activity associated with control of viraemia in primary human immunodeficiency virus type 1 infection, *J. Virol.* **68** (1994), 6103-6110.
10. Boucher, C.A.B., de Wolf, F., Houweling, J.T.M., *et al*: Antibody response to a synthetic peptide covering a LAV-I/HTLV-IIIB neutralization epitope and disease progression, *AIDS* **3** (1989), 71-76.
11. Boufassa, F., Bachmeyer, N., Deveau, C.C., *et al*: Influence of neurologic manifectations of primary human immunodeficiency virus infection on disease progression, *J. Infect. Dis.* **171** (1995), 1190-1195.
12. Brehmer-Andersson, E. and Torssander, J.: The exanthema of acute (primary) HIV infection: identification of a characteristic histopathological picture?, *Acta Derm. Venereol. (Stockh)* **70** (1990), 85-87.
13. Bryson, Y.J., Pang, S., Wei, L.S., Dickover, R., Diagne, A. and Chen, I.S.Y: Clearance of HIV infection in a perinatally infected infant, *New Engl. J. Med.* **332** (1995), 833-838.
14. Burgess-Cassler, A., Barriga-Angulo, G., Wade, S.E., Castillo-Torres, P. and Schramm, W. A.: Field test for the detection of antibodies to human immunodeficiency virus types 1 and 2 in saliva or plasma, *Clin. Diag. Lab. Immunol.* **3** (1996), 480-482.
15. Busch, M.P., Satten, G.A., Sakahara, N.S., Garrett, P.E. and Herman, S.A.: Time course and dynamics of viremia during early stages of HIV seroconversion. In: *Abstracts of the 4th Conference on Retroviruses and Opportunstic Infections*, Washington DC, January, 1997 (abstract 749).
16. Calabresse, L.H., Proffitt, M.R., Levin, K.H., *et al*: Acute infection with the human immunodeficiency virus (HIV) associated with acute brachial neuritis and exanthematous rash, *Ann. Intern. Med.* **107** (1987), 849-851.

17. Carr, A. and Cooper, D.A.: Primary HIV infection, in M. Sande and P. Volberding (eds.), *The Medical Management of AIDS* 1997, W.B. Saunders, pp. 89-106.
18. Carney, W.P., Rubin, R.H., Hoffman, R.A., *et al*: Analysis of T lymphocyte subsets in cytomegalovirus mononucleosis, *J. Immunol.* **126** (1981), 2114-2116.
19. Centers for Disease Control: CDC classification system for human T-lymphotropic virus type III/lymphadenopathy-associated virus infections, *MMWR* **35** (1986), 334-339.
20. Centers for Disease Control: Case-control study of HIV seroconversion in health-care workers after percutaneous exposure to HIV-infected blood-France, United Kingdom, and United States, January 1988-August 1994, *MMWR* **44** (1995), 929-933.
21. Chant, K., Lowe, D., Rubin, G., *et al*: Patient-to-patient transmission of HIV in private surgical consulting rooms, *Lancet* **1544** (1993), 8886-8887.
22. Cheingsong-Popov, R., Panagiotidi, C., Ali, M., *et al*: Antibodies to HIV-1 nef(p27): Prevalence, significance and relationship to seroconversion, *AIDS Res. Human Retrovir.* **6** (1990), 1099-1105.
23. Cheingsong-Popov, R., Panagiotidi, C., Bowcock, S., *et al*: Relation between humoral responses to HIV gag and env proteins at seroconversion and clinical outcome of HIV infection, *Brit. Med. J.* **302** (1991), 23-26.
24. Clark, S.J., Saag, M.S., Decker, W.D., *et al*: High titers of cytopathic virus in plasma of patients with symptomatic primary HIV-I infection, *New Engl. J. Med.* **324** (1991), 954-60.
25. Cocchi, F., de Vico, A.L., Garzino-Demo, A., Arya, S.K., Gallo, R.C. and Lusso, P.: Identification of RANTES, MIP-1α and MIP-1β as the major HIV-suppressive factors produced by $CD8^+$ T cells, *Science* **270** (1995), 1811-1816.
26. Connor, E.M., Sperling, R.S., Gelber, R., *et al:* Reduction of maternal-infant transmission of human immunodeficiency virus type 1 by zidovudine treatment, *New Engl. J. Med.* **331** (1994), 173-180.
27. Cooper, D.A., Gold, J., Maclean, P., *et al*: Acute AIDS retrovirus infection. Definition of a clinical illness associated with seroconversion, *Lancet* **i** (1985), 537-540.
28. Cooper, D.A., Imrie, A.A. and Penny, R.: Antibody response to human immunodeficiency virus following primary infection, *J. Infect. Dis.* **155** (1987), 1113-1118.
29. Cooper, D.A., Tindall, B., Wilson, E.J., *et al*: Characterization of T lymphocyte responses during primary HIV infection, *J. Infect. Dis.* **157** (1988), 889-896.
30. Cossarizza, A., Ortolani, C., Mussini, C., *et al*: Lack of selective Vß deletion in $CD4^+$ or $CD8^+$ lymphocytes and functional integrity of T-cell repertoire during acute HIV syndrome, *AIDS* **11** (1995), 547-553.
31. Daar, E.S., Moudgil, T., Meyer, R.D. and Ho, D.D.: Transient high levels of viremia in patients with primary human immunodeficiency virus type I infection, *New Engl. J. Med.* **324** (1991), 961-964.
32. de Ronde, A., Reiss, P., Dekker, J., *et al*: Seroconversion to HIV-1-negative regulation factor, *Lancet* **ii** (1988), 574.
33. de Waele, M., Thielmans, C., Van Camp, B.K.G.: Characterization of immunoregulatory T cells in EBV-induced infectious mononucleosis by monoclonal antibodies, *New Engl. J. Med.* **304** (1981), 460-462.
34. de Wolf, F., Lange, J.M.A., Bakker, M., *et al*: Influenza-like syndrome in homosexual men: A prospective diagnostic study, J. R. *Coll. Gen. Pract.* **38** (1988), 443-446.
35. Demarest, J.F., Pantaleo, G., Schacker, T., *et al*: The qualitative nature of the primary immune response to primary HIV infection is a prognosticator of disease progression independent of the initial level of pl.asma viremia, *Proc. Natl. Acad. Sci. USA* **94** (1997), 254-258.
36. Erice, A., Mayers, D.L., Strike, D.G., *et al*: Primary infection with zidovudine-resistant human immunodeficiency virus type 1, *New Engl. J. Med.* **328** (1993), 1163-1165.
37. Gaines, H., von Sydow, M., Sonnetborg, A., *et al*: Antibody response in primary human immunodeficiency virus infection, *Lancet* **i** (1987), 1249-1253.
38. Gaines, H., Albert, J., von Sydow, M., *et al*: HIV antigenaemia and virus isolation from plasma during primary HIV infection, *Lancet* i (1987), 1317-1318.
39. Gaines, H., von Sydow, M., Parry, J.V., *et al*: Detection of immunoglobulin M antibody in primary human immunodeficiency virus infection, *AIDS* **2** (1988), 11-15.
40. Gaines, H., von Sydow, M., Pehrson, P.O. and Lundbergh, P.: Clinical picture of primary HIV infection presenting as a glandular-fever-like illness, *Brit. Med. J.* **297** (1988), 1363-1368.
41. Gaines, H., von Sydow, M.A.E., von Stedingk, L.V., *et al*: Immunological changes in primary HIV infection, *AIDS* **4** (1990), 995-999.
42. Georgoulias, V.A., Malliaraki, N.A., Theodoropoulou, M., *et al*: Indeterminate human immunodeficiency virus type 1 western blot may indicate abortive infection in some low risk blood donors, *Transfusion* **37** (1997), 65-72.

43. Goudsmit, J., de Wolf, F., Paul, D.A., *et al*: Expression of human immunodeficiency antigen (HIV-Ag) in serum and cerebrospinal fluid during acute and chronic infection, *Lancet* **ii** (1986), 177-180.
44. Graziosi, C., Gantt, K., Vaccarezza, M., *et al*: Kinetics of cytokine during human immunodeficiency virus infection, *Proc. Natl. Acad. Sci. USA* **93** (1996), 4386-4391.
45. Henderson, D.K., Fahey, B.J., Willy, M., *et al*: Risk for occupational transmission of human immunodeficiency virus type 1 (HIV-1) associated with clinical exposures: a prospective evaluation, *Ann. Intern. Med.* **113** (1990), 740-746.
46. Ho, D.D., Sarngadharan, M.G., Resnick, L., *et al*: Primary human T-lymphoptropic virus type III infection, *Ann. Intern. Med.* **103** (1985), 880-883.
47. Hoen, B., Harzic, M., Fleury, S., *et al*: ANRS053 trial of zidovudine, lamivudine and ritonavir combination in patients with symptomatic primary HIV-1 infection: preliminary results, In: *Abstracts of the 4th Conference on Retroviruses and Opportunstic Infections*, Washington DC, January, 1997 (abstract 232).
48. Imrie, A., Carr, A., Duncombe, C., *et al*: Primary infection with zidovudine-resistant HIV-1 does not adversely affect outcome at one year, *J. Infect. Dis.* **174** (1996), 195-198.
49. Imrie, A., Beveridge, A., Genn, W., Vizzard, J., Cooper, D.A. and the Sydney Primary HIV Infection Study Group: Transmission of human immunodeficiency virus type 1 resistant to nevirapine and zidovudine, *J. Infect. Dis.* **175** (1997), 1502-1507.
50. Isaksson, B., Albert, J., Chiodi, F., *et al*: AIDS two months after primary human immunodeficiency virus infection, *J. Infect. Dis.* **158** (1988), 866-868.
51. Jacquez, J.A., Koopman, J.S., Simon, C.P. and Longing, I.M.: Role of the primary infection in epidemics of HIV infection in gay cohorts, *J. Acq. Immune Def. Syndr. Human Retrovir.* **7** (1994), 1169-1184.
52. Keay, S., Wecksler, W., Wasserman, S.S., Margolick, J. and Farzadegan, H.: Association between anti-CD4 antibodies and a decline in $CD4^+$ lymphocytes in human immunodeficiency virus type 1 serconverters, *J. Infect. Dis.* **171** (1995), 312-319.
53. Keet, I.P.M., Krijnen, P., Koot, M., *et al*: Predictors of rapid progression to AIDS in HIV-1 seroconvertors, *AIDS* **7** (1993), 51-57.
54. Kessler, H.A., Blaauw, B., Spear, J., *et al*: Diagnosis of human immunodeficiency virus infection in seronegative homosexuals presenting with an acute viral syndrome, *JAMA* **258** (1987), 1196-1199.
55. Kinloch-Löes, S., Hirschel, B.J., Hoen, B., *et al*: A controlled trial of zidovudine in primary HIV infection, *New Engl. J. Med.* **333** (1995), 408-413.
56. Kitayporn, D., Kaewkungwal, J., Bejrachandra, S., Rungroung, E., Chandanayingyong, D. and Mastro, T.D.: Estimated rate of HIV-1 infectious but seronegative blood donations in Bangkok, Thailand, *AIDS* **10** (1996), 1157-1162.
57. Koup, R.A., Safrit, J.T., Cao, Y., *et al*: Temporal association of cellular immune responses with the initial control of viremia in primary human immunodeficiency virus type 1 infection, *J. Virol.* **68** (1994), 4650-4655.
58. Lange, J.M.A., Parry, J.V., de Wolf, F., *et al*: Diagnostic value of specific IgM antibodies in primary HIV infection, *Brit. Med. J.* **293** (1986), 1459-1462.
59. Lima, J., Ribera, A., Garcia-Bragado, F., *et al*: Antiplatelet antibodies in primary infection by human immunodeficiency virus, *Ann. Intern. Med.* **106** (1987), 333.
60. Lindback, S., Brostrom, C., Karlsson, A. and Gaines, H.: Does symptomatic primary HIV-1 infection accelerate progression to CDC stage IV disease, CD4 count below 200 x 106/l, AIDS, and death from AIDS?, *Brit. Med. J.* **309** (1994), 1535-1537.
61. Ludlam, C.A., Tucker, J., Steel, C.M., *et al*: Human T-lymphotropic virus type III (HTLV-III) infection in seronegative haemophiliacs after transfusion of factor VIII, *Lancet* **ii** (1985), 233-236.
62. McCune, J.M., Namikawa, R., Shih, C.-C., *et al*: Suppression of HIV infection in AZT-treated SCIDhu mice, *Science* **247** (1990), 564-566.
63. Marion, S.A., Schechter, M.T., Weaver, M.S., *et al*: Evidence that prior immune dysfunction predisposes to human immunodeficiency virus infection in homosexual men, *J. Acq. Immune Def. Syndr. Human Retrovir.* **2** (1989), 178-186.
64. Mariotti, M., Leftere, J.-J., Noel, B., *et al*: DNA amplification of HIV-1 in seropositive individuals and in seronegative at-risk individuals, *AIDS* **4** (1990), 633-637.
65. Markowitz, M., Cao, Y., Vesanen, M., *et al*: Recent HIV infection treated with AZT, 3TC, and a protease inhibitor, In: *Abstracts of the 4th Conference on Retroviruses and Opportunistic Infections*, Washington DC, January, 1997 (abstract LB8).

66. Mellors, J.W., Kingsley, L.A., Rinaldo, C.R., *et al*: Quantitation of HIV-1 RNA in plasma predicts outcome after seroconversion, *Ann. Intern. Med.* **122** (1995), 573-579.

67. Nielsen, C., Pedersen, C., Lundgren, J.D. and Gerstoft, J.: Biological properties of HIV isolates in primary HIV-1 infection: Consequences for the subsequent course of infection, *AIDS* **7** (1993), 1035-1040.

68. Oksenhendler, E., Gerrard, L., Molina, J.M. and Ferchal, F.: Lymphocytosis in primary HIV infection is associated with CMV co-infection, *AIDS* **7** (1993), 1023-1024.

69. Pedersen, C., Dickmeiss, E., Gaub, J., *et al*: T-cell subset alterations and lymphocyte responsiveness to mitogens and antigens during severe primary infection with HIV: A case series of seven consecutive HIV seroconverters, *AIDS* **4** (1990), 523-526.

70. Pedersen, C., Getstoft, J., Lindhardt, B.O. and Sindrup, J.: Candida esophagitis associated with acute human immunodeficiency virus infection, *J. Infect. Dis.* **156** (1987), 529-530.

71. Pedersen, C., Lindhardt, B.O., Jensen, B.L., *et al*: Clinical course of primary HIV infection: Consequences for subsequent course of infection, *Brit. Med. J.* **299** (1989), 154-157.

72. Pedersen, C., Nielsen, C.M., Vestergaard, B.F., *et al*: Temporal relation of antigenaemia and loss of antibodies to core antigens to development of clinical disease in HIV infection, *Brit. Med. J.* **295** (1987), 567-569.

73. Perrin, L., Markowitz, M., Calandra, G., Chung, M. and the MRL Acute HIV Infection Study Group: An open treatment study of acute HIV infection with zidovudine, lamivudine and indinavir sulfate, In: *Abstracts of the 4th Conference on Retroviruses and Opportunistic Infections,* Washington DC, January, 1997 (abstract 238).

74. Phillips, A.N.: Reduction in HIV concentration during acute infection: independence from a specific immune response, *Science* **271** (1996), 497-499.

75. Phillips, A.N., Sabin, C.A., Elford, J., *et al*. Serum beta-2 microglobulin at HIV-1 seroconversion as a predictor of severe immunodeficiency during 10 years of followup, *J. Acq. Immune Def. Syndr. Human Retrovir.* **13** (1996), 262-266.

76. Raffi, F., Boudart, D. and Billaudel, S.: Acute co-infection with human immunodeficiency virus (HIV) and cytomegalovirus, *Ann. Intern. Med.* **112** (1990), 234-235.

77. Reiss, P., de Ronde, A., Dekker, J., *et al*: Seroconversion to HIV-1 rev- and tat-gene-encoded proteins, *AIDS* **3** (1989), 105-106.

78. Reiss, P., de Ronde, A., Lange, J.M.A., *et al*: Antibody response to the viral negative factor (nef) in HIV-1 infection: A correlate of levels of HIV-1 expression, *AIDS* **3** (1989), 227-233.

79. Reiss, P., Lange, J.M.A., de Ronde, A., *et al*: Antibody response to viral proteins U (vpu) and R (vpr) in HIV-1 infected individuals, *J. Acq. Immune Def. Syndr. Human Retrovir.* **3** (1990), 115-122.

80. Richman, D.D. and Bozzette, S.A.. The impact of the syncytium-inducing phenotype of human immunodeficiency virus on disease progression, *J. Infect. Dis.* **169** (1994), 968-974.

81. Ridzon, R., Gallagher, K., Ciesielski, C., *et al*: Simultaneously transmission of human immunodeficiency virus and hepatitis C virus from a needle stick injury, *New Engl. J. Med.* **336** (1997), 919-922.

82. Roos, M.Th.L., Lange, J.M.A., de Goede, R.E.Y., *et al*: Virus phenotype and immune response in primary human immunodeficiency virus type 1 (HIV-1) infection, *J. Infect. Dis.* **165** (1992), 427-432.

83. Rosenberg, E.S., Billingsley, J.M., Caliendo, A.M., *et al*: Vigorous HIV-1-specific $CD4^+$ T cell responses associated with control of viremia, *Science* **278** (1997), 1447-1450.

84. Rouleau, D., Conway, B., Patenaude, P., Schechter, M.T., O'Shaughnessy, M.V. and Montaner, J.S.G.: Combination antiretroviral therapy for the treatment of acute HIV infection, In: *Abstracts of the 3rd National Conference of Human Retroviruses and Related Infections,* Washington DC, January 1996 [abstract 287].

85. Ruprecht, R.M., O'Brien, L.G., Rossoni, L.D., *et al*: Suppression of mouse viraemia and retroviral disease by 3'-azido-3'-deoxythymidine, *Nature* **323** (1986), 467-469.

86. Ruprecht, R.M., Chou, T.-C., Chipty, F., *et al*: Interferon-alpha and 3'-azido-3'-deoxythymidine are highly synergistic in mice and prevent viremia after acute retrovirus exposure, *J. Acq. Immune Def. Syndr. Human Retrovir.* **3** (1990), 591-600.

87. Schacker, T.W., Collier, A., Hughes, J., Shea, T. and Corey, L.: Clinical and epidemiologic features of primary HIV infection, *Ann. Intern. Med.* **125** (1996), 257-264.

88. Schreiber, G.B., Busch, M.P., Kleinman, S.H. and Korelitz, J.J.: The risk of transfusion-transmitted viral infections: the retrovirus epidemiology donor study, *New Engl. J. Med.* **334** (1996), 1685-1690.

89. Sinicco, A., Palestro, G., Caramello, P., *et al*: Acute HIV-1 infection: Clinical and biological study of 12 patients, *J. Acq. Immune Def. Syndr. Human Retrovir.* **3** (1990), 260-265.
90. Sinicco, A., Biglino, A., Sciandra, M., *et al*: Cytokine network and acute HIV-1 infection, *AIDS* **7** (1993), 625-631.
91. Skolnik, P.R., Kosloff, B.R. and Hirsch, M.S.: Bidirectional interaction between human immunodeficiency virus type 1 and cytomegalovirus, *J. Infect. Dis.* **157** (1988), 508-514.
92. Steeper, T.A., Horwitz, C.A., Hanson, M., *et al*: Heterophil-negative mononucleosis-like illnesses with atypical lymphocytosis in patients undergoing seroconversion to the human immunodeficiency virus, *Am. J. Clin. Pathol.* **89** (1988), 169-174.
93. Stramer, S.L., Heller, J.S., Coombs, R.W., Parry, J.V., Ho, D.D. and Allain, J.P.: Markers of HIV infection prior to IgG antibody seropositivity, *JAMA* **262** (1989), 64-649.
94. Stuyver, L., Wyseur, A., Rombout, A., *et al*: Line probe assay for rapid detection of drug-selected mutations in the human immunodeficiency virus type 1 reverse transcriptase gene, *Antimicrob. Agent Chemother.* **41** (1997), 284-291.
95. Teeuwsen, V.J.P., Siebelink, K.H.J., de Wolf, F., *et al*: Impairment of *in-vitro* immune responses occurs within 3 months after HIV-1 seroconversion, *AIDS* **4** (1990), 77-81.
96. Terpstra, F.G., Ai ,B.M.J., Roos, M.Th.L., *et al*: Longitudinal study of leukocyte function in homosexual men seroconverted for HIV: Rapid and persistent loss of B cell function after HIV infection, *Eur. J. Immunol.* **19** (1989), 667-673.
97. Tersmette, M., Lange, J.M.A., de Goede, R.E.Y., *et al*: Association between biological properties of human immunodeficiency virus variants and risk for AIDS and AIDS mortality, *Lancet* **i** (1989), 983-985.
98. Tindall, B., Barker, S., Donovan, B., *et al*: Characterization of the acute clinical illness associated with human immunodeficiency virus infection, *Arch. Intern. Med.* **148** (1988), 945-949.
99. Tindall, B., Carr, A., Goldstein, D. and Cooper, D.A.: Administration of zidovudine during primary HIV-1 infection may be associated with a less vigorous response, *AIDS* **7** (1993), 127-128.
100. Tindall, B. and Cooper, D.A.: Primary HIV infection: Host responses and intervention strategies, *AIDS* **5** (1991), 1-14.
101. Tindall, B., Gaines, H., Imrie, A., *et al*: Zidovudine in the management of primary HIV infection, *AIDS* **5** (1991), 477-484.
102. Tindall, B., Hing, M., Edwards, P., *et al*: Severe clinical manifestations of primary HIV infection, *AIDS* **3** (1989), 747-749.
103. Tsang, M.I., Evans, L.A., McQueen, P., *et al*: Neutralizing antibodies against sequential autologous human immunodeficiency virus type 1 isolates after seroconversion, *J. Infect. Dis.* **170** (1994), 1141-1147.
104. Valle, S.-L.: Febrile pharyngitis as the primary sign of HIV infection in a cluster of cases linked by sexual contact, *Scand. J. Infect. Dis.* **19** (1987), 13-17.
105. Vanhems, P., Allard, R., Cooper, D.A., *et al*: Acute HIV-1 disease as a mononucleosis-like illness: is the diagnosis too restrictive?, *Clin. Infect. Dis.* **24** (1997), 965-970.
106. Vento, S., Di Perri, G., Garofano, T., Concia, E. and Bassetti, D.: Pneumocystis carinii pneumonia during primary HIV-1 infection, *Lancet* **342** (1993), 24-25.
107. von Sydow, M., Gaines, H., Sonnerborg, A., *et al*: Antigen detection in primary HIV infection, *Brit. Med. J.* **296** (1988), 238-240.
108. von Sydow, M., Sonnerborg, A., Gaines, H. and Strannegard, O.: Interferon-alpha and tumor necrosis factor in serum of patients in varying stages of HIV-1 infection, *AIDS Res. Human Retrovir.* **7** (1991), 375-380.
109. Walker, C.M., Moody, D.J., Stites, D.P. and Levy, J.A.: $CD8^+$ lymphocytes can control infection *in vitro* by suppressing virus replication, *Science* **234** (1986), 1563-1566.
110. Ward, J.W., Bush, T.J., Perkins, H.A., *et al*: The natural history of transfusion-associated HIV infection: Factors influencing progression to disease, *New Engl. J. Med.* **321** (1989), 947-952.
111. Weissman, D., Rubbert, A., Comabadiere, C., *et al*: Multifactorial nature of non-cytolytic $CD8^+$ T-cell-mediated suppression of HIV replication: beta-chemokine-dependent and -independent effects, *AIDS Res. Human Retrovir.* **13** (1997), 63-69.
112. Zaunders, J., Carr, A., McNally, L., Penny, R. and Cooper, D.A.: Effects of primary HIV-1 infection on subsets of $CD4^+$ and $CD8^+$ T lymphocytes, *AIDS* **9** (1995), 561-566.
113. Zhu, T., Mo, H., Wang, N., *et al*: Genotypic and phenotypic characterization of HIV-1 in patients with primary infection, *Science* **261** (1993), 1179-1181.

114. Zhu, T., Wang, N., Carr, A., Nam, D.S., Cooper, D.A. and Ho, D.D.: Genetic characterization of human immunodeficiency virus type 1 in blood and genital secretions: evidence for viral compartmentalization and selection during sexual transmission, *J. Virol.* **70** (1996), 3098-107.

115. Zhu, T., Wang, N., Carr, A., Wolinsky, S. and Ho, D.D.: Evidence for coinfection by multiple strains of human immunodeficiency virus type 1 infection subtype B in an acute seroconvertor, *J. Virol.* **69** (1995), 1324-1327.

116. Ziegler, J.B., Cooper, D.A., Johnson, R.O. and Gold, J.: Postnatal transmission of AIDS-associated retrovirus from mother to infant, *Lancet* **i** (1985), 896-898.

CHAPTER 3

BIOLOGICAL VARIABILITY OF HIV-1

JAN ALBERT[1] and MAARTEN KOOT[2]

[1]Department of Clinical Virology, Swedish Institute for Infectious Disease Control, Karolinska Institute, S-105 21 Stockholm, Sweden, and [2]Department of Clinical Viro-Immunology of the CLB and Laboratory for Experimental and Clinical Immunology, Academic Medical Centre, University of Amsterdam, Amsterdam, The Netherlands

1. Introduction

HIV-1 displays one of the highest evolutionary rates detected in any life form with genetic distances in hypervariable regions of the genome reaching as high as 10-15% within single individuals. This extreme variation is the result of an error-prone reverse transcriptase, high population turnover, viral proteins that accept high variation, and strong environmental selective pressures [1-5]. Thus, the reverse transcriptase enzyme, which lacks proof reading capacity, has a misincorporation rate of 10^{-4} to 10^{-5} per base or approximately one misincorporation per genome per replication cycle [1,2]. Furthermore, it has been estimated that up to 10^{10} virus particles are produced every day in an HIV-1-infected individual [3-5]. Consequently, an HIV-1-infected individual harbours a swarm of closely related viruses [6]. Importantly, the genetic variation of HIV-1 translates into biological variation, such as cell tropism, virulence, and sensitivity to neutralizing antibodies and antiviral drugs.

In this chapter we will review the current knowledge about variation in the biological phenotype of HIV-1. These *in vitro* properties of HIV-1 are of importance because they show strong correlation with clinical outcome. The interest in the biological properties of HIV-1 has been further spurred by the recent finding that chemokine receptors act as co-receptors for HIV-1 entry into cells and that differences in the biological phenotype of HIV-1 isolates is largely explained by differences in chemo-kine co-receptor usage.

In the first part of this chapter we will review the somewhat confusing and partially overlapping nomenclature systems for differences in biological phenotype and their correlation to chemokine co-receptor usage. In the second part we will examine the molecular basis for these differences. Finally, we will discuss the importance of differences in biological phenotype on disease progression, transmission and treatment.

H. Schuitemaker and F. Miedema (ed.), Aids Pathogenesis, 41–54.

2. Biological Phenotype of HIV-1

The term biological phenotype refers to at least three different *in vitro* properties of HIV-1 isolates, namely replication kinetics, cytopathicity and cell tropism. One reason why these properties are collectively referred to as biological phenotype is that they for the most part are closely correlated and therefore to a large degree represent different manifestations of the same underlying determinants, namely co-receptor usage. As a consequence, several nomenclature systems, which emphasize different aspects of the biological properties of HIV-1, have been used. These nomenclature systems are summarised in Table 1 and further discussed below.

TABLE 1. Overview of several essentially overlapping classification system for differences in biological phenotype between HIV-1 isolates

Main emphasis	Nomenclature		References
Replication kinetics	slow/low	rapid/high	7-9
Cytopathicity	non-syncytium-inducing (NSI)	synsytium-inducing (SI)	11-15
Cellular host range	macrophage-tropic (M-tropic)	T-cell-tropic (T-tropic)	15-20
MT-2 cell tropism	MT-2-negative	MT-2-positive	26
Co-receptor usage	CXCR4-negative	CXCR4-positive	29-34

2.1. RAPID/HIGH AND SLOW/LOW PHENOTYPES

In 1986, Åsjö *et al.* were the first to describe that the *in vitro* properties of HIV isolates from AIDS patients differed from those obtained from asymptomatic individuals [7,8]. Isolates from patients with advanced disease replicated faster and to higher titres in primary lymphocyte cultures and were therefore called rapid/high in contrast to most viruses from asymptomatic HIV carriers which replicated more slowly and to lower titres, *i.e.* slow/low isolates. Even though this nomenclature stressed differences in replication kinetics, the authors described that rapid/high isolates also showed increased cytopathicity *in vitro*, with extensive syncytium formation. In addition, rapid/high isolates, but not slow/low isolates, could establish continuous infection of several culture established CD4-positive cell lines, including H9, CEM and U937. The replication kinetics in culture are influenced by inoculum size [9], but several studies show that there also exist true strain-specific differences in replicative capacity [10-12].

2.2. SYNCYTIUM-INDUCING (SI) AND NONSYNCYTIUM-INDUCING (NSI) PHENOTYPES

Tersmette *et al.* also studied the biological properties of primary HIV-1 isolates and found that cytopathicity varied between viruses [13]. Isolates from about half of the AIDS patients induced cell-to-cell fusion *in vitro*, causing the formation of multinucleated cells or syncytia. These viruses were designated syncytium-inducing (SI) [13]. The viruses that lacked this capacity were called non-syncytium-inducing (NSI) [13]. All SI variants were rapidly replicating, as were the NSI variants from AIDS patients. In contrast, NSI viruses obtained from asymptomatic individuals always replicated slowly [11,12]. Longitudinal studies indicated that SI variants were generally not present at the moment of seroconversion but emerged during the course of the infection [11,14,15].

2.2.1. *Macrophage (M)-tropic and T-cell-line (T)-tropic phenotypes*

A third classification system is based on cell tropism and, according to this system, HIV-1 isolates are classified as macrophage (M)-tropic and T-cell line (T)-tropic. Cheng-Mayer *et al.* [16] first described that isolates capable of infecting T-cell lines had a reduced capacity to replicate in primary macrophages and other groups have since reported similar findings [15, 17-20]. However, existing data concerning the inverse correlation between macrophage tropism and T-cell-line tropism are conflicting. Some groups have reported that the ability to infect and replicate in primary monocyte/macrophage cultures is general characteristic of all primary HIV-1 isolates [21-23]. Some of the differences are likely to be due to differences in methodology [23], but clearly more experimental data are required to clarify this point. Another cause for confusion is that primary HIV-1 isolates with rapid/high, SI phenotype are sometimes grouped together with isolates which have been adapted to growth in T-cell lines, *i.e.* T-cell-line-adapted (TCLA) isolates. It is clear that adaptation to growth in T-cell lines changes many biological properties of primary HIV-1 isolates and leads to reduced infectivity for macrophages as well as increased sensitivity to neutralization by antibody and soluble receptor [24,25].

2.3. MT-2-NEGATIVE AND MT-2-POSITIVE PHENOTYPES

MT-2 is an HTLV-I-transformed CD4-positive cell line that can readily be infected by certain HIV-1 isolates and infection usually results in extensive syncytia formation. The MT-2 cell assay was described by Koot *et al.* [26] as a convenient way of determining the biological phenotype of HIV-1 isolates. Consequently, the MT-2 cell assay is today the most widely used assay to determine the biological phenotype of HIV-1. The MT-2-positive phenotype correlates very closely with the original SI phenotype, which was evaluated on PBMC phenotype, as well as the rapid/high phenotype [26-28]. Similarly, the MT-2-negative, slow/low and NSI phenotypes are very closely correlated [26-28]. In contrast, and in line with the discussion in the previous paragraph, the correlation between the MT-2 cell assay and macrophage tropism is more controversial. Some authors find that MT-2-positive isolates cannot infect macrophages [15,17-20], whereas others find no such correlation [22,27,28].

2.4. CHEMOKINE CO-RECEPTORS

The differences in biological phenotype of HIV-1 isolates was recently shown to be due to differences in co-receptor usage. Thus, HIV-1 entry into cells requires interactions with certain co-receptors in addition to CD4. Several studies have shown that the ß-chemokine receptor CCR5 acts as the major co-receptor for primary non-T-cell-line-adapted viruses, *i.e.* slow/low, NSI isolates [29-32]. Primary isolates with rapid/high, SI phenotype and T-cell-line-adapted HIV-1 strains instead use the α-chemokine receptor CXCR4 (previously called fusin or Lestr) as co-receptor [33]. However, many primary SI HIV-1 isolates can also enter via CCR5 and are thus dual tropic [34-37]. In agreement with the co-receptor usage, the ß-chemokines RANTES, MIP-1α and MIP-1β block the entry of slow/low, NSI isolates [33, 38], whereas SDF-1, the ligand of CXCR4, blocks the entry of rapid/high, SI and T-cell-line-adapted isolates [39,40]. Most T-cell lines, including MT-2, express high levels of CXCR4, but not CCR5 [41-43]. In agreement with this, there is an almost perfect correlation between ability to replicate in MT-2 cells and ability to infect via the CXCR4 co-receptor, irrespective of the genetic subtype of the isolate [35-37]. In contrast, there appears to be no direct correlation between macrophage-tropism and chemokine co-receptor usage of primary HIV-1 isolates [34,44]. This is probably explained by the findings that monocytes/macrophages express both CCR5 and CXCR4 [41-43].

2.5. COMMENTS ON BIOLOGICAL PHENOTYPE AND A NEW CLASSIFICATION SYSTEM BASED ON CHEMOKINE CO-RECEPTOR USAGE

As reviewed above, several different classification systems that emphasize different aspects of the biological variation of HIV-1 are in use. The systems are largely, but not completely, overlapping. As recently pointed out by Berger *et al.*, all of the biological classification systems have draw-backs [45]. This is especially true for the system that defines primary HIV-1 isolates as macrophage (M)-tropic or T-cell line (T)-tropic, because it disguises the fact that all primary HIV-1 isolates replicate in activated primary $CD4^+$ T lymphocytes. Berger *et al.* also point to the problem that it is often incorrectly perceived by those unfamiliar with the system that M-tropic viruses replicate exclusively in macrophages [45]. An additional problem is that many T-tropic primary isolates replicate well in macrophages. The rapid/high - slow/low and SI - NSI systems are also misleading because replication kinetics depend not only on biological phenotype and because NSI isolates can readily form syncytia in CCR5-positive cells.

Because of the deficiencies of the current classification systems an expert committee recently proposed a new system based on coreceptor usage [45]. This nomenclature is based on whether a particular HIV-1 isolate, or a group of isolates, uses one or both of the major co-receptors, CCR5 and CXCR4, to enter its target cells. According to this nomenclature, isolates which use CCR5 but not CXCR4 are termed R5 viruses, and isolates with the converse property X4 viruses. Isolates able to use both co-receptors with comparable efficiency would be called R5X4. Whether an X4 or R5X4 virus is a cell-line-adapted isolate should also be specified. It should be pointed out, however, that the MT-2 cell assay remains an effective way to determine the biological phenotype of HIV-1 isolates and the results from the MT-2 assay is determined by the CXR4 usage of the isolate [35-37]. Thus, MT-2-negative, NSI isolates use CCR5 only (R5), whereas MT-2-positive, SI isolates use CXCR4 and in addition often CCR5 (X4 or R5X4).

3. Molecular Basis for HIV-1 Biological Phenotypes

As reviewed above, the biological phenotype of HIV-1 isolates is determined by co-receptor usage, *i.e.* CXCR4 usage. In agreement with these recent findings, it was shown already several years ago that cell tropism of HIV-1 is determined by specific regions in the *env* gene [46,47]. Extensive analysis of the V3 domains of many HIV-1 isolates with known biological phenotype showed that in the V3 loop two fixed positions (306 and 320) were usually negatively charged or uncharged in NSI HIV-1 variants. In SI variants, either one or both these positions usually carried a positively charged amino acid [48]. Site-directed mutagenesis indeed confirmed the functional involvement of the amino acids at positions 306 and 320 and in addition position 324 [49]. In line with these findings, the co-receptor usage has been shown to be determined by individual amino acids within the V3 loop [29,50-52]. It has been reported that other regions of the HIV-1 envelope, *i.e.* the V2 loop, may also be involved in the switch from NSI to SI phenotype. SI variants and their co-existing NSI variants were shown to have elongated V2 domains, as well as a relocation or addition of an N-linked glycosylation site [53]. NSI isolates from AIDS patients that had never developed SI variants had short V2 domains [53]. However, others have not been able to confirm these data [54,55], but it has been proposed that V2 elongation is a temporary requirement during the phenotype switch that is no longer needed once a stable SI phenotype has been obtained [56,57].

4. Genetic Subtypes and Biological Phenotype

One manifestation of the genetic diversity of HIV-1 is the existence of genetic subtypes. Thus, HIV-1 has been divided into two genetic groups, called major (M) and outlier (O). The M group has been further subdivided into ten subtypes (A to J) on the basis of sequence diversity in the *env* and *gag* genes [58]. These subtypes show a distinct geographical distribution [58]. Our knowledge about HIV-1 and AIDS, including biological variation of HIV-1, is primarily based on studies on subtype B, which dominates in the U.S. and Europe. At present, little is known about possible subtype-specific differences in biological properties of HIV-1. It has been suggested that subtype E HIV-1 variants may be more efficiently transmitted via the heterosexual route than subtype B variants [59], but this hypothesis was recently challenged [60,61]. However, studies within the WHO Network for HIV Isolation and Characterization showed that both the phenotypic distinction between SI and NSI viruses and the association between biological phenotype and V3 mutations are present among HIV-1 subtypes other than B [62].

Other studies have shown that the tight association between biological phenotype tested in the MT-2 cell assay and CXCR4 usage exists for all genetic subtypes [35,37]. An association between low $CD4^+$ lymphocyte counts and SI phenotype has also been documented, suggesting that the SI phenotype may be linked to accelerated rate of disease progression also in subtypes other than subtype B [37]. Interestingly, the SI, CXCR4-positive phenotype may be rare among subtype-C-infected individuals [37]. It will be interesting to study whether this difference translates into differences in clinical outcome.

5. Clinical Relevance of Biological Phenotypes

5.1. DISEASE PROGRESSION

In transsectional studies, rapid/high, SI variants are more often isolated from individuals with low CD4-cell counts [7,8,13,63,64]. This indicates that either their presence is only possible in severely immuno-compromised individuals or that their emergence is directly responsible for the low $CD4^+$ T-cell numbers. Prospective cohort studies of HIV-1-infected individuals have shown that the MT-2 assay has considerable prognostic power [65,66]. Thus, proportional hazard analysis showed that the isolation of SI variants is associated with increased risk for clinical progression also after controlling for CD4 counts and p24 antigenaemia. Despite this CD4 independence in hazard analysis, it is likely that the association between SI phenotype and increased risk of disease progression is at least partly mediated via accelerated CD4 decline. In line with this, the rate of CD4-cell decline was faster after the switch to SI phenotype than before the switch and also faster than in the NSI control group [65-67]. In addition, it has been reported that individuals who become infected by rapid/high, SI variants show faster disease progression than those infected with slow/low, NSI variants [27,28,68].

The exact mechanism by which SI variants cause accelerated CD4-cell decline has not yet been identified. Plasma HIV-1 RNA load is a very good predictor of clinical outcome [69]. There are indications that individuals with SI virus have higher virus load than individuals with NSI virus [11,70-72], but conclusive evidence is still lacking. However, there are indications from the SCID-hu mouse system and from experimental infection of human lymphoid tissue that SI variants may be more cytopathic than NSI variants *in vivo* [73, 74]. It is important to note that when SI variants are present in patient PBMC they always outgrow the co-existing NSI variants in *in vitro* bulk culture of the patient PBMC with PHA-stimulated target cells. *In vivo*, however, NSI and SI HIV-1 variants generally co-exist several years after an NSI to SI switch has been observed in cultured virus [12,70,75]. Thus, the selective growth advantage for SI HIV-1, which has been observed *in vitro*, is apparently less pronounced *in vivo*. It is likely that the replication advantage of SI variants is counteracted *in vivo* by other selective forces, such as humoral and cellular immunity [76].

5.2. BIOLOGICAL PHENOTYPE AND CCR5 GENOTYPE

It has been reported that a relatively high proportion of Caucasian individuals carry a 32 base pair deletion (Δ32) in the CCR5 gene, *i.e.* in the gene that encodes the co-receptor for NSI isolates. Thus, approximately 19% are heterozygous and 1% homozygous for Δ32 in the Caucasian population. Individuals who are heterozygous for Δ32 have a reduced rate of disease progression [77,78], a finding that is in agreement with the observation that CCR5 is an important co-receptor for all NSI HIV-1 variants and some SI variants. Similarly, Δ32 homozygots are strongly protected against HIV-1 infection [77,78]. Because primary SI HIV-1 often can use both CCR5 and CXCR4 as co-receptor for cell entry, whereas primary NSI only use CCR5, it could be envisioned that the beneficial effect of a heterozygous Δ32 genotype on the clinical course of infection would be most pronounced in individuals carrying NSI variants and findings in line with this have recently been published [79]. However, it appears that a significant beneficial effect of the heterozygous Δ32 genotype remains also after the emergence of SI virus [80].

This may have at least two explanations. First, many SI isolates are dual tropic and can use CCR5 in addition to CXCR4 [34-37]. Secondly, NSI variants usually co-exist with SI variants *in vivo* after a phenotypic switch has been detected *in vitro*, and these NSI variants may significantly contribute to disease progression, just as they do in patients who develop AIDS without experiencing a phenotypic switch [12,70,75].

5.3. IMPORTANCE OF BIOLOGICAL PHENOTYPE DURING TRANSMISSION

In recently infected individuals the virus population is very homogeneous, at least with respect to V3 sequences [81-84]. The basis for this sequence homology is still not understood, but it has been noted that a majority of virus variants in a newly infected individual are slow/low, NSI [82-84]. However, transmission of rapid/high, SI variants have been documented [27,28,68]. The predominance of NSI variants has been interpreted as an indication that such variants are selected for during transmission [82-84]. According to this hypothesis, these variants are selected because they are more macrophage-tropic than SI variants. However, as reviewed above, the association between macrophage-tropism and SI phenotype is controversial [21,23,45].

Thus, it remains a possibility that the predominance for NSI variants during and early after primary infection may have epidemiological rather than virological explanations [28,68,85]. It should also be pointed out that an NSI carrier can only be expected to transmit NSI variants, whereas an SI carrier can transmit both SI and NSI variants [12, 70,75].

5.4. BIOLOGICAL PHENOTYPE AND ANTI-RETROVIRAL TREATMENT

Next to a role for SI HIV-1 in the natural course of infection, also the efficacy of zidovudine treatment appeared to be dependent on the biological phenotype of HIV present in patients receiving treatment [86,87]. Zidovudine treatment significantly delayed clinical progression in individuals that had only NSI at start of the study and who did not develop SI variants under treatment. In addition, development of resistance to zidovudine was more common among individuals harbouring SI variants [87]. Development of resistance was, however, unlikely to be the only explanation for the poor response to zidovudine among individuals with SI variants, because the difference in response was observed before resistance could be demonstated [87].

A possible mechanism for this viral phenotype-dependent efficacy of zidovudine was provided by a study in which the clonal composition of HIV-1 populations was analysed in individuals receiving zidovudine treatment [88]. Individuals who started zidovudine treatment showed a decline in their cellular infectious load which could be attributed mainly to the suppression of NSI HIV-1 variants, whereas the co-existing SI load remained at base line levels. This difference in response was not due to differential kinetics of development of zidovudine resistance since all viruses still had a wild-type genotype.

A reverse phenomenon has been reported in individuals treated with didanosine [89]. Individuals that carried SI variants at the start of the study some times lost their SI HIV-1 which was not due to a decline in viral load below detection level since NSI HIV-1 could still be isolated [90,91]. The mechanism for this virus-phenotype-dependent efficacy of these nucleoside analogues has not been elucidated, but it is possible that cell-type-specific differences exist in phosphorylation of zidovudine and didanosine. Fortunately, modern triple combination therapy with at least one protease inhibitor appears equally effective on SI and NSI variants [72,89].

6. Conclusions

HIV-1 is characterized by an unusually high degree of genetic variation. This genetic variation is of great importance because it translates into variation in a number of biological properties of which the biological phenotype is probably the most important. The biological phenotype is determined by chemokine co-receptor usage, *i.e.* CXCR4 usage, and is expressed in differences in replication kinetics, cytopathicity and cell tropism. The biological phenotype influences the virulence of HIV-1. Thus, individuals who carry CXCR4-positive, SI variants show an accelerated rate of disease progression compared to individuals who carry CXCR4-negative, NSI variants. Furthermore, individuals with SI variants show limited response to single treatment with zidovudine, but there appears to be no difference in response to modern combination therapy between SI and NSI variants. It has been suggested that NSI variants are selected for during transmission, but available data are not conclusive. Several different classification systems which emphasise different aspects of the biological variation have been described. The most commonly used classify isolates as rapid/high - slow/low, SI - NSI or M-tropic - T-tropic, respectively. All of these systems have drawbacks and the realization that the biological phenotype is determined by co-receptor usage has led to the suggestion of a new classification system based on chemokine co-receptor usage.

7. References

1. Preston, B.D., Poiesz, B.J. and Loeb, L.A.: Fidelity of HIV-1 reverse transcriptase, *Science* **242** (1988), 1168-1171.
2. Roberts, J.D., Bebenek, K. and Kunkel, T.A.: The accuracy of reverse transcriptase from HIV-1, *Science* **242** (1988), 1171-1173.
3. Wei, X., Ghosh, S.K., Taylor, M.E., Johnson, V.A., Emini, E.A., Deutsch, P., Lifson, J.D., Bonhoeffer, S., Nowak, M.A., Hahn, B.H., Saag, M.S. and Shaw, G.M.: Viral dynamics in human immunodeficiency virus type 1 infection, *Nature* **373** (1995), 117-122.
4. Ho, D.D., Neumann, A.U., Perelson, A.S., Chen, W., Leonard, J.M. and Markowitz, M.: Rapid turnover of plasma virions and CD4 lymphocytes in HIV-1 infection, *Nature* **373** (1995), 123-126.
5. Perelson, A.S., Neumann, A.U., Markowitz, M., Leonard, J.M. and Ho, D.D.: HIV-1 dynamics *in vivo*: virion clearance rate, infected cell life-span, and viral generation time, *Science* **271** (1996), 1582-1586.
6. Meyerhans, A., Cheynier, R., Albert, J., Seth, M., Kwok, S., Sninski, J., Morfeldt-Månsson, L., Åsjö, B. and Wain-Hobson, S.: Temporal fluctuations in HIV quasispecies *in vivo* are not reflected by sequential HIV isolations, *Cell* **58** (1989), 901-910.
7. Åsjö, B., Morfeldt-Månsson, L., Albert, J., Biberfeld, G., Karlsson, A., Lidman, K. and Fenyö, E.M.: Replicative capacity of human immunodeficiency virus from patients with varying severity of HIV infection, *Lancet* **ii** (1986), 660-662.
8. Fenyö, E.M., Morfeldt-Månsson, L., Chiodi, F., Lind, B., von Gegerfelt, A., Albert, J., Olausson, E. and Åsjö, B.: Distinct replicative and cytopathic characteristics of human immunodeficiency virus isolates, *J. Virol.* **62** (1988), 4414-4419.
9. Lu,W. and Andrieu, J.M.: Similar replication capacities of primary human immunodeficiency virus type 1 isolates derived from a wide range of clinical sources, *J. Virol.* **66** (1992), 334-340.
10. Åsjö, B., Sharma, U.K., Morfeldt-Månsson, L., Magnusson, A., Barkhem, T., Albert, J., Olausson, E., von Gegerfelt, A., Lind, B., Biberfeld, P. and Fenyö, E.M.: Naturally occurring HIV-1 isolates with differences in replicative capacity are distinguished by *in-situ* hybridization of infected cells, *AIDS Res. Human Retrovir.* **6** (1990), 1177-1182.
11. Connor, R.I., Mohri, H., Cao, Y. and Ho, D.D.: Increased viral burden and cytopathicity correlate temporally with CD4$^+$ T-lymphocyte decline and clinical progression in human immunodeficiency virus type-1-infected individuals, *J. Virol.* **67** (1993), 1772-1777.
12. Van 't Wout, A.B., Blaak, H., Ran, L.J., Brouwer, M., Kuiken, C. and Schuitemaker, H.: Evolution of syncytium-inducing and non-syncytium-inducing biological virus clones in relation to replication kinetics during the course of human immunodeficiency virus type-1 infection, *J. Virol.* **72** (1998), 5099-5107.
13. Tersmette, M., de Goede, R.E.Y., Al, B.J.M., Winkler, I.N., Gruters, R.A., Cuypers, H.T.M., Huisman, J.G. and Miedema, F.: Differential syncytium-inducing capacity of human immunodeficiency virus isolates: Frequent detection of syncytium-inducing isolates in patients with acquired immunodeficiency syndrome (AIDS) and AIDS-related complex, *J. Virol.* **62** (1988), 2026-2032.
14. Tersmette, M., Gruters, R.A., de Wolf, F., de Goede, R.E.Y., Lange, J.M.A., Schellekens, P.Th.A., Goudsmit, J., Huisman, J.G. and Miedema, F.: Evidence for a role of virulent human immunodeficiency virus (HIV) variants in the pathogenesis of acquired immunodeficiency syndrome: studies on sequential HIV isolates, *J. Virol.* **63** (1989), 2118-2125.
15. Schuitemaker, H., Kootstra, N.A., de Goede, R.E.Y., de Wolf, F., Miedema, F. and Tersmette, M.: Monocytotropic human immunodeficiency virus type 1 (HIV-1) variants detectable in all stages of HIV-1 infection lack T-cell line tropism and syncytium-inducing ability in primary T-cell culture, *J. Virol.* **65** (1991), 356-363.
16. Cheng-Mayer, C., Homsy, J., Evans, L.A. and Levy, L.A.: Identification of human immunodeficiency virus subtypes with distinct patterns of sensitivity to serum neutralization, *Proc. Natl. Acad. Sci. USA* **85** (1988), 2815-2819.

17. Gendelman, H.E., Orenstein, J.M., Baca, L.M., Weiser, B., Burger, H., Kalter, D. and Meltzer, M.S.: The macrophage in the persistence and pathogenesis of HIV infection, *AIDS* **3** (1989), 475-495.
18. Collman, R., Hassan, N.F., Walker, R., Godfrey, B., Cutilli, J., Hastings, J.C., Friedman, H., Douglas, S.D. and Nathanson, N.: Infection of monocyte-derived macrophages with human immunodeficiency virus type 1 (HIV-1). Monocyte-tropic and lymphocyte-tropic strains of HIV-1 show distinctive patterns of replication in a panel of cell types, *J. Exp. Med.* **170** (1989), 1149-1163.
19. von Briesen, H., Andreesen, R. and Rubsamen-Waigmann, H.: Systematic classification of HIV subtypes on lymphocytes and monocytes/macrophages, *Virology* **178** (1990), 597-602.
20. Kozak, S.L., Platt, E.J., Madani, N., Ferro Jr., F.E., Peden, K. and Kabat, D.: CD4, CXCR-4, and CCR-5 dependencies for infections by primary patient and laboratory-adapted isolates of human immunodeficiency virus type 1, *J. Virol.* **71** (1997), 873-882.
21. Valentin, A., Albert, J., Fredriksson, R., Fenyö, E.M. and Åsjö, B.: Dual tropism for lymphocytes and mononuclear phagocytes is a common feature of primary HIV-1 and HIV-2 isolates, *J. Virol.* **68** (1994), 6684-6689.
22. Connor, R.I. and Ho, D.D.: Human immunodeficiency virus type-1 variants increased replicative capacity develop during the asymptimatic stage before disease progression, *J. Virol.* **68** (1994), 4400-4408.
23. Stent, G., Joo, G.B., Kierulf, P. and Åsjö, B.: Macrophage tropism: fact or fiction?, *J. Leukocyte Biol.* **62** (1997), 4-11.
24. Moore, J.P., Cao, Y., Qing, L., Sattentau, Q.J., Pyati, J., Koduri, R., Robinson, J., Barbas, C.F.3, Burton, D.R. and Ho, D.D.: Primary isolates of human immunodeficiency virus type 1 are relatively resistant to neutralization by monoclonal antibodies to gp120, and their neutralization is not predicted by studies with monomeric gp120, *J. Virol.* **69** (1995), 101-109.
25. Sullivan, N., Sun, Y., Li, J., Hofmann, W. and Sodroski, J.: Replicative function and neutralization sensitivity of envelope glycoproteins from primary and T-cell line-passaged human immunodeficiency virus type 1 isolates, *J. Virol.* **69** (1995), 4413-4422.
26. Koot, M., Vos, A.H., Keet, I.P.M., de Goede, R.E.Y., Dercksen, M.W., Terpstra, F.G., Coutinho, R.A., Miedema, F. and Tersmette, M.: HIV-1 biological phenotype in long-term infected individuals evaluated with an MT-2 cocultivation assay, *AIDS* **6** (1992), 49-54.
27. Scarlatti, G., Hodara, V., Rossi, P., Muggiasca, L., Bucceri, A., Albert, J. and Fenyö, E.M.: Transmission of human immunodeficiency virus type 1 from mother to child correlates with viral phenotype, *Virology* **197** (1993), 624-629.
28. Fiore, J., Björndal, Å., Aperia-Peipke, K., Di Stefano, M., Angarano, G., Pastore, G., Gaines, H., Fenyö, E.M. and Albert, J.: The biological phenotype of HIV-1 is usually preserved during and after sexual transmission, *Virology* **204** (1994), 297-303.
29. Choe, H., Farzan, M., Sun, Y., Sullivan, N., Rollins, B., Ponath, P.D., Wu, L., Mackay, C.R., LaRosa, G., Newman, W., Gerard, N., Gerard, C. and Sodroski, J.: The beta-chemokine receptors CCR3 and CCR5 facilitate infection by primary HIV-1 isolates, *Cell* **85** (1996), 1135-1148.
30. Doranz, B.J., Rucker, J., Yi, Y.J., Smyth, R.J., Samson, M., Peiper, S.C., Parmentier, M., Collman, R.G. and Doms, R.W.: A dual-tropic primary HIV-1 isolate that uses fusin and the beta-chemokine receptors CKR-5, CKR-3, and CKR-2b as fusion cofactors, *Cell* **85** (1996), 1149-1158.
31. Dragic, R., Litwin, V., Allaway, G.P., Martin, S.R., Huang, Y.X., Nagashima, K.A., Cayanan, C., Maddon, P.J., Koup, R.A., Moore, J.P. and Paxton, W.A.: HIV-1 entry into $CD4^+$ cells is mediated by the chemokine receptor CC-CKR-5, *Nature* **381** (1996), 667-673.
32. Alkhatib, G., Combadiere, C., Broder, C.C., Feng, Y., Kennedy, P.E., Murphy, P.M. and Berger, E.A.: CC-CKR5: a RANTES, MIP-1alpha, MIP-1beta receptor as a fusion cofactor for macrophage-tropic HIV-1, *Science* **272** (1996), 1955-1958.
33. Feng, Y., Broder, C.C., Kennedy, P.E. and Berger, E.A.: HIV-1 entry cofactor: functional cDNA cloning of a seven-transmembrane, G protein-coupled receptor, *Science* **272** (1996), 872-877.
34. Simmons, G., Wilkinson, D., Reeves, J.D., Dittmar, M.T., Beddows, S., Weber, J., Carnegie, G., Desselberger, U., Gray, P.W., Weiss, R.A. and Clapham, P.R.: Primary, syncytium-inducing human immunodeficiency virus type 1 isolates are dual-tropic and most can use either lestr or CCR5 as coreceptors for virus entry, *J. Virol.* **70** (1996), 8355-8360.
35. Zhang, L., Huang, Y., He, T., Cao, Y. and Ho, D.D.: HIV-1 subtype and second-receptor use, *Nature* **383** (1996), 768.

36. Björndal, Å., Deng, H., Jansson, M., Fiore, J.R., Colognesi, C., Karlsson, A., Albert, J., Scarlatti, G., Littman, D.R. and Fenyö, E.M.: Co-receptor usage of primary human immunodeficiency virus type 1 isolates varies according to biological phenotype, *J. Virol.* **71** (1997), 7478-7487.
37. Tscherning, C., Alaeus, A., Fredriksson, R., Björndal, Å., Deng, H.K., Littman, D.R., Fenyö, E.M. and Albert, J.: Differences in chemokine coreceptor usage between genetic subtypes of HIV-1, *Virology* **241** (1998), 181-188.
38. Jansson, M., Popovic, M., Karlsson, A., Cocchi, F., Rossi, P., Albert, J. and Wigzell, H.: Sensitivity to inhibition by α-chemokines correlates with biological phenotype of primary human immunodeficiency virus type 1 isolates, *Proc. Natl. Acad. Sci. USA* **93** (1996), 15382-15387.
39. Oberlin, E., Amara, A., Bachelerie, F., Bessia, C., Virelizier, J.L., Arenzanaseisdedos, F., Schwartz, O., Heard, J.M., Clarklewis, I., Legler, D.F., Loetscher, M., Baggiolini, M. and Moser, B.: The CXC chemokine SDF-1 is the ligand for lestr/fusin and prevents infection by T-cell-line-adapted HIV-1, *Nature* **382** (1996), 833-835.
40. Bleul, C.C., Farzan, M., Choe, H., Parolin, C., Clarklewis, I., Sodroski, J. and Springer, T.A.: The lymphocyte chemoattractant SDF-1 is a ligand for lestr/fusin and blocks HIV-1 entry, *Nature* **382** (1996), 829-833.
41. Bleul, C.C., Wu, L., Hoxie, J.A., Springer, T.A. and Mackay, C.R.: The HIV coreceptors CXCR4 and CCR5 are differentially expressed and regulated on human T lymphocytes, *Proc. Natl. Acad. Sci. USA* **94** (1997), 1925-1930.
42. McKnight, A., Wilkinson, D., Simmons, G., Talbot, S., Picard, L., Ahuja, M., Marsh, M., Hoxie, J.A. and Clapham, P.R.: Inhibition of human immunodeficiency virus fusion by a monoclonal antibody to a coreceptor (CXCR4) is both cell-type- and -strain-dependent, *J. Virol.* **71** (1997), 1692-1696.
43. Wu, L., Paxton, W.A., Kessam, N., Ruffing, N., Rottman, J.B., Sullivan, N., Choe, H., Sodroski, J., Newman, W., Koup, R.A. and Mackay, C.R.: CCR5 levels and expression patterns correlate with infectability by macrophage-tropic HIV-1, *in vitro*, *J. Exp. Med.* **185** (1997), 1681-1691.
44. Cheng-Mayer, C., Liu, R., Landau, N.R. and Stamatatos, L.: Macrophage tropism of human immunodeficiency virus type 1 and utilization of the CC-CKR5 coreceptor, *J. Virol.* **71** (1997), 1657-1661.
45. Berger, E.A., Doms, R.W., Fenyö, E.M., Korber, B.T., Littman, D.R., Moore, J.P., Sattentau, Q.J., Schuitemaker, H., Sodroski, J. and Weiss, R.A.: A new classification for HIV-1, *Nature* **391** (1998), 240.
46. O'Brien, W.A., Koyanagi, Y., Namazie, A., Zhao, J.Q., Diagne, A., Idler, K., Zack, J.A. and Chen, I.S.Y.: HIV-1 tropism for mononuclear phagocytes can be determined by regions of gp120 outside the CD4-binding domain, *Nature* **348** (1990), 69-73.
47. Shioda, T., Levy, J.A. and Cheng-Mayer, C.: Macrophage and T cell-line tropisms of HIV-1 are determined by specific regions of the envelope gp120 gene, *Nature* **349** (1991), 167-169.
48. Fouchier, R.A., Groenink, M., Kootstra, N.A., Tersmette, M., Huisman, J.G., Miedema, F. and Schuitemaker, H.: Phenotype-associated sequence variation in the third variable domain of the human immunodeficiency virus type 1 gp120 molecule, *J. Virol.* **66** (1992), 3183-3187.
49. De Jong, J.J., De Ronde, A., Keulen, W., Tersmette, M. and Goudsmit, J.: Minimal requirements for the human immunodeficiency virus type 1 V3 domain to support the syncytium-inducing phenotype: analysis by single amino acid substitution, *J. Virol.* **66** (1992), 6777-6780.
50. Cocchi, F., De Vico, A.L., Garzino-Demo, A., Cara, A., Gallo, R.C. and Lusso, P.: The V3 domain of the HIV-1 gp120 envelope glycoprotein is critical for chemokine-mediated blockade of infection, *Nature Med.* **2** (1996), 1244-1247.
51. Wu, L.J., Gerard, N.P., Wyatt, R., Choe, H., Parolin, C., Ruffing, N., Borsetti, A., Cardoso, A.A., Desjardin, E., Newman, W., Gerard, C. and Sodroski, J.: CD4-induced interaction of primary HIV-1 gp120 glycoproteins with the chemokine receptor CCR-5, *Nature* **384** (1996), 179-183.
52. Speck, R.F., Wehrly, K., Platt, E.J., Atchison, R.E., Charo, I.F., Kabat, D., Chesebro, B. and Goldsmith, M.A.: Selective employment of chemokine receptors as human immunodeficiency virus type 1 coreceptors determined by individual amino acids within the envelope V3 loop, *J. Virol.* **71** (1997), 7136-7139.
53. Groenink, M., Fouchier, R.A.M., Broersen, S., Baker, C.H., Koot, M., van 't Wout, A.B., Huisman, J.G., Miedema, F., Tersmette, M. and Schuitemaker, H.: Relation of phenotype evolution of HIV-1 to envelope V2 configuration, *Science* **260** (1993), 1513-1516.
54. Wang, N., Zhu, T. and Ho, D.D.: Sequence diversity of V1 and V2 domains of gp120 from human immunodeficiency virus type 1: lack of correlation with viral phenotype, *J. Virol.* **69** (1995), 2708-2715.

55. Cornelissen, M., Hogervorst, E., Zorgdrager, F., Hartman, S. and Goudsmit, J.: Maintenance of syncytium-inducing phenotype of HIV type 1 is associated with positively charged residues in the HIV type 1 gp120 V2 domain without fixed positions, elongation, or relocated N-linked glycosylation sites, *AIDS Res.* Human *Retrovir.* **11** (1995), 1169-1175.
56. Fouchier, R.A.M., Broersen, S.M., Brouwer, M., Tersmette, M., Van 't Wout, A.B., Groenink, M. and Schuitemaker, H.: Temporal relationship between elongation of the HIV type-1 glycoprotein 120 V2 domain and the conversion toward a syncytium-inducing phenotype, *AIDS Res. Human Retrovir.* **11** (1995), 1473-1478.
57. Schuitemaker, H., Fouchier R.A.M., Broersen, S.M., Groenink, M., Koot, M., Van 't Wout, A.B., Huisman, J.G., Tersmette, M. and Miedema, F.: Envelope V2 configuration and HIV-1 phenotype: Clarification, *Science* **268** (1995), 115.
58. Myers, G., Korber, B., Foley, B., Jeang, K.-T., Mellors, J.W. and Wain-Hobson, S.: Human retroviruses and AIDS 1996: A compilation and analysis of nucleic acid and amino acid sequences. Theoretical Biology and Biophysics Group T10, Los Alamos National Laboratory, Los Alamos, New Mexico, USA., 1996.
59. Soto-Ramirez, L.E., Renjifo, B., McLane, M.F., Marlink, R., O'Hara, C., Sutthen, R., Wasi, C., Vithayasai, P., Vithayasai, V., Apichartpiyakul, C., Auewarakul, P., Pena Cruz, V., Chui, D.S., Osathanondh, R., Mayer, K., Lee, T.H. and Essex, M.: HIV-1 Langerhans' cell tropism associated with heterosexual transmission of HIV, *Science* **271** (1996), 1291-1293.
60. Pope, M., Frankel, S.S., Mascola, J.R., Trkola, A., Isdell, F., Birx, D.L., Burke, D.S., Ho, D.D. and Moore, J.P.: Human immunodeficiency virus type 1 strains of subtypes B and E replicate in cutaneous dendritic cell T-cell mixtures without displaying subtype-specific tropism, *J. Virol* .**71** (1997), 8001-8007.
61. Dittmar, M.T., Simmons, G., Hibbitts, S., Ohare, M., Louisirirotchanakul, S., Beddows, S., Weber, J., Clapham, P.R. and Weiss, R.A.: Langerhans cell tropism of human immunodeficiency virus type 1 subtype A through F isolates derived from different transmission groups, *J. Virol.* **71** (1997), 8008-8013.
62. de Wolf, F., Hogervorst, E., Goudsmit, J., Fenyö, E.M., Rubsamen-Waigmann, H., Holmes, H., Galvao-Castro, B., Karita, E., Wasi, C., Sempala, S.D.K., Baan, E., Zorgdrager, F., Lukasov, V., Osmanov, S., Kuiken, K., Cornelissen, M., the WHO Network for HIV Isolation.: Syncytium-inducing and non-syncytium-inducing capacity of human immunodeficiency type 1 subtypes other than B: phenotypic and genotypic characteristics, *AIDS Res. Human Retrovir.* **10** (1994), 1387-1400.
63. Tersmette, M., Gruters, R.A., de Wolf, F., de Goede, R.E.Y., Lange, J.M.A., Schellekens, P.Th.A., Goudsmit, J., Huisman, J.G. and Miedema, F.: Evidence for a role of virulent human immunodeficiency virus (HIV) variants in the pathogenesis of acquired immunodeficiency syndrome: studies on sequential HIV isolates, *J. Virol.* **63** (1989), 2118-2125.
64. Cheng-Mayer, C., Seto, D., Tateno, M. and Levy, J.A.: Biological features of HIV-1 that correlate with virulence in the host, *Science* **240** (1988), 80-82.
65. Koot, M., Keet, I.P.M., Vos, A.H., de Goede, R.E.Y., Roos, M.Th.L., Coutinho, R.A., Miedema, F., Schellekens, P.Th.A. and Tersmette, M.: Prognostic value of HIV-1 syncytium-inducing phenotype for rate of $CD4^+$ cell depletion and progression to AIDS, *Ann. Intern. Med.* **118** (1993), 681-688.
66. Karlsson, A., Parsmyr, K., Sandström, E., Fenyö, E.M. and Albert, J.: MT-2 cell tropism as a prognostic marker for disease progression in HIV-1 infection, *J. Clin. Microbiol.* **32** (1994), 364-370.
67. Richman, D.D., and Bozzette, S.A.: The impact of the syncytium-inducing phenotype of human immunodeficency virus on disease progression, *J. Infect.* Dis. **169** (1994), 968-974.
68. Nielsen, C., Pedersen, C., Lundgren, J.D. and Gerstoft, J.: Biological properties of HIV isolates in primary HIV infection: consequences for the subsequent course of infection, *AIDS* **7** (1993), 1035-1040.
69. Mellors, J.W., Rinaldo, C.R.J., Gupta, P., White, R.M., Todd, J.A. and Kingsley, L.A.: Prognosis in HIV-1 infection predicted by the quantity of virus in plasma. *Science* **272** (1996), 1167-1170.
70. Koot, M., van 't Wout, A.B., Kootstra, N.A., de Goede, R.E.Y., Tersmette, M. and Schuitemaker, H.: Relation between changes in cellular load, evolution of viral phenotype, and the clonal composition of virus populations in the course of human immunodeficiency virus type 1 infection, *J. Infect. Dis.* **173** (1996), 349-354.

71. Blaak, H., de Wolf, F., van 't Wout, A.B., Pakker, N.G., Bakker, M., Goudsmit, J. and Schuitemaker, H.: Temporal relationship between human immunodeficiency virus type 1 RNA levels in serum and cellular infectious load peripheral blood, *J. Infect. Dis.* **176** (1997), 1383-1387.
72. Bratt, G., Karlsson, A., Leandersson, A.-C., Albert, J., Wahren, B. and Sandström, E.: Treatment history and baseline viral load, but not viral tropism or CCR5 genotype, influence prolonged efficacy of highly active antiretroviral treatment, *AIDS* **12** (1998), 2193-2202.
73. Kaneshima, H., Su, L., Bonyhadi, M.L., Connor, R.I., Ho, D.D. and McCune, J.M.: Rapid-high, syncytium-inducing isolates of human immunodeficiency virus type 1 induce cytopathicity in the human thymus of the SCID-hu mouse, *J. Virol.* **68** (1994), 8188-8192.
74. Glushakova, S., Baibakov, B., Zimmerberg, J. and Margolis, L.B.: Experimental HIV infection of human lymphoid tissue: correlation of $CD4^+$ T-cell depletion and virus syncytium-inducing/non-syncytium-inducing phenotype in histocultures inoculated with laboratory strains and patient isolates of HIV type 1, *AIDS Res. Human Retrovir.* **13** (1997), 461-471.
75. Schuitemaker, H., Koot, M., Koostra, N.A., Dercksen, M.W., de Goede, R.E.Y., van Steenwijk, R.P., Lange, J.M.A., Eeftinck Schattenkerk, J.K.M., Miedema, F. and Tersmette, M.: Biological phenotype of human immunodeficiency virus type 1 clones at different stages of infection: progression of disease is associated with a shift from monocytotropic to T-cell-tropic virus populations, *J. Virol.* **66** (1992), 1354-1360.
76. Scarlatti, G., Leitner, T., Hodara, V., Jansson, M., Karlsson, A., Wahlberg, J., Rossi, P., Uhlén, M., Fenyö, E.M. and Albert, J.: Interplay of HIV-1 phenotype and neutralizing antibody response in patho-genesis of AIDS, *Immunol. Lett.* **51** (1996), 23-28.
77. Samson, M., Libert, F., Doranz, B.J., Rucker, J., Liesnard, C., Farber, C.M., Saragosti, S., Lapoumeroulie, C., Cognaux, ., Forceille, C., Muyldermans, G., Verhofstede, C., Burtonboy, G., Georges, M., Imai, T., Rana, S., Yi, Y., Smyth, R.J., Collman, R.G. and Doms, R.W.: Resistance to HIV-1 infection in caucasian individuals bearing mutant alleles of the CCR-5 chemokine receptor gene: see comments, *Nature* **382** (1996), 722-725.
78. Dean, M., Carrington, M., Winkler, C., Huttley, G.A., Smith, M.W., Allikmets, R., Goedert, J.J., Buchbinder, S.P., Vittinghoff, E., Gomperts, E., Donfield, S., Vlahov, D., Kaslow, R., Saah, A., Rinaldo, C., Detels, R. and O'Brien, S.J.: Genetic restriction of HIV-1 infection and progression to AIDS by a deletion allele of the CKR5 structural gene. Hemophilia Growth and Development Study, Multicenter AIDS Cohort Study, Multicenter Hemophilia Cohort Study, San Francisco City Cohort, A, *Science* **273** (1996), 1856-1862.
79. Bratt, G., Sandström, E., Albert, J., Samson, M. and Wahren, B.: The influence of MT-2 tropism on the prognostic implications of the 32 basepair deletion in the CCR5 gene, *AIDS* **11** (1997), 1415-1419.
80. de Roda Husman, A.M., Koot, M., Cornelissen, M., Keet, I.P.M., Brouwer, M., Broersen, S.M., Bakker, M., Roos, M.Th.L., Prins, M., de Wolf, F., Coutinho, R.A., Miedema, F., Goudsmit, J. and Schuitemaker, H.: Association between CCR5 genotype and the clinical course of HIV-1 virus infection, *Ann. Intern. Med.* **127** (1997), 882-890.
81. McNearney, T., Westervelt, P., Thielan, B.J., Trowbridge, D.B., Garcia, J., Whittier, R. and Ratner, L.: Limited sequence heterogeneity among biologically distinct human immunodeficiency virus type 1 isolates from individuals involved in a clustered infectious outbreak, *Proc. Natl. Acad. Sci. USA* **87** (1990), 1917-1921.
82. Zhu, T., Mo, H., Wang, N., Nam, D.S., Cao, Y., Koup, R.A. and Ho, D.D.: Genotypic and phenotypic characterization of HIV-1 in patients with primary infection, *Science* **261** (1993), 1179-1181.
83. Zhang, L.Q., MacKenzie, P., Cleland, A., Holmes, E.C., Leigh Brown, A.J. and Simmonds, P.: Selection for specific sequences in the external envelope protein of human immunodeficiency virus type 1 upon primary infection, *J. Virol.* **67** (1993), 3345-3356.
84. Van 't Wout, A.B., Kootstra, N.A., Mulder-Kampinga, G.A., Albrecht-van Lent, N., Scherpbier, H.J., Veenstra, J., Boer, K., Coutinho, R.A., Miedema, F. and Schuitemaker, H.: Macrophage-tropic variants initiate human immunodeficiency virus type-1 infection after sexual, parenteral and vertical transmission, *J. Clin. Invest.* **94** (1994), 2060-2067.
85. Albert, J., Fiore, J., Fenyö, E.M., Pedersen, C., Lundgren, J.D., Gerstoft, J. and Nielsen, C.: Biological phenotype of HIV-1 and transmission (letter), *AIDS* **9** (1995), 822-823.

86. Koot, M., Schellekens, P.Th.A., Mulder, J.W., Lange, J.M.A., Roos, M.Th.L., Coutinho, R.A., Tersmette, M. and Miedema, F.: Viral phenotype and T-cell reactivity in human immunodeficiency virus type 1-infected asymptomatic men treated with zidovudine, *J. Infect. Dis.* **168** (1993), 733-736.

87. Karlsson, A., Parsmyr, K., Aperia, K., Sandström, E. and Albert, J.: MT-2 cell tropism and development of zidovudine and didanosine resistance in HIV-1-infected individuals, *J. Infect. Dis.* **170** (1994), 1367-1375.

88. Van 't Wout, A.B., De Jong, M.D., Kootstra, N.A., Veenstra, J., Lange, J.M.A., Boucher, C.A.B. and Schuitemaker, H.: Changes in cellular virus load and zidovudine resistance of syncytium-inducing and non-syncytium-inducing human immunodeficiency virus populations under zidovudine pressure: a clonal analysis, *J. Infect. Dis.* **174** (1996), 845-849.

89. Van 't Wout, A.B., Ran, L.J., De Jong, M.D., Bakker, M., Van Leeuwen, R., Notermans, D.W., Loeliger, A.E., De Wolf, F., Danner, S.A., Reiss, P., Boucher, C.A.B., Lange, J.M.A. and Schuitemaker, H.: Selective inhibition of syncytium- inducing and non-syncytium-inducing HIV-1 variants in individuals receiving didanosine or zidovudine, respectively, *J. Clin. Invest.* **100** (1997), 2325-2332.

90. Delforge, M.-L., Liesnard, C., Debaisieux, L., Tchetcheroff, M., Farber, C.-M. and Van Vooren, J.-P.: *In-vivo* inhibition of syncytium-inducing variants of HIV in patients treated with didanosine, *AIDS* **9** (1995), 89-101.

91. Zheng, N.N., McQueen, P.W., Hurren, L., Evans, L.A., Law, M.G., Forde, S., Barker, S. and Cooper, D.A.: Changes in biologic phenotype of human immunodeficiency virus during treatment of patients with didanosine, *J. Infect. Dis.* **173** (1995), 1092-1096.

CHAPTER 4

T-CELL DYNAMICS AND RENEWAL IN HIV-1 INFECTION

DAWN R. CLARK and KATJA C. WOLTHERS

Department of Clinical Viro-Immunology of the CLB and Laboratory for Experimental and Clinical Immunology, Academic Medical Centre, University of Amsterdam, Amsterdam, The Netherlands

The mechanism for $CD4^+$ T-cell decline, one of the most prominent features of HIV-1 infection, is still not properly understood. Over the past 10 years of AIDS research, investigators have examined virus-related killing, apoptosis, and disturbed renewal capacity as possible mechanisms for $CD4^+$ T-cell depletion [1-3]. In 1995 these possibilities were linked as two sides of the same coin, namely "turnover" of T cells. Now, the extent of turnover in T cells and its role in HIV-1 pathogenesis is a point of debate [4]. In this chapter, an overview is presented of our current knowledge on T-cell turnover and renewal in HIV-1 infection and the implications for $CD4^+$ T-cell depletion.

1. Virus and $CD4^+$ T-cell dynamics in HIV-1 infection

The T-cell depletion seen in HIV-1 infection is the result of changes in T-cell subsets during the course of infection. Overall $CD4^+$ T cells decline while overall $CD8^+$ T cells increase over time. The increase in the $CD8^+$ T-cell pool is the result of peripheral expansion of memory cells. This subset only begins to decline shortly preceding AIDS diagnosis [5]. The $CD4^+$ memory compartment also initially expands due to peripheral expansion, but memory cells are then progressively lost. Interestingly, in both the $CD4^+$ and the $CD8^+$ subsets, the naïve compartment begins to decline soon after infection [6,7]. Thus, the T-cell depletion observed in HIV-1 infection consists of naïve cells of both $CD4^+$ and $CD8^+$ subsets, and memory cells of the $CD4^+$ subset. The majority of research has focussed on the depletion of $CD4^+$ T cells and the bulk of the discussion which follows will concentrate on this subset.

During the asymptomatic phase, it was believed that HIV was present in low amounts in the body, affecting specific immune functions which would lead to immune deterioration, $CD4^+$ T-cell decline and eventually to AIDS [8,9]. With the development of PCR-based techniques it was shown that even during clinical latency high levels of virus were present [1,10]. However, the first insight into the magnitude of viral replication was provided in 1995 by the studies of Ho *et al.* [11] and Wei *et al.* [12]. Here, new antiretroviral drugs were shown to be highly effective

H. Schuitemaker and F. Miedema (ed.), Aids Pathogenesis, 55–64.

in eliminating virus particles and increasing CD4$^+$ T cells in the first 4 weeks after start of therapy. By calculating the slopes of viral decay and CD4$^+$ T-cell recovery after treatment, they were able to estimate the amount of virus particles that were eliminated and the number of cells that were produced per day. By assuming that during infection a steady state exists, such that the amount of virus produced equals the amount eliminated, virus production was calculated to be on the order of 10^9-10^{10} particles per day [13], which was much higher than previously anticipated. In addition, the virus half-life was determined to be 6 hr. Suddenly, the view on HIV-1 infection had changed into one of a virus infection which constantly produced enormous amounts of virus with high replication rates, which would lead to high mutation rates. Assuming that the CD4$^+$ T cells which arose in the blood during the first 30 days of therapy were newly produced cells, the production rate per day was calculated and extrapolated to the total amount of newly produced CD4$^+$ T cells in the body. Assuming a pretreatment steady state, this production rate would thus balance the destruction of cells by the virus. To better understand this, Ho *et al.* [11] introduced the metaphor of the sink, with the drain (CD4$^+$ T-cell destruction) wide open due to HIV-related killing. The level in the sink is initially balanced by increased flow from the tap (CD4$^+$ T-cell production) but eventually the flow of the tap will be exhausted. The 'turnover rate', which is the amount of cells produced and destroyed per day, was thus calculated to be 2 x 10^9 cells/day. It was easy to imagine that such a high turnover would eventually exhaust the system's capacity to regenerate and in the end would lead to decline of CD4$^+$ T cells.

1.1. T-CELL TURNOVER IN HIV-1 INFECTION

The capacity to generate new T cells, the life-span of T cells and the T-cell expansion in response to antigens are all unknown factors when turnover of cells in a viral infection is addressed. Although the life-span of lymphocytes has been studied in humans [14], little is known about normal turnover rates or turnover rates after virus infections.

1.1.1. *Telomere length as a marker for replication*

Telomere length can be used to determine the amount of divisions a cell or a cell population has gone through. Telomeres are the extreme ends of chromosomes comprised of TTAGGG repeats, approximately 10 kb in length in humans [15]. During cell division, telomeres shorten because of the inability of DNA polymerases to fully replicate the 3' end. Telomeres of normal somatic cells *in vivo* gradually shorten with age; human leukocytes for example lose about 33 bp per year [16]. Telomere shortening upon cell division does not occur in malignant cells or germline cells, which express high levels of telomerase, a ribonucleoprotein polymerase that carries an RNA template to synthesise telomeric repeats onto chromosomal ends. Somatic cells show little or no telomerase activity [17]. Therefore, telomere length of normal somatic cell populations, lymphocytes for example, provides information on replicative history.

Normal to increased telomere lengths were reported in $CD4^+$ T cells of HIV-1-infected individuals compared to healthy persons [18,19], and normal shortening rates were found even in patients who progressed to AIDS [19]. In a small group of HIV-infected individuals with $CD4^+$ T-cell counts <200 mm^{-3}, shorter telomeres were found [20]. $CD8^+$ T cells from HIV-infected patients did lose telomere length at an accelerated rate and therefore $CD8^+$ T-cell telomere length was shorter in HIV-infected patients than in healthy controls [18-21]. This differential telomere length loss of $CD4^+$ and $CD8^+$ T cells in HIV-1 infection could not be explained by altered telomerase activity, since no abnormalities in telomerase activity were found in T lymphocytes from HIV-infected individuals [18,19]. Controversy exists over whether telomere length analysis can be interpreted in terms of T-cell turnover, because telomere length analysis cannot account for cells lost by cell death, and therefore normal telomere length can still be compatible with high turnover rates [20]. A different interpretation is that if increased death rates lead to increased production rates, eventually this would be reflected in shorter mean telomere length of the remaining population [22]. Mathematical interpretation of normal telomere length of $CD4^+$ T cells in HIV-1 infection allows for a maximum 2- to 4-fold increase in production [23].

1.1.2. *Ki-67 as marker for proliferation*

Another parameter for measurement of proliferation of human cells is Ki-67 expression. Ki-67 is a nuclear antigen that is expressed in all cycling cells but not in the G_0 phase [24]. By immunohistological staining and quantitative image analysis, tonsil and lymph-node tissue were analysed for Ki-67 expressing $CD4^+$ T cells. The proportion of proliferating $CD4^+$ T cells was maximally 3-fold increased in HIV-1-infected individuals [25]. The percentage Ki-positive cells increased with decreasing $CD4^+$ T-cell numbers in HIV-1-infected patients [26]. When absolute numbers of Ki-67-positive cells in blood and lymph nodes were calculated, the fraction of proliferating $CD4^+$ T cells was similar in HIV-infected individuals with $CD4^+$ T cells >300 cells mm^{-3} and healthy controls, while the fraction of proliferating $CD8^+$ T cells was increased in the patients [27]. One could argue that the effect of increased destruction rates of proliferating $CD4^+$ T cells in HIV-1 infection will be that the magnitude of proliferation cannot be measured by Ki-67 expression. However, the amount of Ki-67-expressing cells is not increased during the first weeks of therapy, which implies that before therapy the destruction rate of these proliferating cells was small, since otherwise one would have expected an increase in Ki-67-positive cells that directly after therapy would be spared from destruction.

1.1.3. *Labelling of proliferating cells*

Turnover of lymphocytes in mice could be measured by labelling proliferating cells with the DNA precursor bromodeoxyuridine (BrdU) [28]. Recently, labelling SIV-infected and non-infected macaques with BrdU added to the drinking water showed substantial increases in proliferation and death rates of lymphocytes and NK cells in the SIV-infected monkeys with high viral load [29]. Remarkably, the rate of output from thymic or extrathymic sources in uninfected monkeys was calculated to be the same as in mice [28], while the output in infected monkeys was calculated to be 3-

fold higher. This suggests that SIV would not interfere with the capacity to regenerate new cells, while there is evidence that HIV-1 does interfere with renewal capacity.

1.2. INCREASED TURNOVER IN HIV-1 INFECTION: HOW HIGH IS HIGH?

Measurements of proliferation by telomere length and Ki-67 expression suggest a normal to maximally 4-fold increased proliferation rate of $CD4^+$ T cells in HIV-1 infection. These findings seem to contradict the initial statements on high turnover of $CD4^+$ T cells in HIV-1 infection [11,12]. Although no exact definition was given for the magnitude of increased turnover rates, it was suggested that in some patients the turnover rate was 78-fold increased compared to others [11]. In SIV-infected macaques, the exact rate of increase for T-lymphocyte turnover has not been determined, but seems to be in the order of 3-6 fold [29]. In addition to production rates, turnover involves destruction rates. With a high number of virions produced every day [11,12], high destruction of $CD4^+$ T cells is intuitively consistent. However, there is no actual evidence for high destruction rates of $CD4^+$ T cells. Apoptosis does occur in $CD4^+$ T cells of HIV-infected individuals, but to a lesser extent than in $CD8^+$ T cells [3]. Zhang *et al.* reported a 2-fold increase in $CD4^+$ T-cell apoptosis in lymphoid tissue [25]. Haase *et al.* [30] reported low frequencies of productively infected cells and Chun *et al.* [31] showed low frequencies of both resting and activated $CD4^+$ T cells with integrated provirus. This implies that only small numbers of cells are infected, which are then responsible for high virus production. Therefore, turnover of these productively infected cells is only a fraction of the total turnover in $CD4^+$ T cells [25]. In conclusion, turnover rates may be increased in HIV-1 infection, but not to levels as have previously been suggested. However, it has been suggested that the renewal capacity of the immune system is limited and may be unable to keep up with even modestly increased destruction.

2. T-cell renewal from progenitors

The process of T-cell renewal in the maintenance of the T-cell population can involve two separate mechanisms; proliferation of mature cells in the periphery and development of new T cells from a progenitor source. The relative contribution of each of these mechanisms to overall T-cell renewal is dependent upon the age of the individual and the profile of the remaining T-cell pool. The common view is that early T-cell development requires a thymus and involves a high degree of development from progenitor sources but that as the organism ages the T-cell population is maintained primarily through peripheral expansion of dividing mature cells [32-34]. It is also clear that a T-cell-depleted individual cannot completely regenerate the T-cell population in the absence of thymic function [32,35,36].

The impact of HIV-1 infection on T-cell renewal from a progenitor source was originally thought to be minor based on the lack of infection of bone-marrow progenitor cells [37,38]. However, with the advent of potent antiretroviral therapy more interest has focused on parameters of immune reconstitution in infected individuals. For an adult infected with HIV-1 even the low thymic regeneration found in uninfected adults is likely to be lacking.

2.1. T-CELL RENEWAL ASSESSMENT IN HIV-1 INFECTION

Determination of the level of regeneration in humans is complicated by a lack of suitable markers for distinguishing between cells that developed from a progenitor and those arising by proliferation of existing cells. Since the naïve compartment is thought to proliferate little, if at all, without alteration of the CD45 phenotype, increases in this pool are considered to result from development of new cells. By contrast, increases in the memory pool must be via proliferation of either existing memory cells, or of activated naïve cells, which will then change their CD45 phenotype. For newly developed cells, it is difficult to determine whether cells developed within the thymus or outside the thymus. Despite these difficulties, researchers have used a variety of means to try to assess the level of regeneration in HIV-1-infected individuals.

2.1.1. *Regeneration assessment by bone-marrow function*

There are several pieces of evidence that the bone marrow of HIV-1-infected persons display developmental abnormalities. The first observations suggesting diminished development of bone marrow, were the multiple cytopenias experienced by many infected individuals [39]. Enumeration of progenitor subsets by monoclonal antibody staining of surface molecules showed the presence of all progenitor subsets in bone marrow from HIV-1-infected individuals [40], though some reports suggest that certain subsets have been reduced in HIV-1 infection [41,42]. No alteration in the number of very primitive progenitors in infected persons could be detected in assays that support the development of these cells [40]. Therefore, there is no clear evidence that the number of progenitors has been adversely affected by HIV-1 infection.

Bone-marrow stroma from HIV-1-infected persons is able to promote the development of progenitors from an uninfected individual [43]. However, the progenitors from the infected individual were unable to develop on bone-marrow stroma from the uninfected individual. Bone marrow from HIV-1-infected individuals was shown to have diminished capacity to develop cells of the granulocyte, erythrocyte, and megakaryocyte lineages in *in-vitro* colony assays [44-46]. These data add to the accumulating evidence that the bone-marrow progenitors of HIV-1-infected individuals are impaired in their ability to develop to mature haematopoietic cells of multiple lineages.

2.1.2. *Progenitor development capacity measurements*

Progenitor development to the T-cell lineage can be measured by *in-vitro* T-cell development systems, such as fetal thymus organ culture or thymic monolayer cultures, or by development in SCID-hu mice. HIV-1-infected SCID-hu mice show a thymic pathology, similar to that of infected children, with a lack of thymic subsets beyond the very primitive triple negative stage and alterations in stromal architecture [47]. Interpretation of these results in the context of progenitor cell function is difficult as the thymic tissue is also of human origin and a potential target for infection. Additionally, the thymic and liver tissues used are of fetal origin, which may have a different pattern of response than adult thymic tissue and bone marrow.

Measurement of progenitor function was recently measured by fetal thymus organ culture (FTOC). This technique used a murine fetal thymus to support the development of human progenitors into mature T cells. This xenogeneic system eliminates the confounding factor of the thymic tissue as a target for infection. In the initial cross-sectional study, the ability of HIV-1-infected individuals to develop T cells in FTOC was shown to be significantly lower than that of uninfected individuals [48]. Recently, a longitudinal study comparing individuals who progressed to AIDS and long-term non-progressors (LTNP) showed that progressors lost T-cell development capacity very early in infection while LTNP still retained significant capacity after 8 years of infection [49].

2.1.3. *Thymic function assessment*

Another way of measuring the effect of HIV-1 infection on T-cell development is by assessing thymic function, particulary since the naïve $CD4^+$ T-cell compartment has been shown to require a functioning thymus for its maintenance. Thymic mass of HIV-1-infected adults has been measured by CT scan [50]. A number of individuals over 40 years of age were found to have larger thymic mass than the same aged uninfected individuals. The number of naïve cells in the periphery could be correlated to the amount of thymic mass. These results show that the key to maintaining the naïve pool is development of naïve cells which requires a functional thymus.

Recently, another technique has been used to measure thymic function. Excision circle PCR detects T cells that have recently recombined their T-cell receptor (TCR) genes. When a progenitor cell develops into a mature T cell, it must recombine gene segments to form a functional T-cell receptor. During this process in $\alpha\beta$-TCR-bearing cells, the majority of T cells in the body, the entire d locus is excised and forms a stable circle in the nucleus. These excision circles, detected by specific PCR, are only found in $CD45RA^+$ cells [51]. This technique has been described as a measure of thymic function. In fact, it is a measurement of development of T cells irrespective of thymic function as any developing T cell must recombine TCR genes. In the case of $CD4^+$ cells, however, this technique would measure thymic function since development of this subset is thought to occur only in a thymus. Using this technique, Koup *et al.* [51] have shown that the number of cells in the periphery-expressing excision circles declines with HIV-1 infection. Taken together, results on thymic function support the contention that HIV-1 infection alters the T-cell renewal capability of infected individuals.

These data show that HIV-1 infection has a direct inhibitory effect on the already low thymic-dependent development in adults. This translates to a reduced development of $CD4^+$ cells with a naïve phenotype and places even more emphasis on the peripheral expansion of $CD4^+$ cells to maintain this population as it is being depleted by the virus. These data also suggest that the eventual depletion of $CD4^+$

cells is, at least in part, due to a block in the development of T cells from a progenitor source, which prevents the complete regeneration of the naïve CD4$^+$ T-cell compartment. In addition, these data explain the observation that naïve CD8$^+$ T cells are also depleted in HIV-1 infection. Whether the rate of peripheral mechanisms of regeneration are incapable of increasing to make up for the loss of T cells, or whether HIV-1 infection directly interferes with these peripheral mechanisms as well, remains to be determined.

2.1.4. *T-cell renewal after potent antiretroviral therapy*

The effect of antiretroviral therapy on T-cell renewal can provide additional information on the effect of HIV-1 infection on renewal. Patients experience a rise in the number of cells in the periphery after initiation of therapy. Initially, this increase consists of cells with a memory phenotype, however there is a slow, consistent rise in the number of naïve cells in most individuals [52-54]. Data from FTOC has shown that the T-cell development capacity of progenitors increases after therapy [49]. This increase could be correlated with the number of naïve cells in the periphery. In addition, the number of cells bearing TCR excision circles goes up after therapy [51]. In children infected with HIV-1, those with more thymic volume prior to therapy had the largest increase in number of CD4$^+$ T cells, CD4$^+$ CD45RA/RO ratio, and in TCR repertoire [55]. These data show that functional progenitors, in combination with functional thymic tissue, are required for reconstitution of treated individuals. They also support the contention that immune depletion in HIV-1 infection is, at least partially, the result of non-functional progenitors and/or non-functional thymic tissue.

2.2. HOW ARE T CELLS BEING DEPLETED IN HIV-1 INFECTION?

Based on the data reviewed above, a model for CD4$^+$ T-cell depletion can be proposed. During the course of asymptomatic infection, an increasing number of naïve cells will be activated and become memory cells. Data suggest that these cells are not replaced by development of new naïve cells, so the naïve pool will be effectively depleted over time. The memory pool is gradually depleted by activation-induced cell death and, in the case of CD4$^+$ T cells, by HIV-1-related death. Since development is inhibited, few naïve cells would feed into the memory population, which therefore must be maintained by increased proliferation of already existing cells. As discus-sed, the increase in proliferation does not appear to be vast and perhaps, though not directly shown, is insufficient to keep up with cell loss. The result is the eventual loss of the memory CD4$^+$ cells and, concommitant with AIDS diagnosis, the loss of the memory CD8$^+$ cells. Thus CD4$^+$ T-cell depletion is likely to be due to the combination of inhibition of development of new cells and the increase in proliferation of already existing cells.

3. References

1. Ho, D.D., Moudgil, T. and Alam, M.: Quantitation of human immunodeficiency virus type 1 in the blood of infected persons, *New Engl. J. Med.* **321** (1989), 1621-1625.
2. Bonyhadi, M.L., Rabin, L., Salimi, S., Brown, D.A., Kosek, J., McCune, J.M. and Kaneshima, H.: HIV induces thymus depletion *in vivo*, *Nature* **363** (1993), 728-732.
3. Meyaard, L., Otto, S.A., Jonker, R.R., Mijnster, M.J., Keet, I.P.M. and Miedema, F.: Programmed death of T cells in HIV-1 infection, *Science* **257** (1992), 217-219.
4. Clark, D.R., De Boer, R.J., Wolthers, K.C. and Miedema, F.: T-cell dynamics in HIV-1 infection, *Adv. Immunol.* (in press, 1999).
5. Margolick, J.B., Muñoz, A., Donnenberg, A.D., Park, L.P., Galai, N., Giorgi, J.V., O'Gorman, M.R.G., Ferbas, J. and for the Multicenter AIDS Cohort Study: Failure of T-cell homeostasis preceding AIDS in HIV-1 infection, *Nature Medicine* **1** (1995), 674-680.
6. Rabin, R.L., Roederer, M., Maldonado, Y., Petru, A. and Herzenberg, L.A.: Altered representation of naïve and memory CD8 T-cell subsets in HIV-infected children, *J. Clin. Invest.* **95** (1995), 2054-2060.
7. Roederer, M., Gregson Dubs, J., Anderson, M.T., Raju, P.A., Herzenberg, L.A. and Herzenberg, L.: CD8 naïve T-cell counts decrease progressively in HIV-infected adults, *J. Clin. Invest.* **95** (1995), 2061-2066.
8. Schnittman, S.M., Lane, H.C., Greenhouse, J., Justement, J.S., Baseler, M. and Fauci, A.S.: Preferential infection of $CD4^+$ memory T cells by human immunodeficiency virus type 1: evidence for a role in the selective T-cell functional defects observed in infected individuals, *Proc. Natl. Acad. Sci. USA* **87** (1990), 6058-6062.
9. Miedema, F., Petit, A.J.C., Terpstra, F.G., Eeftinck Schattenkerk, J.K.M., Al, E.J.M., Roos, M.Th.L., Lange, J.M.A., Danner, S.A., Goudsmit, J. and Schellekens, P.Th.A.: Immunological abnormalities in human immunodeficiency virus (HIV)-infected asymptomatic homosexual men. HIV-1 affects the immune system before $CD4^+$ T-helper cell depletion occurs, *J. Clin. Invest.* **82** (1988), 1908-1914.
10. Piatak, M., Saag, M.S., Yang, L.C., Clark, S.J., Kappes, J.C., Luk, K.C., Hahn, B.H., Shaw, G.M. and Lifson, J.D.: High levels of HIV-1 in plasma during all stages of infection determined by competitive PCR, *Science* **259** (1993), 1749-1754.
11. Ho, D.D., Neumann, A.U., Perelson, A.S., Chen, W., Leonard, J.M. and Markowitz, M.: Rapid turnover of plasma virions and CD4 lymphocytes in HIV-1 infection, *Nature* **373** (1995), 123-126.
12. Wei, X., Ghosh, S.K., Taylor, M.E., Johnson, V.A., Emini, E.A., Deutsch, P., Lifson, J.D., Bonhoeffer, S., Nowak, M.A., Hahn, B.H., Saag, M.S. and Shaw, G.M.: Viral dynamics in human immunodeficiency virus type 1 infection, *Nature* **373** (1995), 117-122.
13. Perelson, A.S., Neumann, A.U., Markowitz, M., Leonard, J.M. and Ho, D.D.: HIV-1 dynamics *in vivo*: virion clearance rate, infected cell life-span, and viral generation time, *Science* **271** (1996), 1582-1586.
14. Michie, C.A., McLean, A., Alcock, C. and Beverley, P.C.L.: Lifespan of human lymphocyte subsets defined by CD45 isoforms, *Nature* **360** (1992), 264-265.
15. Blackburn, E.H.: Structure and function of telomeres, *Nature* **350** (1991), 569-573.
16. Vaziri, H., Schächter, F., Uchida, I., Wei, L., Zhu, X., Effros, R., Cohen, D. and Harley, C.B.: Loss of telomeric DNA during aging of normal and trisomy 21 human lymphocytes, *Am. J. Human Genet.* **52** (1993), 661-667.
17. Broccoli, D., Young, J.W. and De Lange, T.: Telomerase activity in normal and malignant hematopoietic cells, *Proc. Natl. Acad. Sci. USA* **92** (1995), 9082-9086.
18. Palmer, L.D., Weng, N.P., Levine, B.L., June, C.H., Lane, H.C. and Hodes, R.J.: Telomere length, telomerase activity, and replicative potential in HIV infection: analysis of $CD4^+$ and $CD8^+$ T cells from HIV-discordant monozygotic twins, *J. Exp. Med.* **185** (1997), 1381-1386.
19. Wolthers, K.C., Wisman, G.B.A., Otto, S.A., De Roda Husman, A.-M., Schaft, N., De Wolf, F., Goudsmit, J., Coutinho, R.A., Van der Zee, A.G.J., Meyaard, L. and Miedema, F.: T-cell telomere length in HIV-1 infection: no evidence for increased $CD4^+$ T-cell turnover, *Science* **274** (1996), 1543-1547.

20. Pommier, J.-P., Gaithier, L., Livartowski, J., Galanaud, P., Boué, F., Dulious, A., Marcé, D., Ducray, C., Sabatier, L., Lebeau, J. and Boussin, F.-D.: Immunosenescence in HIV pathogenesis, *Virology* **231** (1997), 148-154.
21. Effros, R.B., Allsopp, R.C., Chiu, C.-P., Haysner, M.A., Hirji, K., Wang, L., Harley, C.B., Villeponteau, B., West, M.D. and Giorgi, J.V.: Shortened telomeres in the expanded CD28-$CD8^+$ cell subset in HIV disease implicate replicative senescence in HIV pathogenesis, *AIDS* **10** (1996), F17-F22.
22. Wolthers, K.C., Schuitemaker, H. and Miedema, F.: Rapid $CD4^+$ T-cell turnover in HIV-1 infection: a paradigm revisited, *Immunol. Today* **19** (1998), 44-48.
23. Wolthers, K.C., Noest, A.J., Otto, S.A., Miedema, F. and De Boer, R.J.: Normal telomere lengths in naïve and memory $CD4^+$ T cells in HIV-1 infection: a mathematical interpretation (submitted, 1998).
24. Gerdes, J., Schlueter, L.-L.C., Duchrow, M., Wohlenberg, C., Gerlach, C., Stahmer, I., Kloth, S., Brandt, E. and Flad, H.-D.: Immunobiochemical and molecular biologic characterization of the cell proliferation-associated nuclear antigen that is defined by monoclonal antibody Ki-67, *Am. J. Pathol.* **138** (1991), 867-873.
25. Zhang, Z.Q., Notermans, D.W., Sedgewick, G., Cavert, W., Wietgrefe, S., Zupancic, M., Gebhard, K., Henry, K., Boies, L., Chen, Z., Jenkins, M., Mills, R., McDade, H., Goodwin, G., Schuwirth, C.M., Danner, S.A. and Haase, A.T.: Kinetics of $CD4^+$ T-cell repopulation of lymphoid tissues after treatment of HIV-1 infection, *Proc. Natl. Acad. Sci. USA* **95** (1998), 1154-1159.
26. Perin, L., in press (1998).
27. Fleury, S., De Boer, R.J., Rizzardi, G.P., Wolthers, K.C., Otto, S.A., Welbon, C.C., Graziosi, C., Knabenhans, C., Soudeyns, H., Bart, P.-A., Gallant, S., Corpataux, J.-M., Gillet, M., Meylan, P., Schnyder, P., Meuwly, J.Y., Spreen, W., Glauser, M.P., Miedema, F. and Pantaleo, G.: Limited $CD4^+$ T-cell renewal in early HIV-1 infection: effect of highly active antiretroviral therapy, *Nature Med.* **4** (1998), 794-801.
28. Tough, D.F. and Sprent, J.: Turnover of naïve- and memory-phenotype T cells, *J. Exp. Med.* **179** (1994), 1127-1135.
29. Mohri, H., Bonhoeffer, S., Monard, S., Perelson, A.S. and Ho, D.D.: Rapid turnover of T lymphocytes in SIV-infected rhesus macaques, *Science* **279** (1998), 1223-1227.
30. Haase, A.T., Henry, K., Zupanc, M., Sedgewick, G., Faust, R.A., Melroe, H., Cavert, W., Gebhard, K., Staskus, K., Zhang, Z., Dailey, P.J., Balfour, H.H., Erice, A. and Perelson, A.S.: Quantitative image analysis of HIV-1 infection in lymphoid tissue, *Science* **274** (1996), 985-989.
31. Chun, T.-W., Carruth, L., Finzi, D., Shen, X., DiGiuseppe, J.A., Taylor, H., Hermankova, M., Chadwick, K., Margolick, J., Quinn, T.C., Kuo, Y.-H., Brookmeyer, R., Zeiger, M.A., Bardltch-Crovo, P. and Siliciano, R.F.: Quantification of latent tissue reservoirs and total body viral load in HIV-1 infection, *Nature* **387** (1997), 183-187.
32. Mackall, C.L. and Gress, R.E.: Thymic aging and T-cell regeneration, *Immunol. Rev.* **160** (1997), 91-102.
33. Rocha, B., Dautigny, N. and Pereira, P.: Peripheral T lymphocytes: expansion potential and homeostatic regulation of pool sizes and CD4/CD8 ratios *in vivo*, *Eur. J. Immunol.* **19** (1989), 905-911.
34. Sprent, J. and Tough, D.: $CD4^+$ cell turnover, *Nature* **375** (1995), 194.
35. Mackall, C.L., Granger, L., Sheard, M.A., Cepeda, R. and Gress, R.E.: T-cell regeneration after bone-marrow transplantation: differential CD45 isoform expression on thymic-derived versus thymic-independent progeny, *Blood* **82** (1993), 2585-2594.
36. Miller, J.F.A.P., Doak, S.M.A. and Cross, A.M.: Role of the thymus in recovery of the immune system, *Proc. Soc. Exp. Biol. Med.* **112** (1963), 785-792.
37. Davis, B.R., Schwartz, D.H., Marx, J.C., Johnson, C.E., Berry, J.M., Lyding, J., Merigan, T.C. and Zander, A.: Absent or rare human immunodeficiency virus infection of bone marrow stem/progenitor cells *in vivo*, *J. Virol.* **65** (1991), 1985-1990.
38. Stanley, S.K., Kessler, S.W., Justement, J.S., Schnittman, S.M., Greenhouse, J.J., Brown, C.C., Musongela, L., Musey, K., Kapita, B. and Fauci, A.S.: $CD34^+$ bone marrow cells are infected with HIV in a subset of seropositive individuals, *J. Immunol.* **149** (1992), 689-697.

39. Scadden, D.T., Zon, L.I. and Groopman, J.E.: Pathophysiology and management of HIV-associated hematologic disorders, *Blood* **74** (1989), 1455-1463.
40. Weichold, F.F., Zella, D., Barabitskaja, O., Maciejewski, J.P., Dunn, D.E., Sloand, E.M. and Young, N.S.: Neither human immunodeficiency virus-1 (HIV-1) nor HIV-2 infects most-primitive human hematopoietic stem cells as assessed in long-term bone marrow cultures, *Blood* **91** (1998), 907-915.
41. Bagnara, G.P., Zauli, G., Giovannini, M., Re, M.C., Furlini, G. and La Placa, M.: Early loss of circulating hemopoietic progenitors in HIV-1-infected subjects, *Exp. Hematol.* **18** (1990), 426-430.
42. Marandin, A., Katz, A., Oksenhendler, E., Tulliez, M., Picard, F., Vainchenker, W. and Louache, F.: Loss of primitive progenitors in patients with human immunodeficiency virus infection, *Blood* **88** (1996), 4568-4578.
43. Sloand, E.M., Young, N.S., Sato, T., Kumar, P., Kim, S., Weichold, F.F. and Maciejewski, J.P.: Secondary colony formation after long-term bone marrow culture using peripheral blood and bone marrow of HIV-infected patients, *AIDS* **11** (1997), 1547-1552.
44. Steinberg, H.N., Crumpacker, C.S. and Chatis, P.A.: *In-vitro* suppression of normal human bone marrow progenitor cells by human immunodeficiency virus, *J. Virol.* **65** (1991), 1765-1769.
45. Zauli, G., Re, M.C., Visani, G., Furlini, G., Mazza, P., Vignoli, M. and La Placa, M.: Evidence for a human immunodeficiency virus type-1-mediated suppression of uninfected hematopoietic ($CD34^+$) cells in AIDS patients, *J. Infect. Dis.* **166** (1992), 710-716.
46. Zauli, G., Vitale, M., Gibellini, M. and Capitani, S.: Inhibition of purified $CD34^+$ hematopoietic progenitor cells by human immunodeficiency virus 1 or gp120 mediated by endogenous transforming growth factor b1, *J. Exp. Med.* **183** (1996), 99-108.
47. Aldrovandi, G.M., Feuer, G., Gao, L., Jamieson, B.D., Kristeva, M., Chen, I.S.Y. and Zack, J.A.: The SCID-hu mouse as a model for HIV-1 infection, *Nature* **363** (1993), 732-736.
48. Clark, D.R., Ampel, N.M., Hallet, C.A., Yedavalli, V., Ahmad, N. and DeLuca, D.: Peripheral blood from HIV-infected patients displays diminished T-cell regeneration capacity, *J. Infect. Dis.* **176** (1997), 649-654.
49. Clark, D.R., Repping, S., Pakker, N.G., Prins, J.M., Notermans, D.W., Wit, F.W.N.M., Reiss, P., Danner, S.A., Coutinho, R.A., Lange, J.M.A. and Miedema, F.: Diminished T-cell renewal in HIV-1 infection contributes to $CD4^+$ T-cell depletion and is reversed by antiretroviral therapy, (submitted, 1998).
50. McCune, J.M., Loftus, R., Schmidt, D.K., Carroll, P., Webster, D., Swor-Yim, L.B., Francis, I.R., Gross, B.H. and Grant, R.M.: High prevalence of thymic tissue in adults with human immunodeficiency virus-1 infection, *J. Clin. Invest.* **101** (1998), 2301-2308.
51. Douek, D.C., McFarland, R.D., Keiser, P.H., Gage, E.A., Massey, J.M., Haynes, B.F., Polis, M.A., Haase, A.T., Feinberg, M.B., Sullivan, J.L., Jamieson, B.D., Zack, J.A., Picker, L.J. and Koup, R.A.: Changes in thymic function with age and during the treatment of HIV infection, *Nature* **396** (1998), 690-695.
52. Gorochov, G., Neumann, A.U., Kereveur, A., Parizot, C., Li, T., Katlama, C., Karmochkine, M., Raguin, G., Autran, B. and Debré, P.: Perturbation of $CD4^+$ and $CD8^+$ T-cell repertoires during progression to AIDS and regulation of the $CD4^+$ repertoire during antiviral therapy, *Nature Med.* **4** (1998), 215-221.
53. Kostense, S., Raaphorst, F.M., Notermans, D.W., Joling, J., Hooibrink, B., Pakker, N.G., Danner, S.A., Teale, J.M. and Miedema, F.: Diversity of the TCRBV repertoire in HIV-1-infected patients reflects the biphasic $CD4^+$ T-cell repopulation kinetics during HAART, *AIDS* **12** (1998), F235-F240.
54. Pakker, N.G., Notermans, D.W., De Boer, R.J., Roos, M.Th.L., De Wolf, F., Hill, A., Leonard, J.M., Danner, S.A., Miedema, F. and Schellekens, P.Th.A.: Biphasic kinetics of peripheral blood T cells after triple combination therapy in HIV-1 infection: a composite of redistribution and proliferation, *Nature Med.* **4** (1998), 208-214.
55. Vigano, A., Clerici, M., Bricalli, D., Saresella, M., Difabio, S., Principi, N. and Vella, S.: Immune reconstitution of the thymus during potent antiretroviral therapy in vertically HIV-1-infected children, 12th World AIDS Conference, Geneva, Switzerland (Abstract, 1998).

CHAPTER 5

LONG-TERM NON-PROGRESSIVE HIV INFECTION

SUSAN BUCHBINDER, M.D.[1] and ERIC VITTINGHOFF, Ph.D.[2]

[1]*Director, HIV Research Section, San Francisco Department of Public Healt, San Francisco, CA 94102 and* [2]*Assistant Adjunct Professor, Department of Epidemiology, University of California, San Francisco, CA 94143*

1. Introduction

By the late 1980s it had become clear that the progression time from infection with HIV to development of AIDS was extremely variable, and that a minority of HIV-infected persons remained AIDS-free without overt signs of disease progression many years after infection. In the past decade, such long-term non-progressors (LTNPs) have been intensively studied to determine their distribution by risk group, geographical area, and gender; to identify useful prognostic markers and correlates of delayed disease progression, and to provide insight into the mechanisms preventing or forestalling the development of overt disease in this group, with a view towards development of effective therapies, vaccines, and other preventive strategies.

2. Definitions

Definitions of non-progression differ, based in part on theoretical principles and in part on the populations accessible for study. Investigators generally select persons who have remained AIDS-free with "normal" CD4 counts for a sufficient number of years after documented HIV infection to ensure that these persons have significantly delayed HIV disease progression. Duration of follow-up is the most variable part of these definitions, ranging from a minimum of 5 years to a decade or more. While many investigators require the most recent or all CD4 counts to remain above a cut-point, usually 500 cells per microliter, others also require a stable or positive CD4 slope, calculated by ordinary least squares with a minimum of three CD4 counts. Because counts may drop precipitously post-seroconversion with a subsequent re-bound, slope calculations should generally exclude CD4 measurements from the peri-seroconversion period. Some definitions also exclude persons with HIV-related signs, such as thrush or hairy leukoplakia. Others also exclude persons who have ever taken any anti-retroviral therapy, although until recently, few persons have had access to highly effective combination therapy. With more wide-spread use of highly

H. Schuitemaker and F. Miedema (ed.), Aids Pathogenesis, 65–75.

active combinations of anti-retroviral agents, identification of LTNP who are anti-retroviral-naive will likely be more difficult. In fact, analysis of existing data will be critical in defining the subgroup of individuals who may be expected not to progress for prolonged periods, because this is the one subgroup who may not require early anti-retroviral therapy to prolong clinical health. Inclusion of new prognostic markers, in particular plasma viral burden, may increase the predictive power of existing definitions of LTNP and thus be useful in identifying this special subgroup not clearly requiring therapy.

The proportion of seropositives defined as LTNPs and the actual persons selected vary with the definition used. In the Tricontinental Seroconverter Study, the proportions defined as LTNPs by five different sets of criteria involving AIDS-free time since seroconversion, CD4 counts, and use of anti-retrovirals differed substantially, from 0.5 to almost 25% [1]. Definitions requiring non-negative CD4 slope or all CD4 counts above 500 were the most stringent, excluding the largest number of potential LTNPs. Proportions defined as LTNPs generally decreased as the minimum AIDS-free follow-up time increased from 7 to 10 years, demonstrating the importance of sufficient follow-up time in defining a subgroup with significantly delayed HIV disease progression. Even for the remaining three definitions by which a similar proportion of LTNPs was identified, there was only partial overlap between the specific individuals selected for inclusion, with 50% classified as LTNPs by all three, 22% by two, and the remaining 28% only by the most inclusive definition. Analogous variability of the proportion defined as LTNPs and lack of overlap between definitions were reported for the Italian Seroconverter Study [2]. Differences in definitions and the populations from which they are drawn may account in part for conflicting results from different studies of LTNPs. After reviewing the existing literature on LTNPs, we make some recommendations for how such studies may best be constructed in the future.

At the outset it was unclear whether LTNPs constitute a distinct subgroup that would never progress to AIDS, or simply the subset expected to have the longest progression times. With additional study it has become apparent that few LTNPs completely lack signs or symptoms of HIV disease progression [3] and that many cease to meet LTNP criteria when follow-up is extended. For instance, in the San Francisco City Clinic Cohort, 11% of 590 men with well-characterised dates of seroconversion were AIDS-free and had never had more than one consecutive CD4 count below 500 after 10 years of infection; however, only an estimated 4% will still meet these criteria after 16 years [4]. Furthermore, under a parametric model for the AIDS incubation period, only 13% of HIV-infected homosexual men will remain AIDS-free after 20 years of infection [5], and among these only a small subset would be expected to meet additional clinical and laboratory criteria for non-progression. Although these data do not rule out the possibility of a smaller subset of individuals who will never progress, it appears that, as commonly defined, LTNPs represent that spectrum of persons with longest progression times, rather than a discrete and homogenous group [6].

Often, this discussion of whether LTNP are a discrete subgroup or represent the "tail" of a distribution of latency times confuses disease-delaying mechanisms with random events. Just as the distribution of height within a population is predicated in large part on genetic and environmental factors, "non-progression" as currently defined is likely to be the result of virologic, immunologic and genetic factors that significantly delay disease progression, even if such progression ultimately occurs. Current data suggest that the delayed or arrested disease progression characterising this heterogeneous group is determined by multiple factors operating alone or in combination [7]. In short, there may be multiple pathways leading to non-progression, and LTNPs may never be completely distinguishable by any single characteristic or cluster of characteristics.

3. Epidemiology

There is little evidence that the prevalence of non-progression differs by geography, risk group, or gender, although few systematic studies have addressed this point directly. While significant differences were found in the proportion of homosexual men with non-progressive disease cohorts in Amsterdam, Vancouver, Sidney, and San Francisco in the Intercontinental Seroconverter Study [1], these differences are likely due to systematic differences in measurement of CD4. Evaluation of non-progressive HIV infection in the Italian Seroconverter Study [2] found a somewhat lower proportion of LTNPs among homosexual men than among injection drug users (IDUs), although most of the differences did not achieve statistical significance. In a cross-sectional study comparing LTNPs to rapid progressors in a large Madrid sample including IDUs, women and homosexual men, non-progressors were more likely to be male and come from the IDU risk group [8]. However, another large study comparing IDUs and homosexual men drawn from 12 cohorts found no persuasive evidence for differences by risk group in the proportions defined as LTNPs by CD4 slope and level or in progression to AIDS and death, after adjusting for age and year of seroconversion [9]. An important difficulty in comparing rates of progression and non-progression by risk group is posed by substantial pre-AIDS mortality among IDUs, since to the extent that such mortality may be associated with progression, the proportion defined as LTNPs among surviving IDUs would be increased.

Using typical definitions of non-progression, 6-8% of haemophiliacs have been classified as LTNPs, in both cases within the range of estimates for other risk groups [10,11]. More definitive estimates of relative prevalence of non-progression by gender will require longer follow-up of women with well-characterised dates of seroconversion, and will be complicated by increasingly widespread use of effective anti-retroviral therapies in the developed world. No information is currently available on non-progression among the large numbers of men and women infected by heterosexual transmission in the developing world, which might reflect differences in viral clade or environmental factors. Only a few studies have addressed non-progression in children infected perinatally. These suggest that while very rapid progression is more common in children, prevalence of LTNP appears to be comparable to that in adults, and that several of the same immunologic and virologic correlates of non-progression are found in both groups [12].

Although older age at seroconversion has been associated with more rapid progression to AIDS [13], small nested-case control studies of LTNPs have not found a significant association with age. This may stem in part from the relatively limited range of age at seroconversion in the cohorts of homosexual and bisexual men from which many of the case-control samples have been drawn, and may more broadly reflect the limited power of small studies to detect relatively subtle effects. Nor have associations with race or ethnicity been reported. One nested case-control study of homosexual LTNPs suggested that several measures of socio-economic status (SES), including income, education and occupation were higher in LTNPs than in matched rapid progressors [14]. However, exact duration of infection was unknown for a majority of these men. If men with lower SES measures were infected earlier, on average, than men with higher measures, it is possible that a spurious association between SES and LTNP would result. No other studies have yet identified an association of SES variables and LTNP.

In addition, studies comparing LTNPs to relatively rapid progressors have found only minor between-group differences in lifestyle cofactors. For instance, in a nested case-control study of 42 LTNPs and 382 progressors in the SFCCC, no differences were found in post-seroconversion exposure to STDs or recreational drugs [15]. Similarly, no differences in markers of high-risk sexual behaviour or recreational drug use were found in a nested case-control study of 61 LTNPs and 142 progressors from a Dutch cohort of homosexual and bisexual men [16]. This study did find slightly lower scores for active problem solving among LTNPs, but no other differences in coping skills. Likewise, in a comparison of 67 LTNPs in the MACS cohort to matched controls groups of intermediate and rapid progressors, no significant differences were found in demographics, sexual behaviour, and STDs [5]. In this case, the results for LTNPs are consistent with survival analyses of larger cohorts in which associations of sexual behaviours, STDs, and recreation drug use with risk of progression have in general not been detected.

4. Virologic Factors

Recently, plasma viral RNA levels were found to be a strong independent predictor of HIV disease progression [17]. Longitudinal studies of HIV-infected persons have found that while peri-seroconversion viral levels may not predict long-term outcome, measures taken within the first year after seroconversion are predictive and LTNP have lower plasma viral levels in early HIV infection than other newly HIV-infected persons who later progress [18,19]. LTNP also appear to have lower levels of multiply spliced RNA [10,21], proviral DNA [21], low levels of viral trapping and preserved lymph-node architecture [22] when compared with individuals with progressive HIV infection.

How then to explain the heterogeneity in plasma viral levels among LTNPs as reported in different studies, and even with the same studies? One possible explanation is that some persons are able to sustain high levels of viral replication for extended periods of time without suffering ill effects. Although it appears that this may be the case in some isolated situations, more recent studies suggest that plasma viral level is also the strongest independent predictor of progression among LTNPs [4], and that higher plasma viral levels are early indicators of disease progression. Thus, while it is important to continue to evaluate persons with well-documented cases of persistently high plasma viral levels without immunologic or clinical progression, LTNPs are best defined as the subgroup of individuals with persistently high CD4 counts and low plasma viral levels in the absence of anti-retroviral therapy many years after HIV infection.

Although plasma viral level is a powerful predictor of disease progression, we must still search for the mechanisms responsible for control of viral replication. Several possible explanations have emerged, none of which appears to account for all cases of non-progression. Some subgroups of LTNP appear to be infected with attenuated quasispecies of virus. Mutations in the nef gene were among the first to be reported in isolated cases of LTNP [23,24,25], but soon after, mutations were also reported in other structural [26] and regulatory genes [23,27,28]. Virus isolated from these individuals appears to display impaired replication kinetics. However, the majority of LTNP have not been found to have major structural or functional mutations in any of the genes studied [22,29-32], but rather to have replication-competent virus. In some instances, in which it appeared that virus had attenuated replication kinetics, this was found to be secondary to immunologic suppression of replication, rather than an inherent defect in the virus itself [33]. In addition, recent data from the Sydney Blood Bank Cohort suggests that, despite infection with attenuated HIV, several members of their cohort have detectable viral loads and declining CD4 counts [34].

5. Immunologic Factors

There is indirect evidence that immunologic mechanisms are responsible for delayed HIV disease progression. For instance, increased genetic diversity in quasispecies in persons with delayed disease progression as compared to individuals with progressive HIV infection [35,36] may be a marker for effective immunologic responses driving intrahost evolution. The bulk of evidence points to the importance of cellular immune responses in controlling viral replication. The early decline in plasma viral levels shortly after HIV infection correlates with the development of HIV-specific CTL and precedes the occurrence of neutralising antibody. LTNP appear to have higher $CD8^+$ counts than other HIV-infected persons [15], even those with relatively early HIV infection [4]. In particular, broadly reactive CTL [37-39], high CTL pre-cursor frequency [40], and CTL recognition of highly conserved or functionally important [41] HIV targets have been reportedly associated with LTNP. In longitudinal studies, it appears that loss of CTL function correlates with disease progression [42] and thus emergence of CTL escape mutants may herald disease progression.

The reason for development of a more effective CTL response in LTNP is unclear. Predominance of specific HLA class-I genotypes among LTNP could account for a more robust CTL response in these individuals. However, in at least some cases, the CTL epitopes recognised are identical to those reported in progressive HIV infection [38]. It has been hypothesised recently that LTNP preserve their $CD4^+$ repertoire in early infection to a greater extent than individuals with progressive HIV disease [43,44]. This may affect the humoral as well as the cellular immune response.

In addition, non-cytolytic $CD8^+$ suppressor cell activity also appears to correlate with delayed disease progression [33,44,46]. The factor responsible for such protection has not been clearly identified. Although the beta-chemokines may play a role in suppressing viral replication [47], others have found enhanced suppression in LTNP without measurable increases in these chemokines, suggesting that other mechanisms or factors may also be involved [10,48,49].

Data supporting the relationship of humoral immunity in delayed HIV disease progression are generally weaker than those for the cellular immune response. While a number of investigators have reported that LTNP have higher concentrations of specific antibodies than progressors [50] and antibody-mediated cytolytic responses [51,52], this may be a consequence of immunosuppression in the progressors rather than demonstration of effective virologic control. Several investigators have reported that LTNPs have broader ability to neutralise laboratory [22] and primary isolate [33]. However, when some of the same investigators attempted to confirm the latter finding, they found no significant neutralisation of the same primary isolates in a different cohort of LTNPs [38]. And, although some investigators report low levels of neutralisation against contemporaneous isolates [53], others have found no such activity [54]. Viruses isolated from LTNP appear to be no more sensitive to neutralisation by heterologous sera than viruses from persons with progressive HIV infection [54]. One longitudinal study of LTNP found that virus neutralisation was poor or undetectable with contemporaneous autologous serum, but improved with later serum samples [55]. This suggests that broad neutralising antibody responses be a result of multiple rounds of neutralisation-escape, but are unlikely to be responsible for chronic control of viral replication.

6. Host Genetic Factors

Investigations of host genetic factors potentially associated with non-progression have focused on co-receptors for HIV-1 and on the major histocompatibility complex (MHC) of human leukocyte antigen (HLA) genes. It has been understood for some time that HIV-1 required co-receptors in addition to CD4 to infect host cells. Following identification of the chemokine receptor molecule CCR5 as the co-receptor for the non-syncytium-inducing (NSI) macrophage-tropic variant in 1996 [56], several papers were published in rapid succession identifying a 32 base-pair deletion in the CCR5 gene and reporting its association with resistance to HIV infection and disease progression [57].

While heterozygotes for the CCR5 deletion are not at reduced risk for infection, they do appear more likely to be non-progressors. Statistically significant elevations in the prevalence of the heterozygote genotype has been found in several studies of LTNPs [58-60]. In the wider population of seropositives, the deletion has also been associated with reduced risk of progression to AIDS and death [57] and slower rates of CD4 depletion [60]. Although the heterozygote CCR5 delta-32 genotype is more common among those with progression times to AIDS greater than 7-10 years, the majority of LTNPs have the wild genotype, again showing that this is a heterogeneous group. Furthermore, the protection against diagnosis of AIDS and death detected in survival analyses may endure only so long as NSI variants predominate [59].

In a related recent finding, a variant of the chemokine receptor molecule CCR2 has been shown to predict slower progression in combined cohorts of homosexual men and haemophiliacs, and is substantially less common in men progressing to AIDS within eight years than in those who remain AIDS-free at least that long [61]. However, prevalence of this variant has not yet been shown to differ in LTNPs as currently defined.

Because of the role that HLA antigens and the transporter protein associated with antigen presentation (TAP) play in antigen processing, presentation and the immune response, it is also reasonable to expect that genes in the major histocompatibility complex may be associated with non-progression. While certain class-I alleles, class-II haplotypes, and combinations involving both HLA and TAP variants have been shown to predict progression in wider populations, these factors have not yet been shown to differ systematically in LTNPs.

7. Conclusions

Studies of non-progressors to date have addressed several critical issues. Without treatment, only 10% of individuals are likely to remain LTNPs for ten years, and plasma viral levels in conjunction with CD4 counts are currently our best measures for identifying who those persons will be. This means that the vast majority of HIV-infected persons, particularly those with CD4 counts < 500 and those with moderate to high plasma viral levels are likely to progress unless effective treatment is initiated.

There is no indication that modifiable cofactors, such as exposure to sexually transmitted diseases or illicit drugs, account for a significant proportion of LTNPs, although these factors may lead to secondary morbidity or increase infectiousness. However, most studies of the prevalence and correlates of non-progression have been done in populations of Caucasian homosexual men with Clade B virus. Extending these studies to include other populations in both developed and developing countries might highlight other factors associated with non-progression and thus lead to clues about other mechanisms of protection. Such studies will become increasingly difficult in the developed world as use of anti-retroviral therapies becomes more widespread, although persons most likely to become LTNPs may also be among the least likely to be treated.

Definitions of LTNPs have relied almost exclusively on CD4 counts and relatively short follow-up times. To ensure that persons are selected with the most significantly delayed disease progression, new definitions should stipulate a minimum of 10 years of documented infection, with longer periods of follow-up being preferable. Adding measures of viral replication, such as plasma viral level, will further define the subset of LTNPs with the least likelihood of progression. Although cohort-based studies may best represent the range of LTNPs and can best evaluate prevalence and common features among LTNPs, referral-based cohorts may identify LTNPs with unusual mechanisms of non-progression. Both types of studies require use of appropriately selected controls, including both rapid progressors and persons with early HIV infection, to identify features unique to LTNPs. Some studies, such as those evaluating the influence of HLA and TAP on non-progression, are likely to require larger numbers than can be acquired in any single cohort.

Studies of LTNPs have already begun to yield clues about correlates of protection from disease progression, and are suggesting new preventive and therapeutic strategies. The identification of mutations in viral regulatory genes that attenuate viral replication in both animal and human studies have led to intensive efforts to understand the mechanism of attenuation and develop safe and effective preventive vaccines. The identification of host genetic polymorphisms, particularly in the CCR5 gene, has led some to speculate that therapies directed at this target might be useful for treating HIV infection. The association of broadly reactive CTL with non-progression has encouraged investigators developing preventive vaccines to focus on those strategies that are most likely to stimulate broad CTL responses. The recent data suggesting that preservation of CD4-mediated help may be the critical factor determining the breadth and durability of the CTL response has also led some investigators to suggest that restoration of the CD4 repertoire be a focus of new therapeutic strategies. Although combinations of beta-chemokines appear to be one mechanism for CD8-mediated non-cytolytic HIV suppression, the appropriate chemokine or combination of chemokines that may result in therapeutic benefit have yet to be identified.

There has been much talk, of late, about turning HIV into a "chronic manageable disease". Further investigation of well-defined populations of persons who appear to have reached this state without the benefit of therapeutic interventions will likely provide more detailed clues about mechanisms of delayed progression and help to both broaden and refine attempts to develop safe, effective and accessible preventive and therapeutic strategies.

8. References

1. Strathdee, S.A., Veugelers, P.J., Page-Shafer, K.A., *et al.*: Lack of consistency between five definitions of non-progression in cohorts of HIV-infected seroconverters, *AIDS* **10** (1996), 959-965.
2. Petrucci, A., Dorrucci, M., Alliegro, M.B., *et al.*: How many HIV-infected individuals may be defined as long-term nonprogressors? A report from the Italian Seroconversion Study. Italian Seroconversion Study Group (ISS), *J. Acq. Immune Def. Syndr. Human Retrovir.* **14** (1997), 243-248.
3. Lefrere, J.J., Morand-Joubert, L., Mariotti, M., *et al.*: Even individuals considered as long-term non-progressors show biological signs of progression after 10 years of human immunodeficiency virus infection, *Blood* **90** (1997), 1133-1140.
4. Buchbinder, S.P., Vittinghoff, E., Park, M.S., *et al.* : Long-term non-progression in the San Francisco City Clinic Cohort. XI International Conference on AIDS, Vancouver, July 1996 (Abstract Tu.C.553).
5. Munoz, A., Kirby, A.J., He, Y.D., *et al.*: Long-term survivors with HIV-1 infections: incubation period and longitudinal patterns of $CD4^+$ lymphocytes, *J. Acq. Immune Def. Syndr. Human Retrovir.* **8** (1995), 496-505.
6. Easterbrook, N.J.: Non-progression in HIV infection, *AIDS* **8** (1994), 1179-1182.
7. Sheppard, H.W., Lang, W., Ascher, M.S., *et al.*: The characterization of non-progressors: long-term HIV-1 infection with stable $CD4^+$ T-cell levels, *AIDS* **7** (1993), 1159-1166.

8. Soriano, V., Martin, R., del Romero, J., *et al.*: Rapid and slow progression of the infection by the type 1 human immunodeficiency virus in a population of seropositive subjects in Madrid, *Medicina Clinica* **107** (1996), 761-766.
9. Prins, M. and Veugelers, P.J.: Comparison of progrression and non-progression in injecting drug users and homosexual men with documented dates of HIV-1 seroconversion. European Seroconverter Study and the Tricontinental Seroconverter Study, *AIDS* **11** (1997), 621-631.
10. Vicenzi, E., Bagnarelli, P., Santagostina, E., *et al.*: Hemophilia and non-progressing human immunodeficiency virus type 1 infection, *Blood* **89** (1997), 191-200.
11. Ujhelyi, E., Krall, G., Zimonyl, I., *et al.*: Longitudinal immunological follow-up of HIV-infected haemophiliacs in Hungary, *Acta Microbilogica et Immunologica Hungarica* **42** (1995), 189-198.
12. Martin, N., Koup, R., Kaslow, R., *et al.*: Workshop on perinatally acquired human immunodeficiency virus infection in long-term surviving children: a collaborative study of factors contributing to slow disease progression. The Long-Term Survivor Project, *AIDS Res. Human Retrovir.* **12** (1996), 1565-1570.
13. Veugelers, P.J., Page, K.A., Tindall, B., *et al.*: Determinants of HIV disease progression among homosexual men registered in the Tricontinental Seroconverter Study, *Am. J. Epidemiol.* **140** (1994), 747-758.
14. Schechter, M.T., Hogg, R.S., Aylward, B., *et al.*: Higher socioeconomic status is associated with slower progression of HIV infection independent of access to health care, *J. Clin. Epidemiol.* **47** (1994), 59-67.
15. Buchbinder, S.P., Katz, M.H., Hessol, N.A., *et al.*: Long-term HIV-1 infection without immuno-logic progression, *AIDS* **8** (1994), 1123-1128.
16. Keet, I.P.M., Krol, A., Klein, M.R., *et al.*: Characteristics of long-term asymptomatic infection with human immunodeficiency virus type 1 in men with normal and low $CD4^+$ cell counts, *J. Infect. Dis.* **169** (1994), 1236-1243.
17. Mellors, J.W., Rinaldo Jr., C.R , Gupta, P., *et al.*: Prognosis in HIV-1 infection predicted by the quantity of virus in plasma, *Science* **272** (1996), 1167-1170.
18. O'Brien, T.R., Blattner, W.A., Waters, D., *et al.*: Serum HIV-1 RNA levels and time to development of AIDS in the Multicenter Hemophilia Cohort Study, *JAMA* **276** (1996), 105-110.
19. Hogervorst, E., Jurriaans, S., de Wolf, F., *et al.*: Predictors for non- and slow progression in human immunodeficiency virus (HIV) type 1 infection: low viral RNA copy numbers in serum and maintenance of high HIV-1 p24-specific but not V3-specific antibody levels, *J. Infect. Dis.* **171** (1995), 811-821.
20. Saksela, K., Stevens, C.E., Rubinstein, P., *et al.*: HIV-1 messenger RNA in peripheral blood mononuclear cells as an early marker of risk for progression to AIDS, *Ann. Intern. Med.* **123** (1995), 641-648.
21. Comar, M., Simonelli, C. and Zanussi, S.: Dynamics of HIV-1 mRNA expression in patients with long-term nonprogressive HIV-1 infection, *J. Clin. Invest.* **100** (1997), 893-903.
22. Pantaleo, G., Menzo, S., Vaccarezza, M., *et al.*: Studies in subjects with long-term nonprogressive human immunodeficiency virus infection, *New Engl. J. Med.* **332** (1995), 209-216.
23. Kirchhoff, F., Greenough, T.C., Brettler, D.B., *et al.*: Brief report: Absence of intact nef sequences in a long-term survivor with nonprogressive HIV-1 infection, *New Engl. J. Med.* **332** (1995), 228-232.
24. Saksena, N.K., Ge, Y.C., Wang, B., *et al.*: An HIV-1-infected long-term non-progressor (LTNP): molecular analysis of HIV-1 strains in the vpr and nef genes, *Ann. Acad. Med. Singapore* **25** (1996), 848-854.
25. Mariani, R., Kirchhoff, F., Greenough, T.C., *et al.*: High frequency of defective nef alleles in a long-term survivor with nonprogressive human immunodeficiency virus type 1 infection, *J. Virol.* **70** (1996), 7752-7764.
26. Connor, R.I., Sheridan, K.E., Lai, C., *et al.*: Characterization of the functional properties of env genes from long-term survivors of human immunodeficiency virus type 1 infection, *J. Virol.* **70** (1996), 5306-5311.
27. Wang, B., Ge, Y.C., Palasanthiran, P., *et al.*: Gene defects clustered at the C-terminus of the vpr gene of HIV-1 in long-term nonprogressing mother and child pair: *in-vivo* evolution of vpr quasispecies in blood and plasma, *Virology* **223** (1996), 224-232.

28. Michael, N.L., Chang, G., d'Arcy, L.A., *et al.*: Defective accessory genes in a human immunodeficiency virus type 1-infected long-term survivor lacking recoverable virus, *J. Virol.* **69** (1995), 4228-4236.
29. Zhang, L., Huang, Y., Yuan, H., *et al.*: Genetic characterization of vif, vpr, and vpu sequences from long-term survivors of human immunodeficiency virus type 1 infection, *Virology* **228** (1997), 340-349.
30. Cornelissen, M., Kuiken, C., Zorgdrager, F., *et al.*: Gross defects in the vpr and vpu genes of HIV type 1 cannot explain the differences in RNA copy number between long-term asymptomatics and progressors, *AIDS Res. Human Retrovir.* **13** (1997), 247-252.
31. Schwartz, D.H., Viscidi, R., Laeyendecker, O., *et al.*: Predominance of defective proviral sequences in an HIV$^+$ long-term non-progressor, *Immunol. Lett.* **51** (1996), 3-6.
32. Huang, Y., Zhang, L. and Ho, D.D.: Biological characterizations of nef in long-term survivors of human immunodeficiency virus type 1 infection, *J. Virol.* **69** (1995), 8142-8146.
33. Learmont, J., Geczy, A., Raynes-Greenow, C., *et al.*: The Syndney Blood Bank Cohort infected with attenuated quasi-species of HIV-1: Long-term non-progression. XII World AIDS Conference, Geneva, Switzerland, June 28-July 3, 1998 (Abstract 13350).
34. Cao, Y., Qin, L., Zhang, L., Safrit, J. and Ho, D.D.: Virologic and immunologic characterization of long-term survivors of human immunodeficiency virus type 1 infection, *New Engl. J. Med*. **332** (1995), 201-208.
35. Wolinsky, S.M., Korber, B.T., Neumann, A.U., *et al.*: Adaptive evolution of human immunodeficiency virus-type 1 during the natural course of infection, *Science* **272** (1996), 537-542.
36. Lukashov, V.V., Kuiken, C.L. and Goudsmit, J.: Intrahost human immunodeficiency virus type 1 evolution is related to length of the immunocompetent period, *J. Virol.* **69** (1995), 6911-6916.
37. Lubaki, M.N., Dhruva, B., Quinn, T.C., *et al.*: Characterization of a polyclonal cytolytic T lymphocyte response to human immunodeficiency virus in individuals without clinical progression. XI International Conference on AIDS. Vancouver, 1996 July (Abstract MoA392).
38. Harrer, T., Harrer, E., Kalams, S.A., *et al.*: Strong cytotoxic T cell and weak neutralizing antibody responses in a subset of persons with stable nonprogressing HIV type 1 infection, *AIDS Res. Human Retrovir.* **12** (1996), 585-592.
39. Harrer, T., Harrer, E., Kalams, S.A., *et al.*: Cytotoxic T lymphocytes in asymptomatic long-term non-progressing HIV-1 infection. Breadth and specificity of the response and relation to *in-vivo* viral quasispecies in a person with prolonged infection and low viral load, *J. Immunol.* **156** (1996), 2616-2623.
40. Kalams, S.A., Harrer, T., Harrer, E., *et al.*: HIV-1-specific cytotoxic T lymphocyte and proliferative responses in peripheral blood mononuclear cells (PBMC) of subjects with stable non-progressing HIV-1 infection. XI International Conference on AIDS. Vancouver, 1996 July (Abstract MoA391).
41. Harrer, E., Harrer, T., Barbosa, P., *et al.*: Recognition of the highly conserved YMDD region in the human immunodeficiency virus type 1 reverse transcriptase by HLA-A2-restricted cytotoxic T lymphocytes from an asymptomatic long-term nonprogressor, *J. Infect. Dis.* **173** (1996), 476-479.
42. Klein, M.R., van Baalen, C.A., Holwerda, A.M., *et al.*: Kinetics of Gag-specific cytotoxic T lymphocyte responses during the clinical course of HIV-1 infection: a longitudinal analysis of rapid progressors and long-term asymptomatics, *J. Exp. Med.* **181** (1995), 1365-1372.
43 Kalams, S.A., Johnson, R.P., Trocha, A.K., *et al.*: Longitudinal analysis of T-cell receptor (TCR) gene usage by human immunodeficiency virus 1 envelope-specific cytotoxic T lymphocyte clones reveals a limited TCR repertoire, *J. Exp. Med.* **179** (1994), 1261-71.
44. Hadida, F., Bonduelle, O., Candotti, D., *et al.*: Immunological studies on the French cohort of 70 HIV-1-infected long-term non-progressors (LTNP). XI International Conference on AIDS. Vancouver, 1996 July (Abstract MoA393).
45. Barker, E., Mackewicz, C.E. and Levy, J.A.: Effects of TH1 and TH2 cytokines on CD8$^+$ cell response against human immunodeficiency virus: implications for long-term survival, *Proc. Natl. Acad. Sci. USA* **92** (1995), 11135-11139.
46. Blackbourn, D.J., Mackewicz, C.D., Barker, E., *et al.*: Suppression of HIV replication by lymphoid tissue CD8$^+$ cells correlates with the clinical state of HIV-infected individuals, *Proc. Natl. Acad. Sci. USA* **93** (1996), 13125-13130.

47. Cocchi, F., DeVico, A.L., Garzino-Demo, A., *et al.*: Identification of RANTES, MIP-1-alpha, and MIP-1-beta as the major HIV-suppressive factors produced by CD8⁺ T cells, *Science* **270** (1995), 1811-1815.
48. Zanussi, S., Simonelli, C., D'Andrea, M., *et al.*: CD8⁺ lymphocyte phenotype and cytokine production in long-term non-progressor and in progressor patients with HIV-1 infection, *Clin. Exp. Immunol.* **105** (1996), 220-224.
49. McKenzie, S.W., Dallalio, G., North, M., *et al.*: Serum chemokine levels in patients with non-progressing HIV infection, *AIDS* **10** (1996), F29-F33.
50. Lifson, A.R., Buchbinder, S.P., Sheppard, H.W., *et al.*: Long-term human immunodeficiency virus infection in asymptomatic homosexual and bisexual men with normal CD4⁺ lymphocyte counts: immunologic and virologic characteristics, *J. Infect. Dis.* **63** (1991), 959-965.
51. Okada, N., Wu, X. and Okada, H.: Presence of IgM antibodies which sensitize HIV-1-infected cells to cytolysis by homologous complement in long-term survivors of HIV infection, *Microbiol. Immunol.* **41** (1997), 331-336.
52. Baum, L.L., Cassutt, K.J., Knigge, K., *et al.*: HIV-1 gp 120-specific antibody-dependent cell-mediated cytotoxicity correlates with rate of disease progression, *J. Immunol.* **157** (1966), 2168-2173.
53. Schonning, K., Nielsen, C., Iversen, J., *et al.*: Neutralizing antibodies in slowly progressing HIV-1 infection, *J. Acq. Immune Def. Syndr. Human Retrovir.* **10** (1995), 400-407.
54. Montefiori, D.C., Pantaleo, G., Fink, L.M., *et al.*: Neutralizing and infection-enhancing antibody responses to human immunodeficiency virus type 1 in long-term nonprogressors, *J. Infect. Dis.* **173** (1996), 60-67.
55. Bradney, A.P., Scheer, S., Crawford, J.M., Buchbinder, S.P. and Monetefiori, D.C.: Neutralization-escape in human immunodeficiency virus type-1-infected long-term nonprogressors, *J. Infect. Dis.*, in press.
56. Feng, Y., Broder, C.C., Kennedy, P.E. and Berger, E.A.: HIV-1 entry cofactor: functional cDNA cloning of a seven-transmembrane G protein-coupled receptor, *Science* **272** (1996), 872-877.
57. Dean, M., Carrington, M., Winkler, C., *et al.*: Genetic restriction of HIV-1 infection and progression to AIDS by a deletion allele of the CKR5 structural gene. Homephilia Growth and Development Study, Multicenter AIDS Cohort Study, Multicenter Hemophilia Cohort Study, San Francisco City Cohort, ALIVE Study, *Science* **273** (1996), 1856-1862.
58 Zimmerman, P.A., Buckler-White, A., Alkhatib, G., *et al.*: Inherited resistance to HIV-1 conferred by an inactivating mutation in CC chemokine receptor 5: studies in populations with contrasting clinical phenotypes, defined racial background, and quantified risk, *Mol. Med.* **3** (1997), 23-36.
59. Michael, N.L., Chang, G., Louie, L.G., *et al.*: The role of viral phenotypes and CCR-5 gene defects in HIV-1 transmission and disease progression, *Nature Medicine* **3** (1997), 338-340.
60. Eugen-Olsen, ., Iversen, A.K., Garred, P., *et al.*: Heterozygosity for a deletion in the CKF-5 gene leads to prolonged AIDS-free survival and slower CD4 T-cell decline in a cohort of HIV-seropositive individuals, *AIDS* **11** (1997), 305-310.
61. Smith, M.W., Dean, M., Carrington, M., *et al.*: Contrasting genetic influence of CCR2 and CCR5 variants on HIV-1 infection and disease progression. Hemophilia Growth and Development Study (HGDS), Multicenter AIDS Cohort Study (MACS), Multicenter Hemophilia Cohort Study (MHCS), San Francisco City Cohort (SFCC), ALIVE Study, *Science* **277** (1997), 959-965.

CHAPTER 6

CYTOTOXIC T LYMPHOCYTES IN HIV-1 INFECTION

M.R. KLEIN, Ph.D.[*,1], S.H. Van der BURG, Ph.D.[2] and B. AUTRAN, Ph.D., M.D.[3]

[*]*Present address: TB Research Programme, Medical Research Council (MRC) Laboratories, Atlantic Road, Fajara, P.O. Box 273, Banjul, The Gambia, West Africa;* [1]*Department of Clinical Viro-Immunology, Central Laboratory of the Netherlands Red Cross Blood Transfusion Service and Laboratory of Experimental and Clinical Immunology, University of Amsterdam, Amsterdam, The Netherlands;* [2]*Department of Immunohaematology and Blood Bank, Leiden University Hospital, Leiden, The Netherlands, and* [3]*Laboratoire d'Immunologie Cellulaire, URA CNRS 625, Bâtiment C.E.R.V.I., Paris, France*

1. Introduction

Cytotoxic T lymphocytes (CTL) are part of the major host defence against trespassing intracellular pathogens. Since it has been shown *in vitro* that HIV-1-specific CTL are able to eliminate virus-infected cells via MHC class-I-restricted killing [1, 2] and can interfere with HIV-1 replication via secretion of various antiviral cytokines [3], it is widely held that antiviral CTL responses to HIV-1 are 'salutary' to the infected host. However, after a decade of intense research on HIV-1-specific CTL [1,2], their precise role during the natural history of HIV-1 infection is still not fully resolved. The majority of published studies to date seem to support the concept that HIV-1-specific CTL contribute to controlling viral replication and thus to delaying the onset of disease. However, several observations have suggested that HIV-1-specific CTL are 'pathogenic' to the infected patient. For example, HIV-1-specific CTL have been detected in broncho-alveolar lavage fluids of some HIV-1-infected patients suffering from lymphocytic alveolitis or interstitial pneumonitis [2,4] and in cerebrospinal fluid of some infected patients suffering from neurological disorders [5]. Zinkernagel has even suggested that severe depletion of $CD4^+$ T cells and progression to AIDS result from killing of huge numbers of infected cells by HIV-1-specific CTL [6,7]. If this were true, it could greatly affect the current rationale of developing AIDS vaccines and treatment of HIV-1-infected patients.

In order to start composing a coherent model for the role of HIV-1-specific CTL in the pathogenesis of AIDS, we need to *a*) fully characterise HIV-1-specific CTL responses during the different clinical stages of disease, with respect to dynamics and epitope specificity; *b*) understand why the virus persists in the face of

H. Schuitemaker and F. Miedema (ed.), Aids Pathogenesis, 77–98.

vigorous and ongoing HIV-1-specific CTL responses, and *c*) discover why antiviral CTL fail to provide life-long protection from progression to AIDS in the vast majority of patients.

2. HIV-1-specific CTL Responses in the Clinical Course of HIV-1 Infection

2.1. CTL DURING ACUTE HIV-1 INFECTION

Upon sexual or parenteral transmission of HIV-1 it usually takes about 2 to 4 weeks before clinical symptoms of acute infection ensue. In about 50 to 70% of infected individuals an acute retroviral syndrome develops with mild influenza-like manifestations from which most patients usually fully recover. To date, HIV-1-specific CTL responses have only been documented in a limited number of patients with an acute retroviral syndrome [8-13]. HIV-1-specific CTL have been observed as early as a few days following the onset of acute symptoms and in general before (neutralising) antibody responses could be detected (Fig. 1). The appearance of HIV-1-specific CTL usually parallels a striking diminution of the viraemia in infected patients. Likewise, elegant studies in rhesus monkeys have shown that upon deliberate infection with Simian Immunodeficiency Virus (SIV), CTL appear as early as 4 to 7 days post-virus inoculation, and coincide with viral clearance from blood and lymph nodes [14,15]. Collectively, these studies indicate that CTL are recruited very early during the encounter with HIV-1 or other related retroviruses, and that in part they may be responsible for the initial control of the viraemia. Quite interesting are observations of some individuals who have remained HIV-1-seronegative despite frequent exposure to the virus, and who seem to harbour HIV-1-specific CTL and T-helper cell responses [16,17]. These responses may be coincidental demonstrating that some degree of viral replication has occurred, but could also suggest that the virus has been 'repelled' because of efficient HIV-1-specific immune responses. Furthermore, in a group of women from The Gambia, who were identified because they seemed to have escaped HIV-1 infection despite several years of high-risk sexual behaviour, three out of six patients had CTL that recognised HIV-1- and HIV-2-crossreactive epitopes. This suggests that cellular immunity to HIV-2 may protect against infection with HIV-1 [18]. Along the same lines are findings in recent animal studies which demonstrated that a live attenuated HIV-2 vaccine could protect cynomolgus macaques against development of AIDS for more than 5 years after infection with a pathogenic strain of SIV [19].

These results are quite encouraging, but the putative role of virus-specific CTL for control and clearance of HIV-1 infection remain to be elucidated further. A critical note has been put forward by Phillips who was able to mathematically model termination of the primary viraemia in the absence of HIV-1-specific immune responses [20]. His theory seems to be supported by lack of detectable HIV-1-specific CTL responses in some individuals reported in the current literature. Clearly more systematic studies of patients with acute HIV-1 infection are needed with respect to other effector functions of the immune system to refute or to support a causal role of HIV-1-specific CTL in the initial control of HIV-1 infection.

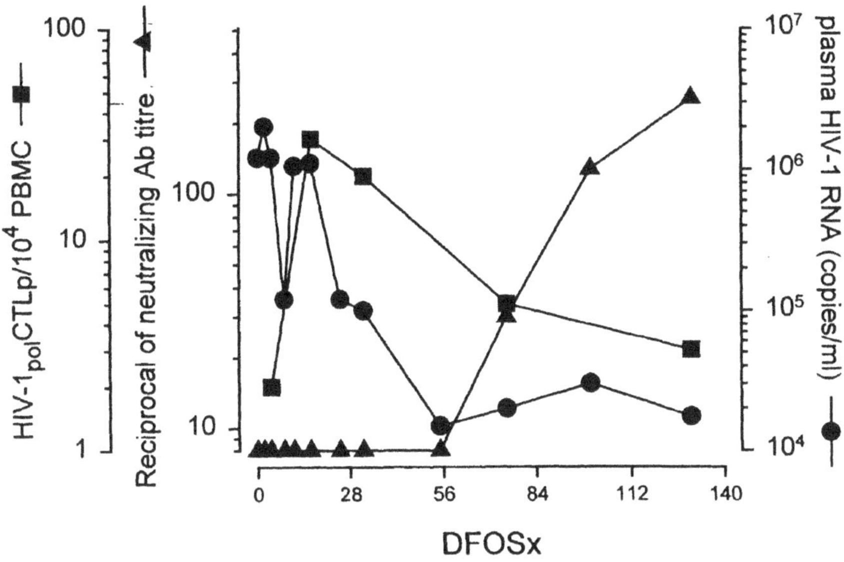

Fig. 1 - Temporal associations of immune responses with initial control of the viraemia in acute HIV-1 infection. Data depicted in this graph were drawn from studies reported by Koup, Safrit and co-workers [8,9,135-138]. Patient AD6 (V), whom they studied in most detail, was a homosexual male who seroconverted for HIV-1 8 days following onset of symptoms (DFOSx) of the acute retroviral syndrome. The viraemia (●) markedly diminished at 28 DFOSx and the virologic setpoint was reached by 56 DFOSx. HIV-1-specific CTL responses were demonstrated as early as 4 DFOSx, with CTL directed against Pol (■) being the most pronounced. Minor responses to Gag and Env were also observed [8,9,135,136]. Neutralising antibody responses (▲) against early autologous HIV-1 isolates were first detected in the plasma at 77 DFOSx [137]. The early virus isolates were macrophage-tropic. Sequence analysis of the gp120-V3 region showed a marked homogeneous virus population, at least until 1 year after the original presentation [9], and without any evidence for viral escape from the HLA-B7-restricted gp120-V3 CTL epitope RPNNNTRKSI [136]. After the acute phase, the $CD4^+$ T-cell returned to relatively normal values and the patient remained asymptomatic during the follow-up of their studies.

2.2. HIV-1-SPECIFIC CTL DURING THE CLINICAL LATENCY PERIOD

Within several months from seroconversion the plasma levels of HIV-1 RNA usually stabilise around the so-called 'virologic setpoint'. Thereafter a variable asymptomatic period follows which appears to correlate with the level of this residual virus replication [21]. The median time from HIV-1 seroconversion to clinical AIDS in adults has been estimated to be about 8 to 10 years. Vigorous HIV-1-specific CTL responses have been observed in most asymptomatic individuals studied to date and involve the recognition of the entire set of HIV-1 proteins [22-24]. The precursor frequencies of HIV-1-specific CTL, as determined in limiting dilution assays, typically range from 10^{-3} to 10^{-5} and are thought to represent mainly memory CTL [25-30]. High levels of circulating HIV-1-specific CTL effectors can often be demonstrated directly *ex vivo* without the need for restimulation and expansion *in vitro* [1,12,31,32]. Measurements of epitope-specific CTL by quantitation of TcR-β mRNA transcripts [33] or flow cytometric analysis of antigen-specific CTL with tetrameric peptide HLA complexes have also indicated extraordinary levels of circulating HIV-1-specific CTL effectors ranging up to several percent of the peripheral blood mononuclear cells (PBMC) [34,147]. Recent studies on viral dynamics have revealed unexpectedly rapid kinetics of HIV-1 replication [35]. The virion half-life has been estimated to be 6 hours and the total number of virions produced per day at 10^{10} [36]. Assuming a minimum burst size of about 100 virions per cell, the total number of productively infected cells is approximately 10^8. Extrapolating these figures would indicate that the *in-vivo* ratios between killer cells and infected target cells approximate those used in *in-vitro* killing assays, in theory suggesting that sufficient CTL can be generated *in vivo* to battle the huge numbers of HIV-1-infected cells [37].

Unresolved to date is the exact relationship between the effector and memory components of the HIV-1-specific CTL responses. It could be hypothesised that the ratio of HIV-1-specific CTL effector to memory frequency reflects how successfully virus replication is contained. For example, if virus replication is well controlled, the number of circulating effector- *c.q.* activated CTL might be low while (resting) memory CTL frequencies are either high or normal. In the case of a high ratio with increased numbers of effector CTL for a prolonged period of time this will probably predict more rapid disease progression. In addition, one could envisage in case of HIV-1-infected patients on highly active anti-retroviral therapy (HAART) that phenotypic analysis of HIV-1-specific CTL could give complementary information to clinicians and patients on the effect of therapy once viral load has become undetectable [38].

Another feature of HIV-1-specific CTL responses is that they typically involve recognition of a large array of epitopes from multiple HIV-1 proteins (Fig. 2). The ever-increasing list of CTL epitopes has frequently been updated in the past [see *e.g.* references 22-24]. Information on HIV-1-derived CTL epitopes is being compiled in the Los Alamos HIV Molecular Immunology database and has been made widely accessible through the computer internet [23]. Characterisation of CTL epitopes, recognised by dominant HIV-1-specific CTL responses during the natural history of HIV-1 infection, is clearly relevant for future vaccine development as it is generally expected that this will involve the induction or boosting of HIV-1-specific CTL responses.

It should be stated that the majority of CTL epitopes identified to date are relatively conserved. This may, however, reflect a bias in the *in-vitro* detection assays of CTL which mainly utilise sequences derived from laboratory strains of HIV-1. This has clearly been demonstrated for some unfortunate labworkers infected with laboratory strain HIV-1_{IIIb} in which case the autologous virus sequences and the *in-vitro* tools for detection were fully matched [39]. It was shown that a significant proportion of CTL responses were directed against sequences unique to this viral isolate [39], suggesting that the breadth and the magnitude of HIV-1-specific CTL responses in general are likely to be underestimated.

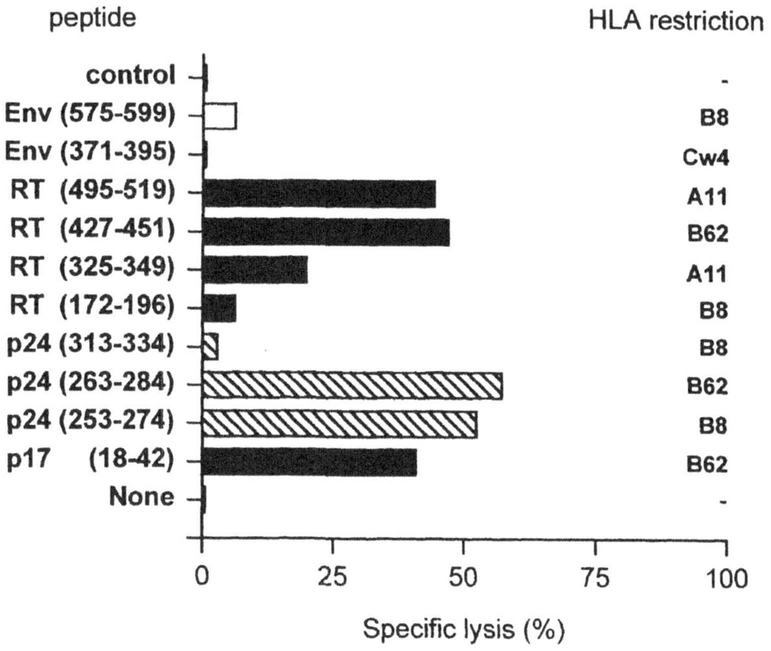

Fig. 2 - Simultaneous recognition of multiple HIV-1-derived CTL epitopes in cultures of unstimulated PBMC. Data depicted in this graph were drawn from studies reported by Walker, Johnson and co-workers [1,23,24,89-91]. The graph represents experiments with freshly isolated and unstimulated PBMC of HIV-1-seropositive individual 63 (010-035i), whom they studied in great detail. PBMC were directly tested for cytotoxicity in a standard 6-hr ^{51}Chromium-release assay using EBV-transformed autologous B-lymphoblastoid cell lines that had been pulsed with the indicated synthetic peptides previously demonstrated to contain CTL epitopes (100 µg/ml) [92]. Numbers in parentheses indicate the position of the first and last amino acid of the peptide. For each peptide, the HLA restriction element is indicated on the right. By analysing the epitope specificity of HIV-1-specific clones from this individual, it was found that up to 14 distinct HIV-1 CTL epitopes were recognised [92]. No extensive clinical follow-up data were reported on this individual: at the time the experiments were carried out, he was asymptomatic and, although the single published determination of his $CD4^+$ T-cell counts (145 cells/µl) and HIV-1 RNA levels (94,800 copies/ml) indicate a significant increased risk for progression to AIDS, as of recently he is still alive and doing well. At this point, he is on extensive anti-retroviral therapy (B.D. Walker, *personal communication*).

2.3. HIV-1-SPECIFIC CTL DURING PROGRESSION TO AIDS

Many studies have documented that HIV-1-specific CTL responses deteriorate during disease progression [25,29,30,32,40,41,144]. From longitudinal studies on HIV-1-specific CTL and viral load it follows that there are at least two scenarios for disease progression. In rapid progressors the viral load generally increases in the face of strong HIV-1-specific CTL responses, suggesting that HIV-1 has escaped anti-viral CTL responses. Another pattern is observed in more typical progressors, in whom the viral load seems to increase only after a substantial loss of HIV-1-specific

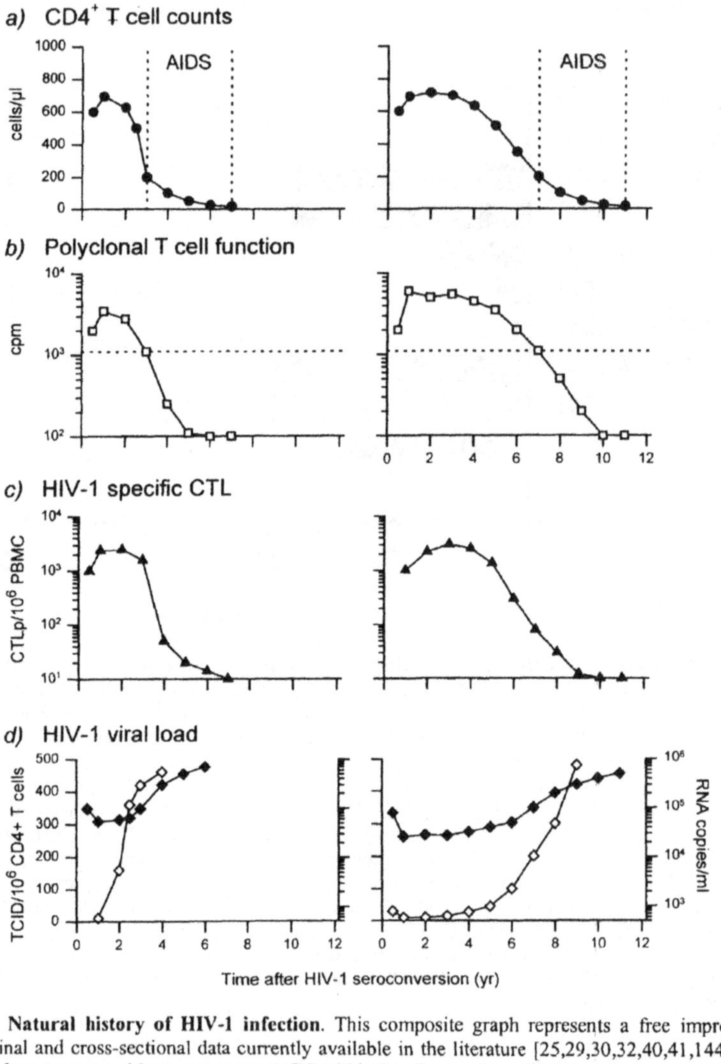

Fig. 3 - Natural history of HIV-1 infection. This composite graph represents a free impression of longitudinal and cross-sectional data currently available in the literature [25,29,30,32,40,41,144]. Panels on the left represent **rapid** progressors to AIDS within 3 to 7 years from HIV-1 seroconversion. Panels on the right represent **typical** progressors who develop AIDS within 7 to 11 years. Panels *a*) show longitudinal $CD4^+$ T-cell counts (●) during the years following HIV-1 seroconversion; *b*) polyclonal T-cell function to CD3 Mab in whole blood lymphocyte cultures (□); *c*) HIV-1-specific CTLp frequencies (▲) determined with limiting dilution analysis, and *d*) HIV-1 viral load representing the number of circulating $CD4^+$ T cells productively infected with HIV-1 (TCID) (◊), and the number of HIV-1 RNA copies/ml (♦). Note *i*) the relatively high virologic setpoint after the acute phase of infection; *ii*) the increase of HIV-1-infected $CD4^+$ T cells in the face of vigorous HIV-1-specific CTL responses for rapid progressors, in violent contrast to typical progressors where viral load significantly increased only after substantial loss of HIV-1-specific CTL responses, and *iii*) the loss of anti-viral CTL responses that parallels deterioration of the polyclonal T-cell function.

CTL has occurred (Fig. 3). The reasons for the apparent failure of CTL to contain HIV-1 replication may be diverse and as of yet are only poorly understood (*see below*).

As mentioned before, it has been proposed by some investigators that HIV-1-specific CTL responses are deleterious to the infected host [6,7]. In general, however, the observations to date do not seem to support this concept. The most obvious counter argument would be the fact that $CD4^+$ T-cell numbers do not recover when HIV-1-specific CTL responses deteriorate during progression to AIDS [25,29,30,32,40,41], whereas it has been shown that most late-stage patients still can regain considerable numbers of $CD4^+$ T cells during highly active anti-retroviral therapy (HAART) [35,42]. Along the same lines, infusions of large amounts of non-specifically expanded autologous $CD8^+$ T lymphocytes, although proportions of HIV-1-specific CTL are largely unknown, have been well-tolerated and without any deleterious effect on $CD4^+$ T-cell counts [43]. More accepted, however, seems the fact that HIV-1-specific CTL, which infiltrate infected tissues such as the lungs and lymphoid organs, are associated with inflammatory side effects and might contribute to the progressive disintegration of lymphoid tissues as observed at late stages of HIV-1 infection [2,4,5,44]. It is generally thought that the latter phenomena are of less significance to the initiation of the disease process.

2.4. CTL AND LONG-TERM SURVIVAL OF HIV-1 INFECTION

In a small proportion of HIV-1-infected individuals which represent the extreme of the right tail of disease progression, an extraordinary benign disease course beyond the median time to AIDS is observed. This group of so-called 'long-term survivors' appears heterogeneous and seems to contain mostly very slow progressors and may be some true non-progressors [45]. It is generally expected that HIV-1-specific CTL from long-term survivors have distinct features which might contribute to prolonged maintenance of the asymptomatic phase. Several studies of long-term survivors have shown robust and persistent HIV-1-specific CTL responses [30,46-49] involving simultaneous recognition of multiple CTL epitopes [49-53]. These CTL responses coincide with relatively low numbers of HIV-1-infected cells and intuitively point toward a situation of efficient viral control.

It has been suggested that qualitative differences already in the initial immune response to HIV-1 may be responsible for distinct clinical outcomes [54, 55]. Pantaleo *et al.* have reported that simultaneous expansions of many Vβ-families during primary HIV-1 infection, probably resembling a broadly directed HIV-1-specific CTL response, at least protects against rapid progression to AIDS [54]. It could imply that the virus is less able to escape or exhaust HIV-1-specific responses (*see below*) if the response mounted was polyclonal. In addition, Van Baalen *et al.* have shown that early CTL responses from long-term survivors were more frequently targeted at epitopes in Tat and Rev compared to those of rapid progressors [55]. This suggests that CTL directed against early viral proteins may be more effective in controlling viral load, as they are thought to kill HIV-1-infected cells before a major release of virions takes place.

Central to the controversy on HIV-1-specific CTL are the findings that some long-term survivors seem to have rather low levels of HIV-1-specific CTL [48]. This condition, however, may be a consequence of infection with attenuated viruses [56,57] or may be caused by genetic defects in HIV-1 co-receptor expression [58,59], which both could result in a lower level of viral replication and hence in a lower level of immune activation.

Even more paradoxical are the findings that patients who progress to AIDS usually also mount strong HIV-1-specific CTL responses during their asymptomatic period [30], frequently involving epitopes recognised by CTL from long-term survivors [32,50,51,60]. However, consistent associations between HLA alleles and time to AIDS (*e.g.* HLA-B8, -B35 and rapid disease progression [61] and HLA-B27 or -B57 and long-term survival [45,50,62-64,145]) indicate that there may be qualitative differences between HIV-1-specific CTL responses. Either structural or functional peculiarities of these HLA molecules, affecting the number or the kind of CTL epitopes presented [65], or the level of cross-talk with other cell types involved in the immune response to HIV-1 (*e.g.* interactions with inhibitory receptors on natural killer cells [66]) could be responsible for the observed associations. It must be emphasised, however, that associations with HLA are never absolute and, consequently, HIV-1-specific CTL responses restricted by so-called 'protective' alleles have also been observed in people with a progressive disease [60]. Inconsistencies may be explained by taking into account the virulence of transmitted viruses which can differ significantly from person to person [67,68].

3. Persistence of HIV-1 Despite Vigorous Anti-viral CTL Responses

Despite seemingly potent CTL responses, HIV-1 is almost never completely eradicated from the body but instead persists for many years at the virologic setpoint [21]. It seems likely that release of virions from HIV-1-infected cells, which are relatively resistant or not readily accessible to CTL-mediated killing or other types of immune responses, will significantly contribute to this level of residual viral replication. The ability of the virus to escape from CTL recognition and the broad cell tropism of HIV-1 (*i.e.* specific target-cell preference) are both considered to confer viral persistence in the face of vigorous HIV-1-specific CTL responses.

3.1. VIRAL ESCAPE FROM CTL RECOGNITION

Given the error rate of HIV-1 reverse transcriptase in the order of 10^{-4} to 10^{-5} per base, the size of the viral genome of approximately 10^4 bp and the huge number of virions produced each day, the numbers of mutant viruses which appear each day are tremendous [36,69]. This high mutation rate results in steady accumulation of random mutations in the viral genome. Only those mutations will persist in the viral quasispecies which do not significantly interfere with functional or structural constraints of viral proteins. At the same time, virus variants are continuously subjected to strong selective forces which will result in accumulation of mutations at specific sites of the viral genome. Mutant viruses may appear in the course of disease which carry mutations in CTL epitopes that affect CTL recognition, either by interfering with intracellular processing and transport of viral peptides, or with peptide/MHC T-cell receptor interactions [70]. Although initially some concern was raised [71,72], there is now an increasing body of experimental evidence showing that escape of HIV-1 from CTL recognition can result in viral persistence [12,13,73-77]. In cases where the HIV-1-specific CTL response is oligoclonal and involves recognition of only a single variable epitope, the virus can easily adapt and dramatic acceleration of progression to AIDS has been observed (Fig. 4) [12,75]. For example, adoptive transfer of an *ex-vivo* expanded autologous nef-specific CTL clone resulted in the emergence of viral escape mutants lacking the specific epitope sequence recognised, while at the same time the patient rapidly progressed to AIDS [75]. The same phenomenon has been observed by Borrow *et al.* who reported two patients suffering from acute HIV-1 infection [12]. In the face of a strong and oligoclonal expansion of $CD8^+$ T cells [10], which reflected the dominant HIV-1 gp160-specific CTL response to a single HLA-B44-restricted CTL epitope [11,12], there was a rapid accumulation of viruses which had escaped this dominant CTL response (Fig. 4). The patients subsequently progressed to AIDS and death within 3 years. The level of reduction of wild-type virus, the kinetics with which mutant viruses appeared, and the genetic pathways by which the virus escaped CTL recognition bear significant similarities to the kinetics with which drug-resistant strains of HIV-1 emerge during anti-retroviral therapy [78]. This indicates that HIV-1-specific CTL can exert selective pressure on the viral quasispecies. As such, these observations are strong evidence that HIV-1-specific CTL are relevant for controlling viral replication *in vivo* (Fig. 4).

Most HIV-1-infected persons, however, mount polyclonal CTL responses to the virus involving simultaneous recognition of many epitopes. This situation will seriously limit the opportunities for the virus to escape *in vivo* [79], as newly emerging variants can still be eliminated by CTL recognising epitopes which have not changed. Recently, it was shown in 4 asymptomatic HIV-1-infected individuals that particular variants dominated the viral quasispecies only for a short period of time and that they seemed to be eliminated by HIV-1-specific CTL later [76]. Depending on the dominance of the CTL response from which the virus escapes, the virus will experience a transient decrease of the cytolytic force of the immune system and subsequently a temporary selective advantage *in vivo*. Once CTL responses with new specificities are mounted, or CTL from subdominant responses are boosted, the viral load is expected to return to more or less steady-state values. This transient escape mechanism may only be operational during the asymptomatic period when HIV-1-specific CTL responses are not yet affected. During disease progression when the immune competence of the infected host is gradually decaying (*i.e.* when the selection pressure by CTL subsides), the need for the virus to adapt to anti-viral immune responses will slowly disappear and the most fit virus strains will be selected. In order to further explore viral dynamics under the selective constraints of broad immune responses, mathematical models may need to be adjusted with more experimental data to see what the exact clinical consequences are of these temporary, and often non-linear virus-host interactions [79,80].

Many HIV-1-derived CTL epitopes are well conserved among the published HIV-1 sequences [81], *i.e.* the virus can hardly escape from CTL recognising these epitopes. Although in these cases the immune pressure exerted by HIV-

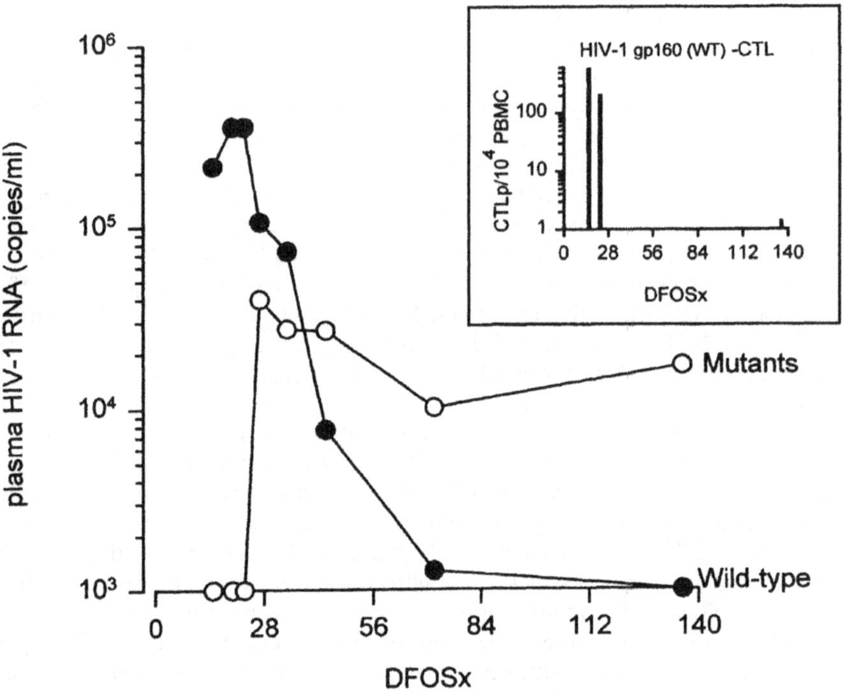

Fig. 4 - Emergence of HIV-1 escape variants from a dominant CTL response. Data depicted in this graph were drawn from studies by Borrow, Shaw and co-workers [10-12,141-143]. Patient #1 (WEAU), whom they studied in most detail, was a homosexual male who presented with symptomatic primary HIV-1 infection 20 days after a single encounter with a patient with AIDS [141]. He seroconverted for HIV-1 at 23 DFOSx of the acute retroviral syndrome. The viraemia reached peak levels at 20 DFOSx and decreased significantly thereafter [142,143]. The first detectable HIV-1-specific CTL responses were observed at 16 DFOSx [10-12] and were exclusively directed against the HLA-B44-restricted wild-type (WT) gp160 CTL epitope AENLWVTVY [12]. This response was exceptional since it comprised at least 5 per cent of the circulating PBMC, although later it declined significantly (*see insert*) [10-12]. From 30 DFOSx on, CTL with other specificities were also observed [10-12]. The early WT isolates (●) were T-cell-tropic and syncytium-inducing [141]. Sequence analysis of the WT immunodominant gp160 CTL epitope AENLWVTVY revealed that escape mutants (○) appeared from 30 DFOSx, which were no longer recognised by the initial CTL response to this epitope. After an initial rebound, CD4$^+$ T-cell counts quickly dropped below 200 cells/μl at 212 DFOSx and the patient died 1601 DFOSx [12].

1-specific CTL is less clear, it does not imply that HIV-1-specific CTL are not involved in effectively eliminating virus-infected cells. On the contrary, it strongly adds to the suggestion that there are additional escape mechanisms which allow the virus to persist for many years in the face of vigorous HIV-1-specific CTL responses [82].

3.2. VIRAL SANCTUARIES

There are many ways for viruses to evade the host immune response [82]. One of the strategies is to create a viral sanctuary where the virus can remain dormant or a site where the virus can continue to replicate, however, without being noticed by the immune system.

Down-regulation of MHC class-I molecules on virus-infected cells or interference with antigen-processing pathways could prevent viral peptides from being presented to T lymphocytes. In HIV-1-infected cells, partial down-regulation of MHC class-I molecules has been observed and ascribed to the viral proteins Tat and Nef [83,84]. Since it appears that only a few peptide-MHC class-I complexes per infected cell are needed to activate the cytolytic machinery of HIV-1-specific CTL, the level of down-regulation of MHC class-I molecules in general may not be sufficient to completely resist CTL-mediated killing [85]. It should be noted that this escape mechanism is rather leaky as down-regulation of MHC class-I molecules would render virus-infected cells susceptible to killing by natural killer (NK) cells.

Another effect of Nef could be the upregulation of Fas ligand (CD95L) expression in virus-infected cells [86], which could also serve as a means to create local sanctuaries. This situation may resemble the proposed FasL-mediated evasion of melanoma cells, which is supposed to induce apoptosis of infiltrating tumour-specific T lymphocytes [87]. However, it has been shown extensively that $CD4^+$ T cells and macrophages infected with HIV-1 can be readily killed by CTL *in vitro* [2, 85], showing that FasL-mediated apoptosis is probably not sufficient to completely abrogate HIV-1-specific CTL responses.

The recent identification of the co-receptors for HIV-1 has substantiated the long-standing observations of HIV-1 strains being variably able to infect different cell types [88]. Containment of virus-infected cells by CTL in various tissues may be complex. Some infected cell types may be efficiently eliminated by CTL, whereas others may be relatively resistant to cell-mediated killing (*e.g.* cells from the macrophage lineage in the brain, bone marrow and epididymis). By analogy with hepatitis-B virus infection (HBV), control of viral replication at sanctuary sites may be dependent on production of anti-viral cytokines [89]. It has been shown that HIV-1-specific CTL are capable of secreting various soluble factors which interfere *in vitro* with HIV-1 replication (*i.e.* Interferon-γ, Tumour Necrosis Factor (TNF)-α, $CD8^+$ T-cell anti-viral factor (CAF), Interleukin (IL)-16, and β-chemokines: RANTES (for regulated on activation normal T cell expressed and secreted), macrophage inflammatory protein 1α (MIP-1α) and MIP-1β (reviewed in [3]). These anti-viral factors appear to extend the action radius of CTL and may to some extent also protect certain non-infected cells, whereas cytolytic processes can act only on cells in which the virus is actively replicating. Elimination of virus from cellular reservoirs resistant to CTL-mediated killing and/or anti-viral factors will be dependent on the natural half-life of the infected cells in which the virus reside as well as on viral cytopathic effects. As such, these infected cells are expected to contribute significantly to the virologic setpoint and subsequently to viral persistence.

Central to the debate on HIV-1-specific CTL is the lack of our current understanding how the different measures of viral load relate to HIV-1-specific CTL responses. To date, the exact relationships still need to be established between the level of residual virus replication as measured by the amount of HIV-1 RNA in the plasma (*i.e.* cell-free virions) and the numbers and different types of cells in tissues or in the circulation able to give viable virus progeny. In addition, infected cells may contain a proviral DNA genome which is defective or which is just temporarily dormant. Latently infected cells which hardly express viral antigens are virtually invisible to the immune system until they start to produce virions upon activation and become targets for CTL recognition [90]. Recently, it was shown for patients on highly active anti-retroviral therapy (HAART) for more than 2 years that the frequency of resting $CD4^+$ T cells harbouring latent HIV-1 was low and did not significantly decrease with increasing time on therapy [91,92]. Therefore, these cells could also constitute an important viral reservoir that significantly contributes to persistence of HIV-1 infection despite high levels of circulating CTL.

An important consequence of the broad cell tropism of HIV-1 is that cells, which are susceptible to infection with the virus, are also crucial to a properly functioning immune system (*see below*). Since HIV-1 attacks the Achilles' heel of the cognate immune system (*i.e.* $CD4^+$ T cells and macrophages), it is likely that the virus eventually will destroy or at least in part will interfere with cells which are involved in the anti-viral immune response [93].

4. Loss of Specific Anti-viral Immune Surveillance

Based on the current literature, at least two non-mutually exclusive scenarios could be proposed for declining HIV-1-specific CTL responses during progression to AIDS. The first scenario would be a physical deletion (*i.e.* exhaustion) of HIV-1-specific CTL. The second would concern a functional depletion of anti-viral CTL due to lack of sufficient levels of specific $CD4^+$ T-helper cells.

4.1. PHYSICAL DELETION OF HIV-1-SPECIFIC CTL

To date, there is substantial experimental evidence to support that HIV-1-specific CTL are preferentially lost during disease progression [29,40,94,95). It was shown both in a cross-sectional study [29], and in a longitudinal study [94], that HIV-1-specific CTL are hardly detectable at low $CD4^+$ T-cell counts, whereas Epstein-Barr virus (EBV)-specific CTL were seemingly unaffected at late stage HIV-1 infection. Similarly, in some AIDS patients broad cytolytic CTL activity was preserved, whereas HIV-1-specific CTL activity was lost [40]. It must be said that there are also observations of impaired CTL responses to other pathogens, such as influenza virus [96], *Mycobacterium tuberculosis* [97], as well as EBV [98] and polyclonal CTL responses [99,100]. These phenomena, however, could well be related to virus-mediated selective depletion of specific $CD4^+$ T-helper cell responses (*see below*).

For mice infected with lymphocytic choriomeningitis virus (LCMV), it has been shown that overwhelming virus infection results in exhaustion of the anti-viral CTL response [101]. A similar situation may arise in HIV-1-infected patients. Long-standing observations of the $CD8^+$ T-cell compartment in HIV-1-infected patients show directly *ex-vivo* detectable CTL effector activity which is uncommon to many other viral infections [1,31,32]. Furthermore, the number of $CD8^+$ T cells are usually increased with many cells carrying an activated phenotype (*e.g.* increased expression of HLA-DR, CD38 and CD57 (reviewed in [102]), and with a high percentage of $CD8^+$ T cells primed for apoptosis [103,104] (*i.e.* upregulation of Fas [105] and down-modulation of Bcl-2 [106]). In addition, telomeres of $CD8^+$ T cells from HIV-1-infected patients are considerably shorter than those of their $CD4^+$ T-cell counterparts, suggesting a prolonged history of peripheral expansion [107,108]. Taken together, these results point toward continuous activation and expansion of specific and non-specific (*bystander*) $CD8^+$ T cells [109], which could possibly lead to exhaustion of anti-viral CTL responses. Recently, Pantaleo *et al.* have observed in a few patients suffering from acute HIV-1 infection that a significant number of HIV-1-specific CTL disappeared in time, which could not be attributed to viral escape from the epitopes recognised by these CTL clones [95]. On the other hand, it has been reported for some patients that HIV-1-specific CTL clones can persist for many years [110,145], so it remains to be established in which situations clonal exhaustion of anti-viral CTL occurs and what the exact clinical effect thereof could be.

It may be hypothesised that CTL responses towards relatively conserved epitopes in HIV-1 are more susceptible to exhaustion as they are constantly being stimulated. In that scenario it would make the immune control of HIV-1 infection more dependent on recognition of variable epitopes, and thus it could contribute to viral escape from CTL recognition.

4.2. FUNCTIONAL ANERGY OF ANTI-VIRAL CTL

Deterioration of HIV-1-specific CTL responses during disease progression can alternatively be explained by the fact that anti-viral CTL become functionally inactive. For example, Moss *et al.* found that frequencies of T-cell receptor sequences of certain HIV-1-specific CTL clones persisted in the peripheral blood despite the fact that CTL reactivity was waning [33]. This phenomenon could result from continuous activation and persistent stimulation of HIV-1-specific CTL which may cause a state of hyporesponsiveness [99,102,111]. Alternatively, functional anergy of CTL could result from lack of sufficient $CD4^+$ T-helper activity, either being reduced absolute $CD4^+$ T-cell numbers or altered functional phenotype of these cells. Of interest are elegant studies in mice experimentally infected with parasites [112], and studies in CD4-deficient mice [113-115], which have shown in cases where T-helper function is altered or absolute $CD4^+$ T-cell counts are very low to absent, that the induction as well as maintenance of anti-viral CTL responses can be affected. For example, BALB/c mice infected with *Schistosoma mansoni* have decreased T-helper type-I responses correlated with diminished virus-specific CTL and subsequently with delayed virus clearance [112]. In HIV-1-seropositive individuals it has been shown that co-infection with intracellular parasite *Toxoplasma gondii* usually results in the severe and life-threatening disease toxoplasmic encephalitis. These patients display impaired IL-2, IFNγ and IL-12 production, whereas IL-6 and Tumour Necrosis Factor α (TNFα) seems not to be affected [116]. Furthermore, there are many long-standing observations which suggest selective impairment of type-I cytokine-mediated immune responses in HIV-1-infected patients [117,118]. Proliferation *in vitro* of T cells in response to ligation of the T-cell receptor (TCR)-CD3 complex is usually impaired and IL-2 production decreased [117,119]. In addition, *in-vivo* delayed-type hypersensitivity (DTH) reactions usually subside during the disease course [120,121]. On the other hand, recent observations from patients treated with genetically modified HIV-1-specific CTL clones have shown that HIV-1-infected patients with low $CD4^+$ T-cell counts and absent DTH responses to several recall antigens were still able to generate primary CTL responses to endogenously expressed foreign antigens [43]. This may suggest that in these cases the induction of MHC class-I-restricted CTL responses was relatively preserved even when the immune system was already significantly affected, and that somehow the problem is more likely to be the regulation of the maintenance of these responses.

Very well documented is the sequential loss of *in-vitro* proliferative responses to HIV-1 proteins [93,122,123], recall antigens (*e.g.* influenza virus, tetanus toxoid and microbial antigens) [124-126], allo-antigens and T-cell mitogens (*e.g.* phytohaemagglutinin and CD3 monoclonal antibodies) [117,119,127,128]. The order of events suggest that this is related to the frequency of encountering various antigenic challenges. $CD4^+$ T cells involved in an anamnestic immune response, either activated as bystanders or stimulated by the foreign antigen, are at risk of becoming infected with HIV-1. In support of this are studies where *in-vivo* antigenic stimulation, *e.g.* by recall antigen vaccination [129-131] or IL-2 therapy [132], has shown to activate and support virus replication in infected individuals.

The conspicuous absence of HIV-1-specific $CD4^+$ T-helper cell responses in most infected patients [93,122,123] suggest that this subset of T helpers are either lost very early in infection or are never really induced in significant numbers. Recently, it has been observed by the group of Walker *et al.* that a long-term survivor, who had been infected with HIV-1 for 18 years without major loss of $CD4^+$ T cells, displayed strong T-helper responses to HIV-1 Gag p24 en Env gp160 proteins [93]. Furthermore, they found an inverse correlation between viral load and proliferative responses to p24. Upon examination of patients who had been put on highly active anti-retroviral therapy (HAART) very early in their infection, even before HIV-1 seroconversion occurred, they observed that this group was able to generate strong proliferative T-cell responses to HIV-1 antigens once their viral loads were reduced to undetectable levels [93]. Early loss of HIV-1-specific T-helper cell responses may help to explain why patients ultimately fail to control the virus, even though the induction of HIV-1-specific CTL responses may still be preserved for some time [43]. Reconstitution of specific immune responses after multiple anti-viral drug therapy is now among the main targets for future studies. It has already been shown that HIV-1-infected individuals, who are successfully treated, to some extent regain their proliferative responses to recall antigens [133]. As it may be envisaged that early and effective treatment may preserve specific T-helper cell responses, the actual boosting of HIV-1-specific CTL responses, however, may not occur optimally if the numbers of HIV-1-infected cells are very low. Boosting of T-helper cells specific for HIV-1 may in the long run allow for sustaining HIV-1-specific CTL responses [146]. As has been suggested by Pantaleo [134], it may prove beneficial to combine highly active anti-retroviral therapy (HAART) with immunization against

HIV-1 antigens to ensure sufficient levels of HIV-1-specific CTL in the case of emerging drug-resistant strains or in situations where patients cannot complete the therapy because of the sometimes serious side-effects of these novel drugs [134].

5. Concluding Remarks

The composite picture from the current literature is in favour of a beneficial role for CTL in HIV-1-infected individuals, and does not support the concept that HIV-1-specific CTL responses are deleterious to the infected host. It must be emphasised that HIV-1-specific CTL per se do not mediate life-long protection from progression to AIDS. HIV-1-specific CTL responses appear variably able to control HIV-1 replication to the virologic setpoint. Therefore, HIV-1-specific CTL may variably contribute to slowing down the rate of disease progression, but as of yet the exact determinants thereof remain to be revealed. The current data strongly point towards multiple mechanisms for viral persistence and for failure of HIV-1-specific CTL responses to control HIV-1 infection indefinitely, of which not the least is the conspicuous absence of specific T-helper cell responses to HIV-1 [93,146].

To further increase our understanding on the biological relevance of HIV-1-specific CTL, research in the next decade may need to focus on the following issues:

- Extensive characterisation of the timing, epitope specificity and phenotype of HIV-1-specific CTL (*i.e.* state of activation, memory *versus* effector type, mode of killing) in patients during the acute phase of HIV-1 infection, because the virologic setpoint early in infection strongly correlates with the rate of disease evolution, in long-term survivors, with the hope to find immune correlates of protection to AIDS, and in (typical) progressors, in order to further explore mechanisms which seem to frustrate anti-viral CTL responses.
- Elucidation of temporal relationships between HIV-1-specific $CD8^+$ CTL, HIV-1-specific $CD4^+$ T-helper cells and different measurements of HIV-1 viral load in situations where the quasi steady-state of virus replication and host immune responses is actively disturbed (*i.e.* immunisation and/or HAART) in order to gain more insight into 'cause and effect' of HIV-1-specific immune responses.
- Continuous updating of the Los Alamos HIV Molecular Immunology Database [23] with newly defined HIV-1-derived CTL epitopes and, more importantly, with relevant clinical and virological data on the patients studied. Viral sequence data and information on methods of screening should be included to judge the value of particular epitopes. Furthermore, serious efforts should be made to include data on the natural history of HIV-1 infection of previously reported patients, since a wealth of relevant unpublished data may still be at our disposal.

By addressing these research topics, the controversies on HIV-1-specific CTL will hopefully be resolved by the next decade. Ideally it would accelerate the development of effective vaccines and immunotherapeutic intervention strategies, especially for those unfortunate people who are not able to bear the sometimes insufferable side-effects of anti-viral therapies, and for all of those individuals in developing countries who do not have readily access to potent anti-retroviral drugs.

6. Acknowledgements

Our studies have been performed as part of the Amsterdam Cohort Studies on HIV-1 infection and AIDS, as well as of the French Immunoco and Alt-cohort studies. Financial support was given by the Dutch AIDS Fund, the Dutch Ministry of Public Health (RGO/VWS), Stichting AIDS Research Amsterdam (SARA), the Netherlands Organization for Scientific Research (NWO), the French Ageny of AIDS Research (ANRS), SIDACTION, and the EEC Biomed Program.

7. References

1. Walker, B.D., Chakrabarti, S., Moss, B. *et al.*: HIV-1-specific cytotoxic T lymphocytes in seropositive individuals, *Nature* **328** (1987), 345-348.
2. Plata, F., Autran, B., Martins, L.P. *et al.*: AIDS virus-specific cytotoxic T lymphocytes in lung disorders, *Nature* **328** (1987), 348-351.
3. Levy, J.A., Mackewicz, C.E. and Barker E.: Controlling HIV-1 pathogenesis: the role of the non-cytotoxic anti-HIV-1 response of $CD8^+$ T cells, *Immunol. Today* **17** (1996), 217-224.
4. Autran, B., Mayaud, C.M., Raphael, M. *et al.*: Evidence for a cytotoxic T-lymphocyte alveolitis in human immunodeficiency virus-infected patients, *AIDS* **2** (1988), 179-183.
5. Jassoy, C., Johnson, R.P., Navia, B.A. *et al.*: Detection of a vigorous HIV-1-specific cytotoxic T-lymphocyte response in cerebrospinal fluid from infected persons with AIDS demential complex, *J. Immunol.* **149** (1992), 3113-3119.
6. Zinkernagel, R.M. and Hengartner, H.: T-cell-mediated immunopathology *versus* direct cytolysis by virus: implications for HIV and AIDS, *Immunol. Today* **15** (1994), 262-268.
7. Zinkernagel, R.M.: Are HIV-specific CTL responses salutary or pathogenic?, *Curr. Opin. Immunol.* **7** (1995), 462-470.
8. Koup, R.A., Safrit, J.T., Cao, Y. *et al.*: Temporal associations of cellular immune responses with the initial control of viremia in primary Human Immunodeficiency Virus type 1 syndrome, *J. Virol.* **68** (1994), 4650-4655.
9. Safrit, J.T. and Koup, R.A.: The immunology of primary HIV infection: which immune responses control HIV replication, *Curr. Opin. Immunol.* **7** (1995), 456-461.
10. Pantaleo, G., Demarest, J.F., Soudeyns, H. *et al.*: Major expansion of CD8+ T cells with a predominant Vβ usage during the primary immune response to HIV-1, *Nature* **370** (1994), 463-467.
11. Borrow, P., Lewicki, H., Hahn, B.H. *et al.*: Virus-specific $CD8^+$ cytotoxic T-lymphocyte activity associated with control of viremia in primary human immunodeficiency virus type 1 infection, *J. Virol.* **68** (1994), 6103-6110.
12. Borrow, P., Lewicki, H., Wei, X. *et al.*: Antiviral pressure exerted by HIV-1-specific cytotoxic T lymphocytes (CTLs) during primary infection demonstrated by rapid selection of CTL escape virus, *Nature Med.* **3** (1997), 205-211.
13. Price, D.A., Goulder, P.J., Klenerman, P. *et al.*: Positive selection of HIV-1 cytotoxic T-lymphocyte escape variants during primary infection, *Proc. Natl. Acad. Sci. USA* **94** (1997), 1890-1895.
14. Yasutomi, Y., Reiman, K., Lord, C. *et al.*: Simian immunodeficiency virus-specific $CD8^+$ lymphocyte response in acutely infected rhesus monkeys, *J. Virol.* **67** (1993), 1707-1711.
15. Reimann, K.A., Tenner-Racz, K., Racz, P. *et al.*: Immunopathogenic events in acute infection of rhesus monkeys with Simian immunodeficiency virus of macaques, *J. Virol.* **68** (1994), 2362-2370.

16. Rowland-Jones, S.L. and McMichael, A.J.: Immune responses in HIV-1-exposed seronegatives: have they repelled the virus?, *Curr. Opin. Immunol.* **7** (1995), 448-455.
17. Langlade-Demoyen, P., Ngo-Giang-Huong, N., Ferchal, F. *et al.*: HIV-1 nef-specific cytotoxic T lymphocytes in noninfected heterosexual contact of HIV-infected patients, *J. Clin. Invest.* **93** (1994), 1293-1297.
18. Rowland-Jones, S., Sutton, J., Ariyoshi, K. *et al.*: HIV-specific cytotoxic T cells in HIV-exposed but uninfected Gambian women, *Nature Med.* **1** (1995), 59-64.
19. Putkonen, P., Walther, L., Zhang, Y.J. *et al.*: Long-term protection against SIV-induced disease in macaques vaccinated with a live attenuated HIV-2 vaccine, *Nature Med.* **1** (1995), 914-918.
20. Phillips, A.N.: Reduction of HIV concentration during acute infection: independence from a specific immune response, *Science* **271** (1996), 497-499.
21. Mellors, J.W., Kingsley, L., Rinaldo Jr., C.R. *et al.*: Quantitation of HIV-1 RNA in plasma predicts outcome after seroconversion, *Ann. Intern. Med.* **122** (1995), 573-579.
22. McMichael, A.J. and Walker, B.D.: Cytotoxic T-lymphocyte epitopes: implications for HIV-1 vaccines, *AIDS* **8** (suppl. 1) (1994), S155-S173.
23. Korber, B.T.M., Brander, C., Walker, B.D. *et al.* (eds.): HIV Molecular Immunology Database, 1995 Los Alamos National Laboratory, Los Alamos, New Mexico: Theoretical Biology and Biophysics, 1995 (http://hiv-web.lanl.gov/immunology).
24. Autran, B.: Cytotoxic T-lymphocyte responses to HIV: from primary infection to AIDS. In: Gupta, S. (ed.) *Immunology of HIV Infection*, New York & London, Plenum Medical Book Company (1996), p. 201-228.
25. Hoffenbach, A., Langlade-Demoyen, P., Dadaglio, G. *et al.*: Unusually high frequencies of HIV-specific cytotoxic T lymphocytes in humans, *J. Immunol.* **142** (1989), 452-462.
26. Joly, P., Guillon, J.-M., Mayaud, C. *et al.*: Cell-mediated suppression of HIV-specific cytotoxic T lymphocytes, *J. Immunol.* **143** (1989), 2193-2201.
27. Gotch, F.M., Nixon, D.F., Alp Nd, R.E.Y. *et al.*: High frequency of memory and effector gag-specific cytotoxic T lymphocytes in HIV-seropositive individuals, *Int. Immunol.* **2** (1990), 707-712.
28. Koup, R.A., Pikora, C.A., Luzuriaga, K. *et al.*: Limiting dilution analysis of cytotoxic T lymphocytes to human immunodeficiency virus gag antigens in infected persons: *in-vitro* quantitation of effector cell populations with p17 and p24 specificities, *J. Exp. Med.* **174** (1991), 1593-1600.
29. Carmichael, A., Jin, X., Sissons, P. *et al.*: Quantitative analysis of the human immunodeficiency virus type 1 (HIV-1)-specific cytotoxic T-lymphocyte (CTL) response at different stages of HIV-1 infection: differential CTL responses to HIV-1 and Epstein-Barr virus in late disease, *J. Exp. Med.* **177** (1993), 249-256.
30. Klein, M.R., Van Baalen, C.A., Holwerda, A.M. *et al.*: Kinetics of gag-specific CTL responses during the clinical course of HIV-1 infection: a longitudinal analysis of rapid progressors and long-term asymptomatics, *J. Exp. Med.* **181** (1995), 1365-1372.
31. Walker, B.D., Flexner, C., Paradis, T.J. *et al.*: HIV-1 reverse transcriptase is a target for cytotoxic T lymphocytes in infected individuals, *Science* **240** (1988), 64-66.
32. Johnson, R.P., Trocha, A., Yang, L. *et al.*: HIV-1 gag-specific cytotoxic T lymphocytes recognize multiple highly conserved epitopes: fine specificity of the gag-specific response defined by using unstimulated peripheral blood mononuclear cells and cloned effector cells, *J. Immunol.* **147** (1991), 1512-1521.
33. Moss, P., Rowland-Jones, S., Frodsham, P. *et al.*: Persistent high frequency of human immunodeficiency virus-specific cytotoxic T cells in peripheral blood of infected donors, *Proc. Natl. Acad. Sci. USA* **92** (1995), 5773-5777.
34. Altman, J.D., Moss, P.A.H., Goulder, P.J.R. *et al.*: Phenotypic analysis of antigen-specific T lymphocytes, *Science* **274** (1996), 94-96.
35. Ho, D.D., Neumann, A.U., Perelson, A.S. *et al.*: Rapid turnover of plasma virions and CD4 lymphocytes in HIV-1 infection, *Nature* **373** (1995), 123-126.
36. Perelson, A.S., Neumann, A.U., Markowitz, M. *et al.*: HIV-1 dynamics *in vivo*: virion clearance rate, infected cell life-span, and viral generation time, *Science* **271** (1996), 1582-1586.
37. Autran, B., Hadida, F. and Haas, G.: Evolution and plasticity of CTL responses against HIV, *Curr. Opin. Immunol.* **4** (1998), 546-553.
38. Hamann, D., Baars, P.A., Rep, M.H.G. *et al.*: Phenotypical and functional separation of memory and effector human $CD8^+$ T cells, *J. Exp. Med.* **186** (1997), 1407-1418.

39. Sipsas, N.V., Kalams, S.A., Trocha, A. *et al.*: Identification of type-specific cytotoxic T-lymphocyte responses to homologous viral proteins in laboratory workers accidentally infected with HIV-1, *J. Clin. Invest.* **99** (1997), 752-762.
40. Pantaleo, G., De Maria, A., Koenig, S. *et al.*: CD8+ T lymphocytes of patients with AIDS maintain normal broad cytolytic function despite the loss of human immunodeficiency virus-specific cytotoxicity, *Proc. Natl. Acad. Sci. USA* **87** (1990), 4818-4822.
41. Greenough, T.C., Brettler, D.B., Somasundaran, M. *et al.*: Human immunodeficiency virus type-1-specific cytotoxic T lymphocytes (CTL), virus load, and CD4 T-cell loss: evidence supporting a protective role for CTL *in vivo*, *J. Infect. Dis.* **176** (1997), 118-125.
42. Danner, S.A., Carr, A., Leonard, J.M. *et al.*: A short-term study of the safety, pharmacokinetics and efficacy of Ritonavir, an inhibitor of HIV-1 protease, *New Engl. J. Med.* **333** (1995), 1528-1533.
43. Riddell, S.R., Elliott, M., Lewinsohn, D.A. *et al.*: T-cell-mediated rejection of gene-modified HIV-specific cytotoxic T lymphocytes in HIV-infected patients, *Nature Med.* **2** (1996), 216-223.
44. Cheynier, R., Henrichwark, S., Hadida, F. *et al.*: HIV and T-cell expansion in splenic white pulps is accompanied by infiltration of HIV-specific cytotoxic T lymphocytes, *Cell* **78** (1994), 373-387.
45. Klein, M.R. and Miedema, F.: Long-term survivors of HIV-1 infection, *Trends Microbiol.* **3** (1995), 386-391.
46. Greenough, T.C., Somasundaran, M., Brettler, D.B. *et al.*: Normal immune function and inability to isolate virus in culture in an individual with long-term human immunodeficiency virus type-1 infection, *AIDS Res. Human Retrovir.* **10** (1994), 395-403.
47. Rinaldo, C.R., Huang, X.L., Fan, Z. *et al.*: High levels of anti-human immunodeficiency virus type 1 (HIV-1) memory cytotoxic T-lymphocyte activity and low viral load are associated with lack of disease in HIV-1-infected long-term nonprogressors, *J. Virol.* **69** (1995), 5838-5842.
48. Ferbas, J., Kaplan, A.H., Hausner, M.A. *et al.*: Virus burden in long-term survivors of HIV infection is a determinant of anti-HIV $CD8^+$ lymphocyte activity, *J. Infect. Dis.* **172** (1995), 329-339.
49. Harrer, T., Harrer, E., Kalams, S.A. *et al.*: Strong cytotoxic T-cell and weak neutralizing antibody responses in a subset of persons with stable non-progressing HIV type-1 infection, *AIDS Res.Human Retrovir.* **12** (1996), 585-592.
50. Goulder, P.J.R., Bunce, M., Krausa, P. *et al.*: Novel, cross-restricted, conserved and immuno-dominant cytotoxic T-lymphocyte epitopes in slow progressors in HIV type-1 infection, *AIDS Res. Human Retrovir.* **12** (1996), 1691-1698.
51. Harrer, T., Harrer, E., Kalams, S.A. *et al.*: Cytotoxic T lymphocytes in asymptomatic long-term non-progressing HIV-1 infection. Breadth and specificity of the response and relation to *in-vivo* viral quasispecies in a person with prolonged survival and low viral load, *J. Immunol.* **156** (1996), 2616-2623.
52. Harrer, E., Harrer, T., Barbosa, P. *et al.*: Recognition of the highly conserved YMDD region in the human immunodeficiency virus type-1 reverse transcriptase by HLA-A2-restricted cytotoxic T lymphocytes from an asymptomatic long-term nonprogressor, *J. Infect. Dis.* **173** (1996), 476-479.
53. Van Baalen, C.A., Klein, M.R., Huisman, R.C. *et al.*: Fine-specificity of cytotoxic T lymphocytes which recognize conserved epitopes of the Gag protein of human immunodeficiency virus type 1, *J. Gen. Virol.* **77** (1996), 1659-1665.
54. Pantaleo, G., Demarest, J.F., Schacker, T. *et al.*: The qualitative nature of the primary immune response to HIV infection is a prognosticator of disease progression independent of the initial level of plasma viremia, *Proc. Natl. Acad. Sci. USA* **94** (1997), 254-258.
55. Van Baalen, C.A., Pontesilli, O., Huisman, R.C. *et al.*: Human immunodeficiency virus type-1 Rev- and Tat-specific cytotoxic T-lymphocyte frequencies inversely correlate with rapid progression to AIDS, *J. Gen. Virol.* **78** (1997), 1913-1918.
56. Kirchhoff, F., Greenough, T.C., Brettler, D.B. *et al.*: Absence of intact nef sequences in a long-term survivor with nonprogressive HIV-1 infection, *New Engl. J. Med.* **332** (1995), 228-232.
57. Deacon, N.J., Tsykin, A., Solomon, A. *et al.*: Genomic structure of an attenuated quasi species of HIV-1 from a blood transfusion donor and recipients, *Science* **270** (1995), 988-991.
58. Huang, Y., Paxton, W.A., Wolinsky, S.M. *et al.*: The role of a mutant CCR5 allele in HIV-1 transmission and disease progression, *Nature Med.* **2** (1996), 1240-1243.
59. De Roda Husman, A.-M., Koot, M., Cornelissen, M. *et al.*: Association between CCR5 genotype and the clinical course of HIV-1 infection, *Ann. Intern. Med.* **127** (1997), 882-890.

60. Van der Burg, S.H., Klein, M.R., Pontesilli, O. *et al.*: HIV-1 reverse transcriptase-specific cytotoxic T lymphocytes targeted at epitopes under structural or functional constraints do not protect against progression to AIDS, *J. Immunol.* **159** (1997), 3648-3654.
61. Keet, I.P.M., Klein, M.R. and Just, J.: The role of host genetics in the natural history of HIV-1 infection: the needles in the haystack, *AIDS* **10** (suppl. A) (1996), S59-S67.
62. Haynes, B.F., Pantaleo, G. and Fauci, A.S.: Toward an understanding of the correlates of protective immunity to HIV infection, *Science* **271** (1996), 324-328.
63. Kaslow, R.A., Carrington, M., Apple, R. *et al.*: Influence of combinations of human major histocompatibility complex genes in the course of HIV-1 infection, *Nature Med.* **2** (1996), 405-411.
64. Theodorou, I., Autran, B., Gourbar, A. *et al.*: HLA phenotypes in long-term asymptomatic HIV-infected adult individuals in France. In: Charron, D. (ed.) *Genetic Diversity of HLA*, Functional and Medical Applications, Paris, France, EDK (1997), p. 698-700.
65. Nelson, G.W., Kaslow, R.A. and Mann, D.L.: Frequency of HLA-allele-specific peptide motifs in HIV-1 proteins correlates with the allele's association with relative rates of disease progression after HIV-1 infection, *Proc. Natl. Acad. Sci. USA* **94** (1997), 9802-9807.
66. Ferris, R.L., Buck, C., Hammond, S.A. *et al.*: Class-I-restricted presentation of an HIV-1 gp41 epitope containing an N-linked glycosylation site. Implications for the mechanism of processing of viral envelope proteins, *J. Immunol.* **156** (1996), 834-840.
67. Van 't Wout, A.B., Kootstra, N.A., Mulder-Kampinga, G.A. *et al.*: Macrophage-tropic variants initiate human immunodeficiency virus type-1 infection after sexual, parenteral and vertical transmission, *J. Clin. Invest.* **94** (1994), 2060-2067.
68. Schuitemaker, H.: Macrophage-tropic HIV-1 variants: initiators of infection and AIDS pathogenesis?, *J. Leukocyte Biol.* **56** (1994), 218-224.
69. Coffin, J.M.: HIV population dynamics *in vivo*: implications for genetic variation, pathogenesis and therapy, *Science* **267** (1995), 483-489.
70. McMichael, A.J. and Phillips, R.E.: Escape of human immunodeficiency virus from immune control, *Ann. Rev. Immunol.* **15** (1997), 271-296.
71. Chen, Z.W., Shen, L., Miller, M.D. *et al.*: Cytotoxic T lymphocytes do not appear to select for mutations in an immunodominant epitope of simian immunodeficiency virus gag, *J. Immunol.* **149** (1992), 4060-4066.
72. Meyerhans, A., Dadaglio, G., Vartanian, J.P. *et al.*: *In-vivo* persistence of an HIV-1-encoded HLA-B27-restricted cytotoxic T-lymphocyte epitope despite specific *in-vitro* reactivity, *Eur. J. Immunol.* **21** (1991), 2637-2640.
73. Phillips, R.E., Rowland-Jones, S., Nixon, D.F. *et al.*: Human immunodeficiency virus genetic variation that can escape cytotoxic T-cell recognition, *Nature* **354** (1991), 453-459.
74. Goulder, P.J.R., Phillips, R.E., Colbert, R.A. *et al.*: Late escape from an immunodominant cytotoxic T-lymphocyte response associated with progression to AIDS, *Nature Med.* **3** (1997), 212-217.
75. Koenig, S., Conley, A.J., Brewah, Y.A. *et al.*: Transfer of HIV-1-specific cytotoxic T lymphocytes to an AIDS patient leads to selection for mutant HIV variants and subsequent disease progression, *Nature Med.* **1** (1995), 330-336.
76. Haas, G., Plikat, U., Debré, P. *et al.*: Dynamics of viral variants in HIV-1 Nef and specific cytotoxic T lymphocytes *in vivo*, *J. Immunol.* **157** (1996), 4212-4221.
77. Wolinsky, S.M., Korber, B.T.M., Neumann, A.U. *et al.*: Adaptive evolution of human immunodeficiency virus type 1 during the natural course of infection, *Science* **272** (1996), 537.
78. Schuurman, R., Nijhuis, M., Van Leeuwen, R. *et al.*: Rapid changes in human immunodeficiency virus type-1 RNA load and appearance of drug-resistant virus populations in persons treated with lamivudine (3TC), *J. Infect. Dis.* **171** (1995), 1411-1419.
79. Nowak, M.A. and Bangham, C.R.M.: Population dynamics of immune responses to persistent viruses, *Science* **272** (1996), 74-78.
80. Nowak, M.A., May, R.M., Phillips, R.E. *et al.*: Antigenic oscillations and shifting immunodominance in HIV-1 infections, *Nature* **375** (1995), 606-611.
81. Mayers, G., Hahn, B.H., Mellors, J.W. *et al.* (eds).: Human retroviruses and AIDS 1995: a compilation and analysis of nucleic acid and amino acid sequences. Los Alamos National Laboratory, Los Alamos, New Mexico: Theoretical Biology and Biophysics, 1995.
82. Koup, R.A.: Virus escape from CTL recognition, *J. Exp. Med.* **180** (1994), 779-782.
83. Howcroft, T.K., Strebel, K., Martin, M.A. *et al.*: Repression of MHC class-I gene promotor activity by two-exon Tat of HIV-1, *Science* **260** (1993), 1320-1322.

84. Schwartz, O., Marechal, V., Le Gall, S. *et al.*: Endocytosis of major histocompatibility complex class-I molecules is induced by the HIV-1 Nef protein, *Nature Med.* **2** (1997), 338-342.
85. Yang, O.O., Kalams, S.A., Rosenzweig, M. *et al.*: Efficient lysis of human immunodeficiency virus type-1-infected cells by cytotoxic T lymphocytes, *J. Virol.* **70** (1996), 5799-5806.
86. Xu, X.N., Screaton, G.R., Gotch, F.M. *et al.*: Evasion of cytotoxic T-lymphocyte (CTL) responses by Nef-dependent induction of Fas ligand (CD95L) expression on Simian immunodeficiency virus-infected cells, *J. Exp. Med.* **186** (1997), 7-16.
87. Strand, S., Hofmann, W.J., Hug, H. *et al.*: Lymphocyte apoptosis induced by CD95 (APO-1/Fas) ligand-expressing tumor cells: a mechanism of immune evasion, *Nature Med.* **2** (1996), 1361-1366.
88. Schuitemaker, H. and Miedema, F.: Viral and cellular determinants of HIV-1 replication in macrophages, *AIDS* **10** (suppl. A) (1996), S25-S32.
89. Guidotti, L.G., Ishikawa, T., Hobbs, M.V. *et al.*: Intracellular inactivation of the hepatitis-B virus by cytotoxic T lymphocytes, *Immunity* **4** (1996), 25-36.
90. Chun, T., Carruth, L., Finzi, D. *et al.*: Quantification of latent tissue reservoirs and total body viral load in HIV-1 infection, *Nature* **387** (1997), 183-187.
91. Wong, J.K., Hezareh, M., Gunthard, H.F. *et al.*: Recovery of replication-competent HIV despite prolonged suppression of plasma viremia, *Science* **278** (1997), 1291-1295.
92. Finzi, D., Hermankova, M., Pierson, T. *et al.*: Identification of a reservoir for HIV-1 in patients on highly active anti-retroviral therapy, *Science* **278** (1997), 1295-1300.
93. Rosenberg, E.S., Billingsley, M.L., Caliendo, A.M. *et al.*: Vigorous HIV-1-specific $CD4^+$ T-cell responses associated with control of viremia, *Science* **278** (1997), 1447-1450.
94. Kersten, M.J., Klein, M.R., Holwerda, A.M. *et al.*: EBV-specific cytotoxic T-cell responses in HIV-1 infection: different kinetics in patients progressing to opportunistic infection or non-Hodgkin's lymphoma, *J. Clin. Invest.* **99** (1997), 1525-1533.
95. Pantaleo, G., Soudeyns, H., Demarest, J.F. *et al.*: Evidence for rapid disappearance of initially expanded HIV-specific $CD8^+$ T-cell clones during primary HIV infection, *Proc. Natl. Acad. Sci. USA* **94** (1997), 9848-9853.
96. Shearer, G.M., Salahuddn, S.Z., Markham, P.D. *et al.*: Prospective study of cytotoxic T-lymphocyte responses to influenza and antibodies to human T-lymphotropic virus III in homosexual men, *J. Clin. Invest.* **76** (1985), 1699-1704.
97. Forte, M., Maartens, G., Rahelu, M. *et al.*: Cytolytic T-cell activity against mycobacterial antigens in HIV, *AIDS* **6** (1992), 407-411.
98. Blumberg, R.S., Parxadis, T., Byington, R. *et al.*: Effects of human immunodeficiency virus on the cellular immune response to Epstein-Barr virus in homosexual men: characterization of the cytotoxic response and lymphokine production, *J. Infect. Dis.* **155** (1987), 877-890.
99. Pantaleo, G., Koenig, S., Baseler, M. *et al.*: Defective clonogenic potential of $CD8^+$ T lymphocytes in patients with AIDS. Expansion *in vivo* of a nonclonogenic $CD3^+CD8^+DR^+CD25^-$ T-cell population, *J. Immunol.* **144** (1990), 1696-1704.
100. Gruters, R.A., Terpstra, F.G., De Goede, R.E.Y. *et al.*: Immunological and virological markers in individuals progressing from seroconversion to AIDS, *AIDS* **5** (1991), 837-844.
101. Moskophidis, D., Lechner, F., Pircher, H. *et al.*: Virus persistence in acutely infected immunocompetent mice by exhaustion of antiviral cytotoxic effector T cells, *Nature* **362** (1993), 758-761.
102. Giorgi, J.V.: Phenotype and function of T cells in HIV disease. In: Gupta, S. (ed.) *Immunology of HIV Infection*, New York & London: Plenum Medical Book Company, 1996, p. 181-199.
103. Meyaard, L., Otto, S.A., Jonker, R.R. *et al.*: Programmed death of T cells in HIV-1 infection, *Science* **257** (1992), 217-219.
104. Boudet, F., Lecoeur, H. and Gougeon, M.: Apoptosis associated with *ex-vivo* down-regulation of Bcl-2 and up-regulation of Fas in potential cytotoxic $CD8^+$ T lymphocytes during HIV-1 infection, *J. Immunol.* **156** (1996), 2282-2293.
105. Katsikis, P.D., Wunderlich, E.S., Smith, C.A. *et al.*: Fas antigen stimulation induces marked apoptosis of T lymphocytes in human immunodeficiency virus-infected individuals, *J. Exp. Med.* **181** (1995), 2029-2036.
106. Bofill, M., Gombert, W., Borthwick, N.J. *et al.*: Presence of $CD3^+CD8^+$Bcl-2low lymphocytes undergoing apoptosis and activated macrophages in lymph nodes of HIV-1^+ patients, *Am. J. Pathol.* **146** (1995), 1542-1555.
107. Effros, R.B., Allsopp, R.C., Chiu, C. *et al.*: Shortened telomeres in the expanded $CD28^-CD8^+$ cell subset in HIV disease implicate replicative senescence in HIV pathogenesis, *AIDS* **10** (1996), F17-F22.

108. Wolthers, K.C., Wisman, G.B.A., Otto, S.A. *et al.*: T-cell telomere length in HIV-1 infection: no evidence for increased CD4+ T-cell turnover, *Science* **274** (1996), 1543-1547.
109. Tough, D.F. and Sprent, J.: Viruses and T-cell turnover: evidence for bystander proliferation, *Immunol. Rev.* **150** (1996), 129-142.
110. Kalams, S.A., Johnson, R.P., Trocha, A.K. *et al.*: Longitudinal analysis of T-cell receptor (TCR) gene usage by human immunodeficiency virus 1 envelope-specific cytotoxic T-lymphocyte clones reveals a limited TCR repertoire, *J. Exp. Med.* **179** (1994), 1261-1271.
111. Borthwick, N.J., Bofill, M., Gombert, W.M. *et al.*: Lymphocyte activation in HIV-1 infection. Functional defects of CD28- T cells, *AIDS* **8** (1994), 431-441.
112. Actor, J.K., Shirai, M., Kullberg, M.C. *et al.*: Helminth infection results in decreased virus-specific CD8+ cytotoxic T-cell and Th1 cytokine responses as well as delayed virus clearance, *Proc. Natl. Acad. Sci. USA* **90** (1993), 948-952.
113. Matloubian, M., Concepcion, R.J. and Ahmed, R.: CD4+ T cells are required to sustain CD8+ cytotoxic T-cell responses during chronic viral infection, *J. Virol.* **68** (1994), 8056-8063.
114. Von Herrath, M.G., Yokoyama, M., Dockter, J. *et al.*: CD4-deficient mice have reduced levels of memory cytotoxic T lymphocytes after immunization and show diminished resistance to subsequent virus challenge, *J. Virol.* **70** (1996), 1072-1079.
115. Cardin, R.D., Brooks, J.W., Sarawar, S.R. *et al.*: Progressive loss of the CD8+ T-cell-mediated control of a g-Herpes virus in the absence of CD4+ T cells, *J. Exp. Med.* **184** (1996), 863-871.
116. Gazzinelli, R.T., Bala, S., Stevens, R. *et al.*: HIV infection suppresses type-1 lymphokine and IL-12 responses to *Toxoplasma gondii* but fails to inhibit the synthesis of other parasite-induced monokines, *J. Immunol.* **155** (1995), 1565-1574.
117. Miedema, F., Petit, A.J.C., Terpstra, F.G. *et al.*: Immunological abnormalities in human immunodeficiency virus (HIV)-infected asymptomatic homosexual men. HIV-1 affects the immune system before CD4+ T-helper-cell depletion occurs, *J. Clin. Invest.* **82** (1988), 1908-1914.
118. Miedema, F., Tersmette, M. and Van Lier, R.A.W.: AIDS pathogenesis: a dynamic interaction between HIV and the immune system, *Immunol. Today* **11** (1990), 293-297.
119. Gruters, R.A., Terpstra, F.G., De Jong, R. *et al.*: Selective loss of T-cell functions in different stages of HIV infection, *Eur. J. Immunol.* **20** (1990), 1039-1044.
120. Blatt, S.P., Hendrix, C.W., Butzin, C.A. *et al.*: Delayed-type hypersensitivity skin testing predicts progression to AIDS in HIV-infected patients, *Ann. Intern. Med.* **119** (1993), 177-184.
121. Markowitz, N., Hansen, N.I., Wilcosky, T.C. *et al.*: Tuberculin and anergy testing in HIV-seropositive and HIV-seronegative persons, *Ann. Intern. Med.* **119** (1993), 185-193.
122. Miedema, F.: Immunological abnormalities in the natural history of HIV infection: mechanisms and clinical relevance, *Immunodef. Rev.* **3** (1992), 173-193.
123. Pontesilli, O., Carlesimo, M., Varani, A.R. *et al.*: HIV-specific lymphoproliferative responses in asymptomatic HIV-infected individuals, *Clin. Exp. Immunol.* **100** (1995), 419-424.
124. Shearer, G.M., Bernstein, D.C., Tung, K.S.K. *et al.*: A model for the selective loss of major histocompatibility complex self-restricted T-cell immune responses during the development of acquired immune deficiency syndrome (AIDS), *J. Immunol.* **137** (1986), 2514-2521.
125. Giorgi, J.V., Fahey, J.L., Smith, D.C. *et al.*: Early effects of HIV on CD4 lymphocytes *in vivo*, *J. Immunol.* **138** (1987), 3725-3730.
126. Ballet, J.J., Couderc, L.J., Rabian-Herzog, C. *et al.*: Impaired T-lymphocyte-dependent immune responses to microbial antigens in patients with HIV-1-associated persistent generalized lymphadenopathy, *AIDS* **2** (1988), 291-297.
127. Clerici, M., Stocks, N., Zajac, R.A. *et al.*: Detection of three different patterns of T-helper cell dysfunction in asymptomatic, human immunodeficiency virus-seropositive patients, *J. Clin. Invest.* **84** (1989), 1892-1899.
128. Roos, M.Th.L., Miedema, F., Koot, M. *et al.*: T-cell function *in vitro* is an independent progression marker for AIDS in human immunodeficiency virus (HIV)-infected asymptomatic individuals, *J. Infect. Dis.* **171** (1995), 531-536.
129. O'Brien, W.A., Grovit-Ferbas, K., Namazi, A. *et al.*: Human immunodeficiency virus type-1 replication can be increased in peripheral blood of seropositive patients after influenza vaccination, *Blood* **86** (1995), 1082-1089.
130. Staprans, S.I., Hamilton, B.L., Follansbee, S.E. *et al.*: Activation of virus replication after vaccination of HIV-1-infected individuals, *J. Exp. Med.* **182** (1995), 1727-1737.
131. Stanley, S.K., Ostrowski, M.A., Justement, J.S. *et al.*: Effect of immunization with a common recall antigen on viral expression in patients infected with human immunodeficiency virus type 1, *New Engl. J. Med.* **334** (1996), 1222-1230.

132. Kovacs, J.A., Baseler, M., Dewar, R.J. *et al.*: Increases in CD4 T lymphocytes with intermittent courses of interleukin-2 in patients with human immunodeficiency virus infection, *New Engl. J. Med.* **332** (1995), 567-575.
133. Autran, B., Carcelain, G.L., Li, T.S. *et al.*: Positive effects of combined anti-retroviral therapy on CD4$^+$ T-cell homeostasis and function in advanced HIV disease, *Science* **277** (1997), 112-116.
134. Pantaleo, G.: How immune-based interventions can change HIV therapy, *Nature Med.* **3** (1997), 483-486.
135. Koup, R.A. and Ho, D.D.: Shutting down HIV, *Nature* **370** (1994), 416.
136. Safrit, J.T., Lee, A.Y., Andrews, C.A. *et al.*: A region in the third variable loop of HIV-1 gp120 is recognized by HLA-B7-restricted CTLs from two acute seroconversion patients, *J. Immunol.* **153** (1994), 3822-2830.
137. Moore, J.P., Cao, Y., Ho, D.D. *et al.*: Development of the anti-gp120 antibody response during seroconversion to human immunodeficiency virus type 1, *J. Virol.* **68** (1994), 5142-5155.
138. Zhu, T., Mo, H., Wang, N. *et al.*: Genotypic and phenotypic characterization of HIV-1 in patients with primary infection, *Science* **261** (1993), 1179-1181.
139. Hadida, F., Parrot, A., Kieny, M.-P. *et al.*: Carboxyl-terminal and central regions of human immunodeficiency virus-1 NEF recognized by cytotoxic T lymphocytes from lymphoid organs. An *in-vitro* limiting dilution analysis, *J. Clin. Invest.* **89** (1992), 53-60.
140. Johnson, R.P. and Walker, B.D.: Cytotoxic T lymphocytes in human immunodeficiency virus infection: responses to structural proteins. In: *Current Topics in Microbiology and Immunology*, 189th edition, Berlin-Heidelberg: Springer-Verlag, 1994, p. 35-63.
141. Clark, S.J., Saag, M.S., Decker, W.D. *et al.*: High titers of cytopathic virus in plasma of patients with symptomatic primary HIV-1 infection, *New Engl. J. Med.* **324** (1991), 954-960.
142. Piatak, M., Saag, M.S., Yang, L.C. *et al.*: High levels of HIV-1 in plasma during all stages of infection determined by competitive PCR, *Science* **259** (1993), 1749-1754.
143. Graziosi, C., Pantaleo, G., Butini, L. *et al.*: Kinetics of human immunodeficiency virus type-1 (HIV-1) DNA and RNA synthesis during primary HIV-1 infection, *Proc. Natl. Acad. Sci. USA* **90** (1993), 6405-6409.
144. Pontesilli, O., Klein, M.R., Kerkhof-Garde, S.R. *et al.*: Longitudinal analysis of HIV-1-specific CTL responses: a predominant Gag-specific response is associated with non-progressive infection, *J. Infect. Dis.* **178** (1998), 1008-1018.
145. Klein, M.R., Van der Burg, S., Hovenkamp, E. *et al.*: Characterization of HLA-B57-restricted HIV-1 Gag- and RT-specific CTL responses, *J. Gen. Virol.* **79** (1998), 2191-2201.
146. Van der Burg, S., Kwappenberg, K.M.C., Geluk, A. *et al.*: Identification of a conserved universal T-helper epitope in HIV-1 reverse transcriptase that is processed and presented to HIV-specific CD4$^+$ T cells by at least 4 unrelated HLA-DR molecules, *J. Immunol.* **194** (1999, in press).
147. Ogg, S., Jin, X., Bonhoffer, S. *et al.*: Quantitation of HIV-1-specific CTL and plasma load of viral RNA, *Science* **278** (1998), 2103-2106.

CHAPTER 7

MECHANISMS AND IN-VIVO SIGNIFICANCE OF HIV-1 NEUTRALISATION

PAUL W.H.I. PARREN[1], DENNIS R. BURTON[1] and
QUENTIN J. SATTENTAU[2,*]

[1]*Departments of Immunology and Molecular Biology, The Scripps Research Institute, La Jolla, CA 92037*
[2]*The Centre d'Immunologie de Marseille-Luminy, Marseille, France.*
[*]*Current address: The Sir William Dunn School of Pathology, University of Oxford, Oxford*

1. Introduction

Great emphasis is currently being put on studying the immune response to HIV-1 infection, with the ultimate aim of developing a vaccine to halt the global pandemic. Recent work suggests that cellular immunity is more important than humoral immunity in controlling an established HIV-1 infection, but antibodies may still be required to prevent, or reduce viral transmission. The antibody response to natural HIV-1 infection is not of sufficient quality or potency to efficiently neutralise *in vitro* most autologous and heterologous primary isolates. One reason for this may be that the majority of viral antigen present during infection is in the form of viral debris, which elicits mainly non-neutralising antibodies. Another reason is that the native, functional form of the envelope glycoprotein spike present on the virus is very poorly immunogenic. We, therefore, need alternative strategies to improve on the type of viral antigen present during natural infection, and its presentation to the immune system. Several studies suggest that high plasma levels of neutralising antibody can protect from HIV-1 or SIV infection in animal models, but at present we are unable to induce specific antibody responses of this magnitude.

Recent advances in our understanding of the structure of the HIV-1-envelope glycoproteins will help in designing more appropriate vaccine antigens. This, in association with novel strategies for more effective presentation of the chosen antigen(s), may yet lead to success in vaccination against this virus.

H. Schuitemaker and F. Miedema (ed.), Aids Pathogenesis, 99–132.

2. Anti-viral Effects of Antibodies in vivo

2.1. PASSIVE ANTIBODY TRANSFER STUDIES

Antibodies to a number of epitopes on the HIV-1-envelope glycoproteins (Env) can neutralise HIV-1 *in vitro* (see below). Defining questions, however, are whether such antibodies can (i) protect against transmission of relevant HIV-1 isolates *in vivo* and (ii) interrupt or suppress continued HIV-1 replication in an established infection. The most direct route to demonstrate the ability of antibodies to convey protection against a pathogen is in challenge experiments with that pathogen after passive transfer of monoclonal antibodies (mAb) or polyclonal antisera from vaccinated or convalescent donors. Antibodies capable of *in-vitro* neutralising activity can protect against challenge with neutralisation-sensitive T-cell-line-adapted (TCLA) HIV-1 strains in chimpanzees (Emini et al., 1992) and hu-PBL-SCID mice (Safrit et al., 1993; Parren et al., 1995; Gauduin et al., 1995; Andrus et al., 1998). Protection from infection with primary isolates (viruses that have been passaged only a limited number of times in primary cell cultures) has been less well studied and is more difficult, correlating with the relative resistance of these isolates to *in-vitro* neutralisation (Daar et al., 1990; Moore and Ho, 1995; Moore et al., 1995b). Nevertheless, complete protection from challenge with primary isolates was achieved in hu-PBL-SCID mice by administering a potent neutralising mAb (b12) prior to exposure (Gauduin et al., 1997). Protection was observed even if the antibody was given post-exposure. The effective time interval in which complete protection could still be achieved varied from 6-24 h and was dependent on the primary isolate tested (Gauduin et al., 1997). The only other study that assessed protection against a primary isolate by passive antibody transfer was performed in chimpanzees, using the neutralising anti-gp41 mAb 2F5 (Conley et al., 1996). Here protection was not seen, but the peak of viral RNA in plasma was either delayed or it did not reach levels comparable to that in control animals, and seroconversion was less rapid.

A fusion protein between CD4 and IgG2 (CD4-IgG2) has been prepared in which the Fv portions of both IgG2 heavy and the light chains have been replaced by CD4 (Allaway et al., 1995). This molecule was found to be highly effective at neutralising a wide range of HIV-1 primary isolates (Trkola et al., 1995; Allaway et al., 1995). CD4-IgG2 partially protected hu-PBL-SCID mice from challenge with the same primary isolates used in the mAb b12 study described above, under conditions in which b12 was protective (Gauduin et al., 1998). The slightly lower efficacy of CD4-IgG2 compared to mAb b12 *in vivo*, despite its somewhat greater efficacy against primary isolates *in vitro*, may be due to its relatively short half-life *in vivo* (Allaway et al., 1995). Protection in post-exposure prophylaxis with CD4-IgG2 has not been reported.

In humans, passive immunisation studies with potent neutralising mAb have not yet been performed, with the exception of a very small trial with mAb 2F5. The antibody serum concentrations achieved, however, were relatively low and no obvious effects on viral load were observed (Katinger et al., 1995). A number of studies have investigated the effects of passive transfer to infected individuals of pooled immune globulin from HIV-1-seropositive individuals (HIVIG). Although many of these results are conflicting, the general consensus is that infusion of HIVIG is ineffective. In a study of 28 mother-infant pairs, no changes in viral load were observed (Lambert et al., 1997). Other studies which reported beneficial effects on the basis of clinical rather than virological parameters (*e.g.* (Levy et al., 1994; Vittecoq et al., 1995)) may therefore have measured the effects of passively transferred immunoglobulin specific for opportunistic pathogens present in the preparation, and not specific effects on HIV-1 (Parren et al., 1997b). HIVIG protected chimpanzees against low-dose challenge with the TCLA virus HIV-1_{IIIB} (Prince et al., 1991) but failed to protect against high-dose challenge with the same virus (Prince et al., 1988). Protection related to neutralisation as this HIVIG, prepared from plasma from a group of selected seropositive donors with high TCLA virus-neutralising serum titers, had neutralising activity against HIV-1_{IIIB}. Significantly, HIVIG failed to protect hu-PBL-SCID mice from primary isolate challenge under conditions where mAb b12 was protective (Gauduin et al., 1997). In agreement with these data, HIVIG preparations have generally been found to neutralise primary isolates poorly, even when used at high concentration (Burton et al., 1994; Mascola et al., 1997). These observations suggest that currently available preparations of HIVIG will be ineffective in the prophylaxis of HIV-1 infection in humans, and of little use in preventing maternal-fetal transmission (Lambert et al., 1997; Moore, 1998).

Passive antibody transfer studies to prevent SIV infection in macaques have been reported to confer protection or benefit in some studies (*e.g.* (Putkonen et al., 1991; Lewis et al., 1993; Haigwood et al., 1996)) but not in others (*e.g.* (Kent et al., 1994; Gardner et al., 1995)). A major problem with interpreting some of these studies and understanding the apparently contradictory results is that neutralisation assays *in vitro* and challenge studies *in vivo* were frequently not carried out with the same virus (Montefiori et al., 1996a; Burton, 1997). Moreover, antibodies against cellular molecules derived from the infected cell surface and acquired by the virion during virus budding have been shown to be active in viral inactivation, and may bias the results. For example, mAb against ICAM-1 molecules present on HIV-1 neutralise the virus effectively *in vitro* (Rizzuto and Sodroski, 1997). Passive transfer of antisera raised against SIV virions grown in human cells, or immunised with human HLA class II, protects macaques from challenge with SIV grown in human but not in monkey cells (Gardner et al., 1995; Arthur et al., 1995). By contrast with the majority of HIV-based studies, passive transfer of maternal anti-FIV antibodies protects neonatal kittens from homologous FIV challenge; protection in this study correlated with the level of neutralising antibodies transferred to the kittens (Pu et al., 1995).

In summary, passive immunisation studies with neutralising mAb to HIV-1 or SIV in animal models do provide direct evidence that antibodies can protect from HIV-1 infection *in vivo*. However, a prerequisite is that the antibody preparation used should strongly bind to, and neutralise, the challenge virus. Further caution is to be observed in translating neutralisation potency from *in-vitro* assays to protective antibody serum levels, as in hu-PBL-SCID mice protective antibody serum titers are typically one to two orders of magnitude greater than the antibody concentration necessary to neutralise >90% of the virus *in vitro* in a PHA-activated T-cell-based assay (Parren et al., 1997c). In agreement with this, protection of macaques was only achieved in the presence of serum-neutralising antibody levels sufficient to inactivate all of the challenge virus (Shibata et al., 1999). The failure of some studies to observe protection with a neutralising antibody preparation (*e.g.* (Schutten et al., 1996; Schutten et al., 1996; Conley et al., 1996)) may therefore be caused by a failure to achieve adequate neutralising antibody concentrations in serum. How well these animal studies will translate to human infection with HIV-1 is unclear. These studies do, however, indicate that prospects for passive antibody in prophylaxis are promising, provided sufficient levels of serum antibodies effective against a wide range of primary isolates can be achieved.

2.2. ACTIVE IMMUNISATION STUDIES

Vaccination studies in chimpanzees have shown that recombinant HIV-1 envelope proteins can induce, in some animals, protection or partial protection against challenge with a TCLA or otherwise neutralisation-sensitive HIV-1 challenge (Berman et al., 1990; Girard et al., 1991; Fultz et al., 1992; Bruck et al., 1994; Girard et al., 1995; Lubeck et al., 1997; Boyer et al., 1997). Protection correlated with neutralising antibody titres in some studies (Berman et al., 1990; Bruck et al., 1994; Lubeck et al., 1997). Vaccination studies using DNA immunisation (Boyer et al., 1997), or a prime-boost with adenovirus-expressed TCLA-gp160 followed by gp120 (Lubeck et al., 1997), have demonstrated protection of chimpanzees against challenge with HIV-1_{SF2}, despite relatively weak CTL and neutralising antibody responses. These poor immune responses may have been sufficient since HIV-1_{SF2} is non-pathogenic for chimpanzees and replicates only marginally in this animal model (El-Amad et al., 1995; Berman et al., 1996). In the latter study (Lubeck et al., 1997), neutralising antibody titres against a primary isolate were assessed but were weak even when measured with the exceptionally neutralisation-sensitive isolate HIV-1_{BZ167} (Zolla-Pazner and Sharpe, 1995; Zhou and Montefiori, 1997; VanCott et al., 1997; Moore and Montefiori, 1997). Neutralising antibody titres have therefore only been found to be a correlate of protection in vaccination studies in which viral challenge consisted of highly neutralisation-sensitive isolates, which is probably a reflection of the poor neutralisation titres achieved. Indeed, in general, vaccination studies in chimpanzees have not been very useful in determining a role for antibody in protection from HIV-1 because all such studies to date have demonstrated an inability to elicit strong neutralising anti-body titres against representative primary isolates in relevant assays (Haynes, 1996; Lubeck et al., 1997).

2.3. THE SHIV/MACAQUE MODEL FOR HIV-1 INFECTION AND AIDS

Although HIV-1 can infect apes such as chimpanzees and gibbons, infection of these species typically results in low levels of viral replication and infection does not result in AIDS (Gardner and Luciw, 1989; Schultz and Hu, 1993). These animal models therefore do not provide a rigorous test of passive or active immunisation against HIV-1 (Letvin, 1998). Infection of rhesus and cynomolgus monkeys with several strains of SIV resembles the course of HIV-1 infection in man more closely (Daniel et al., 1985; Daniel et al., 1987; Kestler et al., 1990; Hulskotte et al., 1998). Chimeric SIV/HIV challenge viruses, termed SHIVs, have been developed for this model in which tat, rev, and env genes, and in some cases other SIV genes, have been replaced by HIV-1 homologues from TCLA (*e.g.* HXBc2, NL432) and primary viruses (*e.g.* SF33, SF162 and 89.6) (Luciw et al., 1995; Li et al., 1995; Reimann et al., 1996b). Initial SHIV constructs did not induce any disease, although long-term persistence was achieved. More pathogenic SHIVs have recently been obtained, however, by serial *in-vivo* passage in macaques (Joag et al., 1996; Reimann et al., 1996a; Joag et al., 1997). Challenge of macaques with SHIV is a highly relevant model to study effects on protection and pathogenesis after passive and active immunisation with candidate HIV-1 vaccines. In a recent and significant study, neutralising antibody responses were analysed in sera from macaques infected with a TCLA HIV-1-derived SHIV (SHIV-HXB2), a HIV-1 primary isolate-derived SHIV (SHIV-89.6) and a highly pathogenic isolate, SHIV-89.6PD, derived after four serial passages of SHIV-89.6 through rhesus macaques (Montefiori et al., 1998b). High titres of type-specific neutralising antibodies were observed in some monkeys, although neutralisation of heterologous SHIVs was mostly low to undetectable. Some monkeys infected with TCLA SHIV-HXB2, but none infected with the primary isolate-derived SHIVs, developed neutralising antibodies against HIV-1 primary isolates; detectable antibody titres only developed very late after infection (>40 weeks) (Montefiori et al., 1998b). These results indicate that infection-induced antibodies presumably raised against mature oligomeric envelope from single (primary) isolates of HIV-1 elicit strong neutralising antibody responses against the homologous primary virus. For vaccine design, it may be necessary to immunise with a cocktail of oligomeric, or otherwise appropriately immunogenic envelope glycoprotein preparations in order to obtain broad neutralising responses.

2.4. NEUTRALISATION ESCAPE

The ability of antibodies to interfere with viral replication has been suggested from the emergence of neutralisation escape mutants during the course of HIV-1 and SIV infection (Norrby and Matthews, 1993; Burns et al., 1993; Nyambi et al., 1997). However, envelope glycoprotein (Env) variation should not automatically be attributed to antibody selection pressure; evolution for replication in different cell types (Mcknight and Clapham, 1995), and usage of different co-receptors, also involve changes in Env. Detailed analysis of the temporal emergence of viruses resistant to neutralisation by autologous serum in a chimpanzee naturally infected with SIV_{cpz} indicated that neutralisation escape mutants emerged on average after 15 months, followed by a neutralising antibody response affecting the resistant isolate 8 months later in this animal (Nyambi et al., 1997). This observation is in agreement with the observed lag time for serum to neutralise autologous virus in human primary infection (Pilgrim et al., 1997), and with the neutralisation profile of sequential virus isolates with consecutive autologous serum samples from a health care worker infected by a needle-stick injury (Pratt et al., 1995; Lathey et al., 1997). The isolate obtained from this patient 10 months after infection was resistant to neutralisation by earlier sera but not the contemporary serum, whereas escape was not apparent in isolates obtained after 1 and $2^1/_2$ months of infection (Lathey et al., 1997). This chronology in neutralisation by serum antibodies and the emergence of neutralisation-resistant isolates may indicate cause and effect in the interplay between virus and host; alternatively, it could be explained by the notion that escape is an unavoidable consequence of viral variation resulting from the presence of a great variety of viral quasispecies (Pelletier and Wain-Hobson, 1996). The observed emergence of escape variants is very slow compared to the complete viral population turnover which occurs in a period of 14-28 days in plasma of patients treated with a potent anti-retroviral drug (Wei et al., 1995), and the very rapid escape observed in HIV-1-infected hu-PBL-SCID mice treated with a cocktail of potent neutralising antibodies (Poignard et al., 1999). Some delay is to be expected, since viral turn-over and variation can be measured in hours to days (Wei et al., 1995; Ho et al., 1995; Perelson et al., 1996), whereas antibody maturation is typically measured in weeks. It remains unclear, however, whether the slow emergence of neutralisation-escape variants in natural infection is an effect of a relatively slow adaptation of the antibody response to viral variation, or to an ability of HIV-1 to flourish despite antibody pressure.

In support of the first notion, it has been suggested that the antibody response to HIV-1 Env follows the kinetics of the slow turn-over of envelope proteins trapped on follicular dendritic cells (Binley et al., 1997). In any case, the selective pressure provided by antibody against human HIV-1 infection is very weak as serum titers neutralising both autologous and heterologous primary isolates are generally low to non-existent (Moore et al., 1996; Pilgrim et al., 1997; Moog et al., 1997). Recent studies in hu-PBL-SCID mice indicate that HIV-1 primary isolates can replicate vigorously in the presence of high concentrations of potent neutralising antibodies while remaining sensitive to the antibodies used (Poignard et al., 1999). In addition, post-exposure immunoprophylaxis with the b12 antibody in hu-PBL-SCID mice was unsuccessful when the mAb was given after 24-48 h, and end-point viral loads in the spleen were unaffected, indicating that the antibody had little effect on established infection (Gauduin et al., 1998). The absence of a therapeutic effect of neutralising antibody has furthermore been observed in passive transfer experiment in macaques infected with the pathogenic $SHIV_{KU\text{-}2}$ (Foresman et al., 1998). This may indicate that the principal route of virus transmission *in vivo* is by cell-to-cell spread, within lymphoid tissues, against which antibody is relatively ineffective (Pantaleo et al., 1995).

The continued emergence of viruses that escape autologous neutralisation may lead to the observed broadening of the neutralising antibody responses in long-term non-progressors (LTNP) (Montefiori et al., 1996b; Pilgrim et al., 1997). It has to be stressed, however, that even in LTNP, neutralisation potency against primary isolates in general is still only weak and sporadic; indeed, in some of these studies undiluted serum was used because neutralisation of primary isolates is rarely observed at serum dilutions of 1:16 or greater (Harrer et al., 1996; Montefiori et al., 1996b; Pilgrim et al., 1997). Potent neutralisation, defined as a greater than 90% reduction in infectivity by antibody concentrations readily achievable *in vivo* (<50 μg/ml for a mAb) of a broad range of primary isolates, is therefore very rare.

2.5. THE ANTIBODY RESPONSE TO HIV-1 IN NATURAL INFECTION

The apparent inability of the humoral immune system to mount a good neutralising response against primary isolate HIV-1 during natural infection is paradoxical in view of the very strong and sustained antibody response against gp120 monomer present in most infected individuals (Moore and Ho, 1995; Binley et al., 1997). On the basis of the affinities for different forms of HIV-1 Env of a large panel of anti-Env mAb obtained from antibody phage-display libraries prepared from bone-marrow of HIV-1-infected individuals (Burton et al., 1991; Barbas et al., 1993; Parren et al., 1996; Parren and Burton, 1997; Ditzel et al., 1997), we have argued that the basis of this problem is that most Env-specific antibodies are elicited by viral debris rather than intact virions (Parren et al., 1997a; Parren et al., 1997c). Viral debris contains forms of HIV-1 Env that are antigenically distinct from the mature, oligomeric Env complex on the surface of infected cells or virions. The debris includes unprocessed gp160 precursor, "shed" monomeric gp120 dissociated from the virion/infected cell surface, and inactive gp41 exposed on the virion/ infected cell surface after shedding of gp120. The vast majority of anti-Env antibodies from natural infection thereby have high to very high affinities for unprocessed envelope, inactive gp41 and monomeric gp120, but display low to undetectable binding to mature oligomeric HIV-1 Env (Parren et al., 1997a; Parren et al., 1997c). The hypothesis that the antibody response in HIV-1 infection is directed against viral debris rather than virions is supported by analysis of the response against the V3 loop. At the level of V3 loop or gp120-monomer binding assays, it is easy to distinguish between isolates of the genetic subtypes B and E, or sera from individuals infected with, for example, B or E viruses. These isolates form discrete antibody- binding serotypes (Moore et al., 1994a; Weber et al., 1996; Moore et al., 1996; Louisirirotchanakul et al., 1998; Cheingsong-Popov et al., 1998). Genetic subtypes B and E can even be identified as distinct neutralisation serotypes for TCLA viruses, reflecting the dominance of anti-V3 antibodies in sera in TCLA neutralisation assays (Louisirirotchanakul et al., 1998; Cheingsong-Popov et al., 1998). Neutralisation serotypes cannot, however, be identified for HIV-1 primary isolates (Weber et al., 1996; Nyambi et al., 1996; Moore et al., 1996; Kostrikis et al., 1996; Mascola et al., 1996a). Serological evidence therefore supports the idea of a strong response against the V3 loop on gp120 monomers or unprocessed gp160 with poor cross-reactivity to this epitope on primary isolate virions.

Thus antibodies reactive with Env derived from HIV-1-infected individuals are generally described as antibodies against HIV-1. Strictly this is incorrect, however, as most antibodies appear to be elicited and matured by viral debris rather than the virus. These antibodies cross-react with the virus but their affinity for funtional, oligomeric, virion-associated Env is low. This low affinity results directly in antibodies that are functionally inert or only weakly active against HIV-1 *in vivo*.

3. How Antibodies Neutralise HIV-1

3.1 STRUCTURE-FUNCTION RELATIONSHIPS IN THE VIRAL ENVELOPE GLYCOPROTEINS

HIV-1 Env is organized, in its mature functional form, into a trimer of trans-membrane (gp41) and surface glycoprotein (gp120) heterodimers (Weissenhorn et al., 1997; Chan et al., 1997). These trimers are arranged into 72 glycoprotein 'spikes' at the virion surface (Gelderblom *et al,* 1989) and it is to these that antibodies must bind if they are to inactivate the virus. The antigenic topology of both monomeric gp120 and oligomeric Env has been probed using panels of mAbs (McKeating et al., 1992a; McKeating et al., 1993; Moore et al., 1993a; Moore et al., 1993b; Moore et al., 1994b; Shotton et al., 1995; Moore and Sodroski, 1996; Ditzel et al., 1997), and reviewed in (Moore and Ho, 1995; Poignard et al., 1996a; Burton and Montefiori, 1997; Burton, 1997; Parren et al., 1997c). Data obtained from these studies has been integrated into the recently obtained crystal structure of the HXBc2 gp120 core. The structure obtained at 2.5Å resolution was of a trimolecular complex containing a truncated form of deglycosylated gp120 lacking the COOH- and NH2-termini and theV1V2 and V3 loops, complexed with domains 1 and 2 of CD4 and a Fab fragment of a neutralising mAb (Kwong et al., 1998). Thanks to this extraordinary leap for-ward in our understanding of the gp120 structure, we now have a relatively complete model of the antigenic surfaces of gp120, and how they relate to neutralisation (Wyatt et al., 1998; Kwong et al., 1998). Since the implications of the gp120 structure for HIV-1 neutralisation, neutralisation escape and the potential use of gp120 as a vaccine immunogen have been discussed in several recent reviews (Wyatt and Sodroski, 1998; Sattentau, 1998; Moore and Binley, 1998), only a brief summary is provided here.

3.1.1 *Structure of the surface unit gp120*

The CD4-binding site (CD4bs) on gp120 is located within a cleft, occupies ~800Å^2 on gp120, and is at the interface of the three structural elements making up the gp120 core: an outer domain, an inner domain and a bridging minidomain. Two major cavities are evident within the gp120-CD4 interface: a shallow one filled with water molecules and a deep hydrophobic depression which extends ~10 Å into the gp120 interior. The region encompassing the CD4bs is devoid of glycosylation and is relatively well conserved between HIV-1 isolates, although it does contain islands of variability that have implications for neutralisation escape. The conserved chemokine receptor binding surface (Rizzuto et al., 1998) is a weakly basic surface located at approximately 90° to the CD4bs and is comprised principally of the bridging sheet with contributions from the base of the V1V2 loop structure. The binding of a gp120 spike to a trimer of membrane-anchored CD4 molecules would orientate the conserved co-receptor binding surface towards the cell membrane where, in association with the V3 loop, it could bind the appropriate chemokine receptor. Despite the fact that several antigenic regions of gp120 are missing in the structure either because they were deleted from the gp120 core (the V1V2 and V3 loops) or because of inadequate resolution (the V4 loop), experimental results coupled with the orientation of the bases of these loops allows modelling of their approximate positions in the intact molecule. Gp120 appears to be a conformationally unstable molecule that oscillates between two or more metastable states. Within an Env oligomer this molecular vibration is even more apparent: it is thought that gp120 probably fluctuates between CD4bs 'open' and 'closed' conformations as a result of movement of the V1V2 loops (Wyatt et al., 1995). Serologic and mutagenesis studies suggest that the V1V2 loops partially mask both the CD4bs and the conserved chemokine receptor binding site (Wyatt and Sodroski, 1998), and that multivalent binding between an 'open' gp120 oligomer and a cluster of CD4 molecules further displaces the V1V2 and V3 loops, inducing the co-receptor binding site and weakening the gp120 association with gp41 (Sattentau and Moore, 1991; Thali et al., 1993; Sattentau et al., 1993). The CD4 molecule contains flexible segments (Wu et al., 1997), allowing gp120 to pivot towards the chemokine receptor, bringing the virus and cell membranes into sufficiently close proximity for the fusion peptide of gp41 to insert into the cell membrane. Further conformational changes then take place in gp41 that trigger the formation of a supercoiled structure reminiscent of the activated form of influen-za HA2, and lead to virus-cell membrane coalescence and entry of the virion capsid into the viral cytoplasm (reviewed in (Chan and Kim, 1998)).

3.1.2 *Neutralising epitopes on the transmembrane unit gp41*

The presence of a neutralisation epitope-containing region within the cytoplasmic domain of gp41 (Chanh et al., 1986; Dalgleish et al., 1988; Evans et al., 1989) has not been confirmed by other laboratories (D'Souza et al., 1994; D'Souza et al., 1995). Indeed, it seems unlikely that this epitope is even exposed on virion or infected cell membranes (Sattentau et al., 1995). The only confirmed gp41 neutralisation epitope is defined by a linear stretch of membrane-proximal amino acids (residues 662 to 667) (Muster et al., 1993; Muster et al., 1994; Muster et al., 1995) that probably form a helix in its native conformation (Gallaher et al., 1989; Weissenhorn et al., 1997; Chan et al., 1997). This epitope, defined by the mAb 2F5, appears to be one of the only exposed regions of gp41 in the virion-associated, pre-activation form of the envelope glycoprotein complex (Sattentau et al., 1995; Binley and Moore, 1997), potentially explaining its neutralising activity. MAbs recognising other regions of gp41 (Xu et al., 1991; Binley et al., 1996) may bind only to 'bald' Env spikes that have shed their gp120 and are functionally inactive (Sattentau et al., 1995; Binley and Moore, 1997).

3.1.3 *Neutralising epitopes on mature oligomeric envelope*

Several regions of gp120, particularly the CD4bs, the variable loops V1, V2 and V3, but also regions at the amino- and carboxy-termini are accessible on monomeric gp120 for antibody binding and induction (Moore et al., 1994b; Moore and Sodroski, 1996). However, when many of these same antibodies were tested for reactivity with virion-associated gp120, binding was much reduced or absent for the majority (Moore et al., 1994b; Sattentau and Moore, 1995), demonstrating that the expression of gp120 in its soluble, monomeric form results in the exposure of many novel and relatively immunogenic epitopes that are not present on the mature oligomeric form. It is not surprising, therefore, that immunisation with soluble gp120 has been shown to induce mostly monomer-specific antibodies that do not react with, and hence do not neutralise, primary HIV-1 isolates (*i.e.* viruses that have been passaged only in primary cultures of T cells or macrophages) (VanCott et al., 1997) and see above. Analysis of mAb binding to oligomeric gp120 of HIV-1 adapted to growth in T-cell lines (TCLA viruses), reveals that only a limited number of epitopes are exposed, including the CD4bs and the C4 region, the V1, V2 and V3 loops, the CD4i epitope that corresponds to the conserved co-receptor binding site, and the unique 2G12 epitope (Trkola et al., 1995; Trkola et al., 1996b). These epitopes are considered to occupy a limited space on the 'neutralising face' of gp120 (Moore and Sodroski, 1996; Wyatt et al., 1998) and reviewed in (Sattentau, 1996; Poignard et al., 1996a; Burton and Montefiori, 1997; Wyatt and Sodroski, 1998). In comparison to TCLA viruses, trimeric gp120 on primary isolates exposes even less surface to antibody attack. Generally speaking, the exposed neutralisation-relevant surfaces accessible on the primary virus Env oligomer are a unique CD4bs-associated epitope (b12 epitope) (Burton et al., 1994), parts of the V3 loop (Bou-Habib et al., 1994; Stamatatos et al., 1997), the 2G12 epitope (Trkola et al., 1995; Trkola et al., 1996b) and the 2F5 epitope (Muster et al., 1993; Muster et al., 1994; Muster et al., 1995). In general, antibody affinity for these epitopes on primary isolates appears to be lower than that observed for the same regions on TCLA viruses, further reducing neutra-lisation efficiency (Bou-Habib et al., 1994; Moore et al., 1995b; Stamatatos et al., 1997; Fouts et al., 1997; D'Souza et al., 1997; Burton, 1997). Differences in quater-nary structure between the virions of primary isolates and TCLA strains probably influence the reduction in antibody affinity. The use of oligomeric forms of gp120 as vaccine antigens has been proposed as a way of improving the quality of the neutra-lising antibody response. The major challenge of this approach is to preserve the integrity of the quaternary structure of the mature envelope oligomer, while minimi-sing the creation of irrelevant immunogenic epitopes. Several strategies have been adopted with this in mind. The inactivation of HIV or SIV particles by formaldehyde fixation is a classical approach (Murphey-Corb et al., 1989; Desrosiers et al., 1989), but may influence the conformation of several of the most important neutralisation epitopes (Sattentau, 1995). More recently, HIV-1 has been inactivated by modifica-tion of the zinc finger regions of the nucleocapsid protein, a procedure that preserves the functional and conformational integrity of the Env structure (Rossio et al., 1998). An alternative strategy is to prepare soluble, recombinant oligomers of un-cleaved gp160, that may potentially be cleaved by exogenous proteases such as furin after purification. Uncleaved oligomers have been produced by several laboratories, and their antigenic properties are under assessment (Parren et al., 1996). One major problem with cleaved oligomers is that of spontaneous dissociation of gp120 from gp41; although uncleaved gp160 can be stabilized into trimers with disulfide bonds (Farzan *et al.*, 1998), problems associated with cleavage and gp120-gp41 stability remain to be overcome.

The availability of the gp120 core structure has allowed relatively precise mapping of three mAb epitope clusters. These epitope groups (including the CD4bs and the chemokine receptor binding site) are located in what has been termed the 'neutralisation face' of gp120, a region spanning the inner, outer and bridging domains. The neutralisation face is flanked by the 'non-neutralising' face that is implicated in oligomerization and the 'immunosilent face' that is mostly occluded by glycosylation (see below) (Moore and Sodroski, 1996; Wyatt et al., 1998). CD4bs-specific mAb epitopes overlap the CD4-gp120 binding surface, but unlike CD4 itself, they also make contact with a number of more variable residues, many of which are in line with the hydrophilic cavity or surround the hydrophobic cavity (Wyatt et al., 1998; Kwong et al., 1998). The CD4-induced (CD4i)-epitope-specific mAbs, defined in the structure by the Fab of 17b, bind epitopes that overlap the chemokine receptor binding surface to varying degrees, and may also contact residues in the V3 loop (Rizzuto et al., 1998). Finally, the 2G12 human mAb epitope is unique in that it is on the neutralising face and located within the outer domain. The neutralisation of range of primary isolates by 2G12 is remarkable considering the variability of this region. The sensitivity of mAb 2G12 binding to mutations of a few gp120 glycosylation sites indicates that carbohydrate may be involved in its epitope (Trkola et al., 1996b). Recognition of a carbohydrate structure that is relatively well conserved compared to the underlying primary amino acid sequences in this region may explain the conservation of the 2G12 epitope between virus subtypes (Trkola et al., 1995; Wyatt and Sodroski, 1998).

3.1.4 *The V3 loop*

One of the antigenic structures not present on the crystalized gp120 core is the V3 loop. For many years it was considered that the V3 loop was the principal neutralisation domain of HIV-1 (reviewed in Moore and Nara (1991)). However, it is becoming increasingly clear that the V3 loop of HIV-1 primary isolates is not an important target for neutralising antibodies generated during natural infection or by candidate HIV-1 vaccines. This contrasts markedly with its dominant role in the neutralisation of TCLA strains by sera from HIV-1-infected individuals and gp120 vaccine recipients (Montefiori et al., 1993; Spear et al., 1994; Moore and Ho, 1995; VanCott et al., 1995a; Mascola et al., 1996b; Burton and Montefiori, 1997; Spenlehauser et al., 1998; Beddows et al., 1998). The lesser role of the V3 loop in primary virus neutralisation appears to be a consequence of its decreased accessibility in mature oligomeric envelope of primary isolates, leading to a lower affinity of antibodies to this region (Bou-Habib et al., 1994; Stamatatos and Cheng-Mayer, 1995; Burton and Montefiori, 1997). Studies aimed at determining whether classical neutralisation serotypes exist, and whether these are related to the known genetic subtypes, further suggested the V3 loop is not a primary virus neutralisation epitope. Thus, neutralisation serotypes, in the conventional sense of the phrase, were not readily identified from extensive neutralisation checkerboards involving HIV-1^{+} sera and primary isolates from genetic subtypes A through E. In particular, no general correlation was found between patterns of primary isolate neutralisation and the genetic subtypes of the sera and isolates tested (Weber et al., 1996; Nyambi et al., 1996; Moore et al., 1996; Kostrikis et al., 1996). The existence of neutralisation serotypes based on specific subtypes is controversial: the conclusion that sera and isolates from genetic subtypes B and E form distinguishable neutralisation serotypes relevant to the HIV-1 epidemic in Thailand was drawn from two related studies (Mascola et al., 1994; Mascola et al., 1996a), but not others (Weber et al., 1996; Moore et al., 1996; Kostrikis et al., 1996; Louisirirotchanakul et al., 1998). Additional studies argue against the V3 loop being important for primary isolate neutralisation. Thus, depletion of V3-specific antibodies with V3 loop peptides does not significantly diminish the primary virus neutralisation capacity of HIV-1-positive sera (VanCott et al., 1995a; Stamatos et al., 1998; Spenlehauser et al., 1998; Beddows et al., 1998). Also, V3-peptide-reactive antibodies purified from HIV-1$^+$ sera did not significantly neutralise primary isolates (Stamatos et al., 1998). In contrast, TCLA virus neutralisation titres are usually very strongly reduced by V3 peptide depletion (Vogel et al., 1994; VanCott et al., 1995a; Beddows et al., 1998). Of note is that substantial primary virus-neutralising activity remained undepleted from a rare, strongly neutralising HIV-1$^+$ serum after removal of gp120-reactive antibodies on affinity columns (Stamatos et al., 1998). The identity of these antibodies is not yet clear, but is an important issue in HIV-1-vaccine development. Immune responses to gp120 monomer vaccines are strongly biased towards the V3 loop, probably accounting for the reasonable performance of vaccinee sera against TCLA viruses, but their poor performance against representative primary isolates (Montefiori et al., 1993; VanCott et al., 1995a; VanCott et al., 1995b; Graham et al., 1996; Mascola et al., 1996a; Connor et al., 1998). Finally, some V3 loop mAbs can have extremely potent neutralising activity against TCLA viruses (Scott et al., 1990; Javaherian et al., 1990; Gorny et al., 1992; White-Scharf et al., 1993; Gorny et al., 1993; Gorny et al., 1998), and in some reports, primary isolates (Conley et al., 1994; Zolla-Pazner and Sharpe, 1995). However, the ability of these mAbs to neutralise typical primary isolates is generally extremely limited, or non-existent, when these reagents are tested in independent studies (D'Souza et al., 1997; Beddows et al., 1998).

3.2. STRUCTURE-FUNCTION RELATIONSHIPS - CONCLUSIONS

Based on the model of HIV-1-cell attachment and initiation of virion-cell fusion described above, and the location of antibody epitopes on gp120, it appears that the potential of HIV-1 for viral evasion and escape from neutralising antibodies is great. Major defense mechanisms involve the occlusion of functionally critical gp120 regions by overlying hypervariable loops which can support significant structural polymorphism and glycosylation of conserved surfaces that would otherwise be sensitive to antibody attack (see below).

In summary, the CD4bs has the following characteristics that reduce recognition by antibody: *i*) It is partially masked by the hypervariable V1V2 loops and their associated glycosylation; *ii*) It is recessed to an extent that would impede direct access to an antibody variable region, and flanking residues are variable or shielded by glycosylation; *iii*) Despite being relatively well conserved, it contains clusters of residues that do not contact CD4 and are subject to mutation, such as those lining the water-filled cavity; *iv*) Many gp120 residues contact CD4 via main-chain atoms, allowing for variability within the corresponding side chains that would contribute to antibody epitopes in the CD4-unbound state; *v*) The quaternary structure of the assembled trimer may restrict access of antibodies to CD4bs epitopes on the individual monomer subunits, especially for primary isolates (although suggested by many lines of evidence, this is unproven at the structural level).

Similarly, the conserved chemokine receptor binding site is masked from antibody binding in the CD4-unbound state by the variable loops V1V2 and V3, and may undergo further allosteric rearrangement to create the final binding surface (Kwong et al., 1998). As an additional evasion strategy, it has been suggested that virion binding to cellular CD4 brings the gp120 oligomer sufficiently close to the cell membrane to sterically prohibit the subsequent attachment of antibodies to the chemokine receptor-binding surface (Kwong et al., 1998). This is an important consideration if that site is only fully formed during the fusion process itself; an effective antibody intervention at this site may be severely compromised by both temporal and spatial constraints.

4. Host and Viral Factors Influencing Neutralisation

4.1 EFFECT OF THE PRODUCER CELL

One of the most striking influences on HIV-1 neutralisation is that of the origin of the virus producer cell (reviewed in (Moore and Ho, 1995; Poignard et al., 1996a)). The adaptation of HIV to growth in immortalised CD4$^+$ cell lines selects for TCLA variants that preferentially utilise CXCR4 as a co-receptor (reviewed in (D'Souza and Harden, 1996; Moore et al., 1997; Berger, 1997)), are strongly dependent on heparan sulfate proteoglycans for binding and infection (Roderiquez et al., 1995; Mondor et al., 1998a), and may also select viruses with a high affinity for CD4 (Platt et al., 1997; Ugolini et al., 1999). TCLA viruses which tend to have a strongly basic V3 loop (Fouchier et al., 1992) and a high affinity for CD4, are readily neutralised by sCD4 and a large spectrum of different mAbs (reviewed by (Moore and Ho, 1995)). By contrast, viruses that have been passaged only in primary cultures of activated peripheral blood mononuclear cells (PBMC) may use CXCR4 (now termed X4 viruses), CCR5 (R5 viruses) or CXCR4 in combination with CCR5 (R5X4 viruses) (Berger et al., 1998). They generally have a low affinity for sCD4 and neutralising mAbs, and are relatively resistant to neutralisation by these ligands ((Fouts et al., 1997) and reviewed by (Moore and Ho, 1995)). It seems likely, based on the gp120 structure, that the CD4bs on the primary isolate Env trimer is more completely masked by the V1V2 loops than that of TCLA viruses. The idea that gp120 vibrates between 'closed' and 'open' states is consistent with the dichotomy of primary and TCLA viruses: thus primary isolate gp120 would prefer a 'closed' conformation, whereas TCLA gp120 would be biased towards 'open'. In this way, the virus *in vivo* would sacrifice some efficiency in receptor binding for resistance to antibody attack, whereas cell line-passaged virus would dispense with antibody resistance mechanisms and evolve more efficient receptor interactions instead. This notion appears to be generally applicable to lentiviruses, in that SIV, FIV and EIAV adapt to passage in cell lines in the same way as HIV-1 does (Baldinotti et al., 1994; Cook et al., 1995; Moore et al., 1995b; Means et al., 1997; Montefiori et al., 1998b). In addition, other antibody resistance mechanisms may exist (see below).

4.2 CO-RECEPTOR USAGE

Co-receptor usage is not a major determiant of neutralisation resistance, since X4, R5 and R5X4 primary viruses exhibit similar neutralisation sensitivities (Trkola et al., 1998; LaCasse et al., 1998; Montefiori et al., 1998a). These findings confirm previous reports that the virus groupings based on syncytium induction in MT2 cells NSI (non-syncytium-inducing) and SI (syncytium-inducing) do not correspond to neutralisation-resistant or -sensitive phenotypes (Hogervorst et al., 1995; Groenink et al., 1995; Cornelisen et al., 1995). The factors relating to neutralisation sensitivity that are virus producer cell-determined remain to be fully elucidated; however, several factors may influence viral infectivity and hence neutralisation susceptibility. Nascent virus incorporates cellular adhesion molecules into its membrane that will, to some extent, influence its infectivity for a given target cell. A well-studied example of this is the interaction of the LFA-1-ICAM-1 pair of adhesion molecules, which can increase infectivity by up to 10-fold (Rizzuto and Sodroski, 1997; Fortin et al., 1997), and can substantially decrease sensitivity to antibody neutralization (Gomez and Hildreth, 1995; Rizzuto and Sodroski, 1997). Other interactions between envelope and cell-derived molecules probably also impact on neutralisation in a similar way. For example, gp120 V3 loop binding to cell surface heparans has been shown to increase HIV-1 binding to its target cells (Roderiquez et al., 1995), in some cases dramatically (Mondor et al., 1998a), and may well increase neutralisation resistance by enhancing virion interactions with CD4-co-receptor complexes. HIV-1-heparan sulfate interactions are sensitive to neutralising antibodies specific for various regions of gp120, including the V3 loop (Roderiquez et al., 1995) and the V2 loop and the CD4-induced epitope (Mondor et al., 1998a).

4.3 GLYCOSYLATION

Glycosylation is an important strategy used by HIV-1 to evade the neutralising antibody response. HIV-1 gp120 is heavily glycosylated with about 50% of its molecular weight being carbohydrate. The number of glycosylation sites varies between isolates, but up to 8 O-linked and 24 N-linked glycosylation sites may be utilised (Leonard et al., 1990; Bernstein et al., 1994). N-linked glycans are essential in the correct folding and processing of gp120 during synthesis, so HIV-1 replication in the presence of inhibitors of glycosylation is associated with a decreased affinity of gp120 for CD4 and a reduction in viral infectivity (Pal et al., 1989; Lee et al., 1992; Li et al., 1993). Removal of carbohydrate from mature oligomeric Env does not, however, affect binding to soluble CD4 (Li et al., 1993). Most, but not all, individual glycosylation sites can further be eliminated without affecting the infectivity of HIV-1 (Lee et al., 1992) and SIV (Reitter and Desrosiers, 1998; Ohgimoto et al., 1998). It has been known for some time that the presence or absence of carbohydrate may have a dramatic influence on the antigenicity of viral glycoproteins; carbohydrate has been suggested to play a critical role in the masking of neutralising epitopes on the protein backbone and diversion of the humoral response (Alexander and Elder, 1984). Viruses that have been passaged in primary macrophage cultures have gp120 of a higher molecular weight as a result of modified glycosylation, and are even less susceptible to neutralisation than PBL-grown viruses (Willey et al., 1996). It has furthermore been shown that removal of N-glycan in the V3 loop of HIV-1 gp120 increases the sensitivity of the mutant virus to neutralisation by V3-loop and CD4bs mAb (Back et al., 1994; Schønning et al., 1996b; Schønning et al., 1998). Such mutant viruses cultured in the presence of neutralising V3-loop antibody rapidly escaped the antibody by re-acquiring N-linked carbohydrate (Schønning et al., 1996a). In SIV infection, amino-acid changes have been observed in V1 during disease progression that allow the virus to escape from neutralisation (Overbaugh et al., 1991; Chackerian et al., 1997; Rudensey et al., 1998). One such escape mutant was completely resistant to neutralisation by homologous serum that neutralised the parental strain, and differed at only four amino acids in V1 (Rudensey et al., 1998). All these changes were to serine and threonine residues which biochemical analyses showed were utilised for the addition of N- and O-linked carbohydrate (Chackerian et al., 1997; Rudensey et al., 1998). This indicates that the V1 domain is a major target of a neutralising antibody response from which SIV may escape by acquiring carbohydrate. The role of V1 in HIV-1 neutralisation remains obscure, however.

Glycosylation is therefore used by HIV and SIV to mask epitopes recognized by neutralising antibodies elicited during infection. This raises the possibility that immunisation with glycan-deficient gp120 might induce a more potent neutralising antibody response. A study with HIV-1 gp160, mutated to eliminate carbohydrate in the CD4-binding domain, indicated that the neutralising responses obtained were highly type-specific rather than more broadly neutralising, although as a caveat the neutralising serum titres obtained were fairly poor (Bolmstedt et al., 1996). Recently, Reitter and Desrosiers constructed a number of replication-competent variants of SIV_{mac239} mutated to lack N-linked carbohydrate at specific sites in and around the V1 and V2 loops (Reitter and Desrosiers, 1998). Macaques infected with some of these variant strains induced antibody responses, which strongly neutralised the glycan-deficient challenge virus (Reitter et al., 1998). Significantly, these sera neutralised the fully glycosylated parent virus better than homologous sera did. This demonstrates that carbohydrate at certain sites may be involved in decreasing the immunogenicity of epitopes that are exposed on the viral surface, thereby diverting the antibody response away from neutralising epitopes.

4.4 PLASMA VIRUS

Few studies have examined the neutralisation properties of plasma virus (*i.e.* virus present in the blood of infected individuals which has not been passaged *in vitro*). Such viruses appear in general to be relatively resistant to neutralisation, behaving similarly to primary isolates expanded in activated PBMC (Gauduin et al., 1996), although plasma virus appears to be more sensitive to antibody-dependent complement-mediated lysis. Neutralisation profiles of plasma virus and corresponding virus stocks obtained after a single passage in PBMC did not, however, correlate in all cases studied. This indicates that neutralisation sensitivities may change with passage, or that small subpopulations of virus with distinct neutralisation sensitivities were expanded in culture (Gauduin et al., 1996). Of note is that the YU2 envelope glycoprotein complex, which was cloned directly from the brain, functions more efficiently in the presence of low concentrations of anti-gp120 mAbs (Sullivan et al., 1995; Sullivan et al., 1998). It is possible that antibody-dependent fusion enhancement is selected against by viral isolation on activated PBMC *in vitro*, but is more common *in vivo* (Moore, 1995). Further studies are warranted to determine in greater detail the phenotype of plasma virus.

4.5 NEUTRALISATION AT MUCOSAL SURFACES

The first line of immunological defense against natural HIV-1 infection is formed by antibodies present at mucosal surfaces. Some progress has been made in understanding the inactivation of HIV-1 at the mucosa. The nature of the mucosal epithelium appears to be an important determinant in the mechanism of infection and neutralisation of HIV-1 at these surfaces. A recent study has shown that transcytosis of primary HIV-1 isolates over tight epithelium, which covers the endocervix, rectum and gastrointestinal tract, could be blocked by dimeric immune IgA and IgM (Bomsel et al., 1998). It was shown that basolaterally-internalised antibodies met transcytosing virions at an intracellular compartment, thereby redirecting the virions back to the mucosal compartment (Bomsel et al., 1998). Although the affinity for virions of the polyclonal dimeric IgA and IgM, purified from serum from infected individuals, is likely to be low, efficient binding may be occurring by a multivalent interaction at relatively high concentrations within the intracellular compartment. In mucosa with pluristratified epithelium, which for example covers the vagina, exocervix and anus, dendritic cells may be the first target of HIV-1. A recent study has shown that some mAb, known to be potent neutralisers of HIV-1 infection of $CD4^+$ T cells, also effectively blocked the entry of HIV-1 into dendritic cells and the transmission of HIV-1 from dendritic cells to resting T cells (Frankel et al., 1998). Neutralising antibody inhibited the establishment of a productive infection even when dendritic cells pulsed with a primary HIV-1 isolate were added 24 h later to antibody and T cells, or were left in culture for 24-48 h prior to addition of antibody and T cells (Frankel et al., 1998). This may be an *in-vitro* model relevant to the delivery of HIV-1 from the mucosa to draining lymph nodes by dendritic cells; antibodies in the circulation could therefore play a role in interrupting this mode of HIV-1 sexual transmission.

5. Mechanisms of Neutralisation

A number of mechanisms have been proposed for the neutralisation of HIV-1. Agglutination (Dimmock, 1993) is unlikely to play a major role: monovalent ligands are able to neutralise as efficiently as bivalent ones, virus-aggregates could not be recovered in neutralising antibody-treated HIV-1 preparations (McDougal et al., 1996), and the bell-shaped curve associated with this mechanism of neutralisation has not been described for HIV-1. Conceptually the simplest mechanism is that neutralisation arises from inhibition of virus-receptor binding, manifested as inhibition of virus attachment to its target cells. Antibodies that fall into the CD4bs specificity were proposed to mediate neutralisation by this mechanism, based on their ability to interfere with soluble gp120-CD4 binding (Sun et al., 1989; Tilley et al., 1991; Thali et al., 1991). This view was challenged by the suggestion that the Fab fragment of a CD4bs-specific antibody may neutralise at a post-attachment step (McInerney et al., 1997). Recent experiments with a large number of CD4bs antibodies and Fabs, including the one used in the latter study, indicate that post-attachment neutralisation is unlikely to be an important mechanism for CD4bs antibodies (Ugolini et al., 1997; Parren et al., 1998). A second major cluster of antibodies bind the V3 loop (reviewed in (Moore and Nara, 1991)), although they usually only neutralise TCLA viruses (Bou-Habib et al., 1994; Moore et al., 1995a; Moore et al., 1995b). Finally, several other epitope clusters had undefined activity in terms of HIV-1 neutralisation, including the V2 loop, the CD4i and the 2G12 epitopes on gp120, and 2F5 on gp41. The role of the V2 loop as a neutralisation epitope for primary isolates is not completely clear. Two monoclonal antibodies, one of human origin and one of chimpanzee, have been reported to neutralise primary isolates (Gorny et al., 1994; Vijh-Warrier et al., 1996; Honnen et al., 1996), although the range of isolates is very limited (Pinter et al., 1998).

Recent studies, including the elucidation of the gp120 crystal structure, have provided some answers concerning the potential mechanisms of neutralisation of the anti-gp120 mAbs discussed above. Thus the CD4i-epitope is known to overlap the conserved co-receptor binding surface, and mAbs binding to this region prevent the gp120-co-receptor interaction (Wu et al., 1996; Trkola et al., 1996a; Mondor et al., 1998b). The idea that inhibition of virus attachment is an unusual form of neutralisation (Dimmock, 1993) is now known not to apply to HIV-1 (Ugolini et al., 1997). There is a direct correlation between antibody occupancy of its binding site on the virus, irrespective of the particular epitope recognised, and infectivity neutralisation (Parren et al., 1998). Of note is that this includes mAbs to the V3 loop. V3 loop-specific mAbs were once thought to interfere with late stages of the HIV-1-cell fusion process, since they inhibited soluble gp120-CD4 binding weakly or not at all (Skinner et al., 1988; Linsley et al., 1988; Moore and Nara, 1991), and a few studies had suggested that they mediate post-attachment neutralisation (Lu et al., 1992; Pelchen-Matthews et al., 1995; Armstrong et al., 1996). However, this now appears to be of secondary importance, since antibodies to the V3 loop have now been shown to effectively inhibit HIV-1 attachment (Valenzuela et al., 1997; Ugolini et al., 1997). These studies were carried out with X4 TCLA viruses binding to a $CD4^+$ T- cell line, but we have confirmed that inhibition of attachment also applies as a neutralisation mechanism for R5 isolates (QJS, unpublished data). Neutralisation by gp41 mAb 2F5 is an exception (Ugolini et al., 1997), being mediated by another mecha-nism which is discussed below.

Inhibition of HIV-1 attachment is probably mediated by coating of the virus envelope with antibody, that obstructs the close approach of virion and target cell (Parren et al., 1998). Inhibition of HIV-1 attachment by CD4bs-specific mAbs is understood to be by competition for CD4 binding, but how do mAbs that bind to epitopes spatially distinct from the CD4bs interfere with HIV-1 attachment to cells? One potential explanation may be that since the face of gp120 exposed to antibody attack is very limited in size (Wyatt et al., 1998; Kwong et al., 1998), steric and geometric constraints physically hinder CD4 binding even when the epitope is relatively 'distant' from the CD4bs. A similar steric hindrance model has been suggested from molecular analyses of structures solved for complexes of neutralising Fab with virions of rhinovirus and foot-and-mouth disease virus (Smith et al., 1996; Hewat et al., 1997). The finding that interference with virion attachment to cells is a major mechanism of HIV-1 neutralisation (Ugolini et al., 1997) is consistent with our demonstration that antibody affinity for the oligomer is the single most important determinant of neutralisation (Fouts et al., 1997; Parren et al., 1998), since epitope occupancy by a minimum number of antibody molecules corresponds with blocking of a critical number of receptor-binding sites on the virus. The critical number of receptor-binding sites to be blocked may be dependent on the number of envelope spikes per virion (Klasse and Moore, 1996). However, we do not rule out the possibility that an antibody might have more than one activity, and that an immunoglobulin molecule bound to a virion at a concentration insufficient to prevent attachment might exert other antiviral effects or even enhance infectivity for certain viruses (Sullivan et al., 1998).

A second, related mechanism is based on the idea that for efficient virus attachment HIV-1 requires both CD4 and co-receptor binding. This notion is supported by the observation that virion-associated gp120 on most primary isolates has a lower affinity for CD4 than predicted by the measurement of the monomeric gp120-CD4 interaction (Moore et al., 1992; Kozak et al., 1997). The affinity between monomeric gp120 and the relevant co-receptor is relatively low, estimated to be ~80nM for CXCR4 (Hesselgesser et al., 1997) and >500nM for CCR5 (Wu et al., 1996), but is increased by up to 1,000-fold in the case of CCR5 by complexing with CD4 (Wu et al., 1996; Trkola et al., 1996a). For certain primary isolates, the virion-CCR5 association may therefore be stronger than the virion-CD4 association. Differences between the affinity of gp120 for CD4 and for the relevant co-receptor could lead to differences in the neutralisation efficiency of antibodies against CD4bs or co-receptor binding-site epitopes, respectively. In support of a role for inhibition of receptor binding in neutralisation of HIV-1, all neutralising mAbs tested are able to inhibit either the gp120-CD4 interaction, or the CD4-dependent, gp120-co-receptor interaction (Wu et al., 1996; Trkola et al., 1996a; Mondor et al., 1998b).

The potent neutralising antibody to gp41, mAb 2F5, is exceptional in that it does not inhibit HIV-1-cell binding (Ugolini et al., 1997). This antibody binds to a membrane proximal epitope exposed on the ectodomain of mature oligomeric envelope (Muster et al., 1993; Sattentau et al., 1995). MAb 2F5 probably neutralises at a stage subsequent to virus attachment by interfering with the assembly of the leucine zipper that occurs during gp41 activation, or with the insertion of the fusion peptide into the cell membrane (Binley and Moore, 1997). There is evidence for the existence of a pre-hairpin intermediate in the conformational transition of gp41 from its native non-fusogenic state to its post-fusion state (reviewed by (Chan and Kim, 1998)). These conformational changes are initiated within 1-4 min of receptor binding and are complete within 20 min (Jones et al., 1998). It may be that the 2F5 epitope is well exposed during this transition and that 2F5 acts by interfering with the completion of this process. The membrane proximal positioning of the 2F5 epitope on gp41 in its native state may result in binding of the antibody at an angle more parallel than perpendicular to the viral surface. This would result in a considerably lower molecular profile than an anti-gp120 antibody, explaining the lack of interference with virion attachment. Another indication that potent neutralising epitopes exist on the fusogenic state of the envelope glycoprotein comes from vaccination studies with whole cell vaccines in which fusogenic complexes were fixed using formaldehyde. Sera from mice immunised with this vaccine were found to be capable of neutralising a broad range of primary HIV-1 isolates (LaCasse et al., 1999).

Another factor that may influence HIV-1 neutralisation is the induction of conformational changes in gp120 that can lead to gp120 dissociation from gp41, as has been demonstrated for sCD4 (Moore et al., 1990; Sattentau and Moore, 1991; Moore et al., 1991; Hart et al., 1991). Some mAbs specific for the V3 loop, the C4 region and the CD4i epitope are also able to induce gp120 'shedding' from virions of the TCLA Hx10 clone of HIV-1 (Poignard et al., 1996b). We have recently shown that the MN strain is much less susceptible to antibody-mediated gp120 shedding, and that only CD4i mAbs are able to induce this effect (QJS unpublished data). Moreover, mAb-induced shedding does not always correlate with neutralisation (Stamatatos et al., 1997). However, even if mAb-induced shedding is not a phenomenon generally applicable to HIV-1, it suggests that mAbs, like CD4, are able to induce conformational changes in the envelope glycoproteins that may potentiate neutralisation. Shedding induced by CD4i antibodies is of interest in this respect as these mAbs may mimic co-receptor binding, thereby inducing fusion-relevant confor-mational changes in the gp120 quaternary structure. MAb-induced conformational changes may even cause infectivity enhancement if the fusion mechanism is appropriately activated. In this respect, the infectivity of virus carrying the envelope glycoproteins of the primary virus clone YU2 is enhanced by sCD4 and mAbs at sub-neutralising concentrations (see above) (Sullivan et al., 1995; Schutten et al., 1995; Sullivan et al., 1998). The region of gp120 that determines the 'enhancement' phenotype is, at least in this case, the V3 loop (Sullivan et al., 1998). Another study has also indicated the conformation of the V3 loop to be important in this type of antibody-dependent enhancement of infection (Schutten et al., 1997). The underlying mechanism is unclear, but may result from mAb-induced exposure of the co-receptor binding site leading to more efficient virion attachment or fusion. Whether antibody enhancement of HIV-1 infection has any role *in vivo* remains to be established. This mechanism of enhancement of fusogenicity should not be confused with Fc-mediated enhancement of HIV-1 infectivity.

Induction of conformational changes in gp120, potentially without causing shedding, would be consistent with the notion of synergistic neutralisation induced by pairs of ligands to different gp120 epitopes. Thus CD4bs and V3 loop mAbs added together have been shown to neutralise to a greater extent than the sum of their neutralisation activities separately (Tilley et al., 1992; McKeating et al., 1992b; Potts et al., 1993; Laal et al., 1994). The same has also been seen with double and triple combinations of V2 loop mAb with V3 loop and CD4bs mAbs (Vijh-Warrier et al., 1996; Li et al., 1998); with sCD4-antibody combinations (Kennedy et al., 1991); with a combination of V3 loop mAb, 2F5 and 2G12 (Li et al., 1998); and with combinations of 2F5, 2G12 and HIVIG (Mascola et al., 1997; Li et al., 1998). It should be noted, however, that the extent of synergy with the antibodies tested is generally weak. The *in-vivo* relevance is unknown, but effective synergy in the context of a polyclonal response would be highly significant to HIV-1 vaccine design, so further research is appropriate.

6. Conclusions

A number of shared neutralising epitopes have been identified on TCLA HIV-1 gp120, and one on gp41. Of these, only three epitopes appear to be well recognised on a broad spectrum of primary isolates. Antibodies produced in natural infection or by vaccination strategies studied to date, however, have been found to bind weakly to primary viruses and are likely to be functionally ineffective against virus *in vivo*. The immunogenicity of the mature oligomer is low in comparison to other antigenic forms of envelope present in viral debris. A strong antibody response to viral debris may even prevent antibodies from being elicited and matured against mature oligomer because of original antigenic sin (Parren et al., 1997a; Nara and Garrity, 1998). Antibody-based approaches to vaccine design should therefore aim to present a form of mature oligomeric envelope to the immune system and not forms of envelope with an epitope exposure similar to viral debris. HIV-1 has evolved to express an envelope structure of minimal antigenicity and immunogenicity in its oligomeric form, a strategy that enables it to evade humoral immunity. The virus also has an ability to rapidly escape any minimal neutralising antibody response that does develop. Strong CTL responses as well as broadly neutralising antibody responses will probably be necessary in an effective vaccine against HIV-1 (Burton and Moore, 1998). The view that both arms of the immune system are required is supported by studies on the control of retroviruses and other RNA viruses in mice (Planz et al., 1997; Hasenkrug and Chesebro, 1997; Baldridge et al., 1997; Dittmer et al., 1998; Dittmer et al., 1999). Neutralising antibodies may play a role in the critical early phase of infection, since by reducing the infectivity of the viral inoculum they may buy cell-mediated immunity time to mature and to clear cells that do become infected (Burton and Moore, 1998).

7. Acknowledgements

The authors thank John Moore for advice and many helpful discussions. We thank David Montefiori, Pascal Poignard and Alexandra Trkola for their comments on the manuscript. PWHIP is supported by NIH grants AI40377, AI42653 and AI44293 and by a Scholarship Award from the Pediatric AIDS Foundation (PFR-77348). DRB is supported by NIH grants AI33292 and HL59727. QJS is supported by funding from the Centre National de la Recherche Scientifique (CNRS), the Institute National de la Santé et la Recherche Médicale (INSERM), the Agence Nationale de Recherches sur le SIDA (ANRS), the Fondation pour la Recherche Médicale (SIDACTION) and the European Community Biomed II Shared-Cost Action « Antibody-mediated enhancement and neutralisation of lentivirus infections: role in immune pathogenesis and vaccine development ».

8. References

Alexander,S. and Elder, J.H. (1984). Carbohydrate dramatically influences immune reactivity of antisera to viral glycoprotein antigens, *Science* **226**, 1328-1330.

Allaway, G.P., Davis-Bruno, K.L., Beaudry, G.A., Garcia, E.B., Wong, E.L., Ryder, A.M., Hasel, K.W., Gauduin, M.-C., Koup, R.A., McDougal, J.S. and Maddon, P.J. (1995). Expression and characterization of CD4-IgG2, a novel heterotetramer that neutralizes primary HIV type-1 isolates, *AIDS Res. Human Retrovir.* **11**, 533-539.

Andrus, L., Prince, A.M., Bernal, I., McCormack, P., Lee, D.H., Gorny, M.K. and Zolla-Pazner, S. (1998). Passive immunization with human immunodeficiency virus type-1-neutralizing monoclonal antibody in hu-PBL-SCID mice: isolation of a neutralization escape variant, *J. Infect. Dis.* **177**, 889-897.

Armstrong, S.J., McInerney, T.L., McLain, L., Wahren, B., Hinkula, J., Levi, M. and Dimmock, N.J. (1996). Two neutralising anti-V3 monoclonal antibodies act by affecting different functions of human immunodeficiency virus type., *J. Gen. Virol.* **77**, 2931-2941.

Arthur, L.O., Bess Jr., J.W., Urban, R.G., Strominger, J.L., Morton, W.R., Mann, D.L., Henderson, L.E. and Benveniste, R.E. (1995). Macaques immunized with HLA-DR are protected from challenge with simian immunodeficiency virus, *J. Virol.* **69**, 3117-3124.

Back, N.K.T., Smit, L., De Jong, J.-J., Keulen, W., Schutten, M., Goudsmit, J. and Tersmette, M. (1994). An N-glycan within the human immunodeficiency virus type-1 gp120 V3 loop affects virus neutralization, *Virology* **199**, 431-438.

Baldinotti, F., Matteucci, D., Mazzetti, P., Gianelli, C., Bandecchi, P., Tozzini, F. and Bendinelli, M. (1994). Serum neutralization of feline immunodeficiency virus is markedly dependent on passage history of the virus and host system, *J. Virol.* **68**, 4572-4579.

Baldridge, J.R., McGraw, T.S., Paoletti, A. and Buchmeier, M.J. (1997). Antibody prevents the establishment of persistent arenavirus infection in synergy with endogenous T cells, *J. Virol.* **71**, 755-758.

Barbas, C.F., Collet, T.A., Amberg, W., Roben, P., Binley, J.M., Hoekstra, D., Cababa, D., Jones, T.M., Williamson, R.A., Pilkington, G.R., Haigwood, N.L., Cabezas, E., Satterthwait, A.C., Sanz, I. and Burton, D.R. (1993). Molecular profile of an antibody response to HIV-1 as probed by combinatorial libraries, *J. Mol. Biol.* **230**, 812-823.

Beddows, S., Louisirirotchanakul, S., Cheingsong-Popov, R., Easterbrook, P.J., Simmonds, P. and Weber, J. (1998). Neutralization of primary and T-cell line-adapted isolated of human immunodeficiency virus type 1: role of V3-specific antibodies, *J. Gen. Virol.* **79**, 77-82.

Berger, E.A. (1997). HIV entry and tropism: the chemokine receptor connection, *AIDS* **11** (suppl. A), S3-S16.

Berger, E.A., Doms, R.W., Fenyö, E.-M., Korber, B.T.M., Littman, D.R., Moore, J.P., Sattentau, Q.J., Schuitemaker, H., Sodroski, J. and Weiss, R.A. (1998). A new classification for HIV-1, *Nature* **391**, 240.

Berman, P.W., Gregory, T.J., Riddle, L., Nakamura, G.R., Champe, M.A., Porter, J.P., Wurm, F.M., Hershberg, R.D., Cobb, E.K. and Eichberg, J.W. (1990). Protection of chimpanzees from infection by HIV-1 after vaccination with recombinant glycoprotein gp120 but not gp160, *Nature* **345**, 622-625.

Berman, P.W., Murthy, K.K., Wrin, T., Vennari, J.C., Cobb, E.K., Eastman, D.J., Champe, M., Nakamura, G.R., Davison, D., Powell, M.F., Bussiere, J., Francis, D.P., Matthews, T., Gregory, T.J. and Obijeski, J.F. (1996). Protection of MN-rgp120-immunized chimpanzees from heterologous infection with a primary isolate of human immunodeficiency virus type 1, *J. Infect. Dis.* **173**, 52-59.

Bernstein, H.B., Tucker, S.P., Hunter, E., Schutzbach, J.S. and Compans, R.W. (1994). Human immunodeficiency virus type-1 envelope glycoprotein is modified by 0-linked oligosaccharides, *J. Virol.* **68**, 463-468.

Binley, J.M., Ditzel, H.J., Barbas, C.F., Sullivan, N., Sodroski, J., Parren, P.W.H.I. and Burton, D.R. (1996). Human antibody responses to HIV type-1 glycoprotein 41 cloned in phage display libraries suggest three major epitopes are recognized and give evidence for conserved antibody motifs in antigen binding, *AIDS Res. Human Retrovir.* **12**, 911-924.

Binley, J.M., Klasse, P.J., Cao, Y., Jones, I., Markowitz, M., Ho, D.D. and Moore, J.P. (1997). Differential regulation of the antibody responses to gag and env proteins of human immunodeficiency virus type 1, *J. Virol.* **71**, 2799-2809.

Binley, J.M. and Moore, J.P. (1997). The viral mousetrap, *Nature* **387**, 346-348.

Bolmstedt, A., Sjolander, S., Hansen, J.-E.S., Akerblom, L., Hemming, A., Hu, S.-L., Morein, B. and Olofsson, S. (1996). Influence of N-linked glycans in V4-5 region of human immunodeficiency virus type-1 glycoprotein gp160 on induction of a virus-neutralizing humoral response, *J. Acq. Immune Def. Syndr. Human Retrovir.* **12**, 213-220.

Bomsel, M., Heyman, M., Hocini, H., Lagaye, S., Belec, L., Dupont, C. and Desgranges, C. (1998). Intracellular neutralization of HIV transcytosis across tight epithelial barriers by anti-HIV envelope protein dIgA or IgM, *Immunity* **9**, 277-287.

Bou-Habib, D.C., Roderiquez, G., Oravecz, T., Berman, P.W., Lusso, P. and Norcross, M.A. (1994). Cryptic nature of envelope V3 region epitopes protects primary human immunodeficiency virus type 1 from antibody neutralization, *J. Virol.* **68**, 6006-6013.

Boyer, J.D., Ugen, K.E., Wang, B., Agadjanyan, M., Gilbert, L., Bagarazzi, M.L., Chattergoon, M., Frost, P., Javadian, A., Williams, W.V., Refaeli, Y., Ciccarelli, R.B., McCallus, D., Coney, L. and Weiner, D.B. (1997). Protection of chimpanzees from high-doser heterologous HIV-1 challenge by DNA vaccination, *Nature Med.* **3**, 526-532.

Bruck, C., Thiriart, C., Fabry, L., Francotte, M., Pala, P., Van Opstal, O., Culp, J., Rosenberg, M., DeWilde, M., Heidt, P. and Heeney, J. (1994). HIV-1 envelope-elicited neutralizing antibody titres correlate with protection and virus load in chimpanzees, *Vaccine* **12**, 1141-1148.

Burns, D.P.W., Collignon, C. and Desrosiers, R.C. (1993). Simian immunodeficiency virus mutants resistant to serum neutralization arise during persistent infection of rhesus monkeys, *J. Virol.* **67**, 4104-4113.

Burton, D.R., Barbas, C.F., Persson, M.A.A., Koenig, S., Chanock, R.M. and Lerner, R.A. (1991). A large array of human monoclonal antibodies to type-1 human immunodeficiency virus from combinatorial libraries of asymptomatic seropositive individuals, *Proc. Natl. Acad. Sci. USA* **88**, 10134-10137.

Burton, D.R., Pyati, J., Koduri, R., Sharp, S.J., Thornton, G.B., Parren, P.W.H.I., Sawyer, L.S.W., Hendry, R.M., Dunlop, N., Nara, P.L., Lamacchia, M., Garratty, E., Stiehm, E.R., Bryson, Y.J., Cao, Y., Moore, J.P., Ho, D.D. and Barbas, C.F. (1994). Efficient neutralization of primary isolates of HIV-1 by a recombinant human monoclonal antibody, *Science* **266**, 1024-1027.

Burton, D.R. and Montefiori, D.C. (1997). The antibody response in HIV-1 infection, *AIDS* **11** (Suppl. A), S87-S98.

Burton, D.R. (1997). A vaccine for HIV type 1: the antibody perspective, *Proc. Natl. Acad. Sci. USA* **94**, 10018-10023.

Burton, D.R. and Moore, J.P. (1998). Why do we not have an HIV vaccine and how can we make one?, *Nature Med.* **4**, 495-498.

Chackerian, B., Rudensky, L.M. and Overbaugh, J. (1997). Specific N-linked and 0-linked glycosylation modification in the envelope V1 domain of simian immunodeficiency virus variants that evolve in the host alter recognition by neutralizing antibodies, *J. Virol.* **71**, 7719-7727.

Chan, D.C., Fass, D., Berger, J.M. and Kim, P.S. (1997). Core structure of gp41 from the HIV envelope glycoprotein, *Cell* **89**, 263-273.

Chan, D.C. and Kim, P.S. (1998). HIV entry and its inhibition, *Cell* **93**, 681-684.

Chanh, T.C., Dreesman, G.R., Kanda, P., Linette, G.P., Sparrow, J.T., Ho, D.D. and Kennedy, R.C. (1986). Induction of anti-HIV-neutralizing antibodies by synthetic peptides, *EMBO J.* **5**, 3065-3071.

Cheingsong-Popov, R., Osmanov, S., Pau, C.-P., Schochetman, G., Barin, F., Holmes, H., Francis, G., Ruppach, H., Dietrich, U., Lister, S., Weber, J. and UNAIDS Network for HIV-1 Isolation and Characterization (1998). Serotyping of HIV type-1 infections: definition, relationship to viral genetic subtypes and assay definition, *AIDS Res. Human Retrovir.* **14**, 311-318.

Conley, A.J., Gorny, M.K., Kessler, J.A.I., Boots, L.J., Ossorio-Castro, M., Koenig, S., Lineberger, D.W., Emini, E.A., Williams, C. and Zolla-Pazner, S. (1994). Neutralization of primary human immunodeficiency virus type-1 isolates by the broadly reactive anti-V3 monoclonal antibody, 447-52D, *J. Virol.* **68**, 6994-7000.

Conley, A.J., Kessler, J.A.I., Boots, L.J., McKenna, P.M., Schleif, W.A., Emini, E.A., Mark, G.E.I., Katinger, H., Cobb, E.K., Lunceford, S.M., Rouse, S.R. and Murthy, K.K. (1996). The consequence of passive administration of an anti-human immunodeficiency virus type-1-neutralizing monoclonal antibody before challenge of chimpanzees with a primary virus isolate, *J. Virol.* **70**, 6751-6758.

Connor, R.I., Korber, B.T.M., Graham, B.S., Hahn, B.H., Ho, D.D., Walker, B.D., Neumann, A.U.,

Vermund, S.H., Mestecky, J., Jackson, S., Fenamore, E., Cao, Y., Gao, F., Kalams, S., Kunstman, K.J., McDonald, D., McWilliams, N., Trkola, A., Moore, J.P. and Wolinsky, S.M. (1998). Immunological and virological analyses of persons infected by human immunodeficiency virus type 1 while participating in trials of recombinant gp120 subunit vaccines, *J. Virol.* **72**, 1552-1576.

Cook, R.F., Berger, S.L., Rushlow, K.E., McManus, J.E., Cook, S.J., Harrold, S., Raabe, M.L., Montelaro, R.C. and Issel, C.J. (1995). Enhanced sensitivity to neutralizing antibodies in a variant of equine infectious anemia virus is linked to amino acid substitution in the surface unit envelope glycoprotein, *J. Virol.* **69**, 1493-1499.

Cornelissen, M., Mulder-Mampinga, G., Veenstra, J., Zorgdrager, F., Kuiken, C., Hartman, S., Dekker, J., Van der Hoek, L., Sol, C., Coutinho, R.A. and Goudsmit, J. (1995). Syncytium-inducing (SI) phenotype suppression at seroconversion after intramuscular inoculation of a non-syncytium-inducing/SI phenotypically mixed human immunodeficiency virus population, *J. Virol.* **69**, 1810-1818.

D'Souza, M.P., Geyer, S.J., Hanson, C.V., Hendry, R.M. and Milman, G. (1994). Evaluation of monoclonal antibodies to HIV-1 envelope by neutralization and binding assays: an international collaboration, *AIDS* **8**, 169-181.

D'Souza, M.P., Milman, G., Bradac, J.A., McPhee, D., Hanson, C.V., Hendry, R.M. and Collaborating Investigators (1995). Neutralisation of primary HIV-1 isolates by anti-envelope monoclonal antibodies, *AIDS* **9**, 867-874.

D'Souza, M.P. and Harden, V.A. (1996). Chemokines and HIV-1 second receptors, *Nature Med.* **2**, 1293-1300.

D'Souza, M.P., Livnat, D., Bradac, J.A., Bridges, S. and The AIDS Clinical Trials Group Antibody Selection Working Group, and Collaborating Investigators (1997). Evaluation of monoclonal antibodies to HIV-1 primary isolates by neutralization assays: performance criteria for selecting candidate antibodies for clinical trials, *J. Infect. Dis.* **175**, 1056-1062.

Daar, E.S., Li, X.L., Moudgil, T. and Ho, D.D. (1990). High concentrations of recombinant soluble CD4 are required to neutralize primary human immunodeficiency virus type-1 isolates, *Proc. Natl. Acad. Sci. USA* **87**, 6574-6578.

Dalgleish, A.G., Chanh, T.C., Kennedy, R.C., Kanda, P., Clapham, P.R. and Weiss, R.A. (1988). Neutralization of diverse HIV-1 strains by monoclonal antibodies raised against a gp41 synthetic peptide, *Virology* **165**, 209-215.

Daniel, M.D., Letvin, N.L., King, N.W., Kannagi, M., Sehgal, P.K., Hunt, R.D., Kanki, P.J., Essex, M. and Desrosiers, R.C. (1985). Isolation of T-cell-tropic HTLV-III-like retrovirus from macaques, *Science* **228**, 1201-1204.

Daniel, M.D., Letvin, N.L., Sehgal, P.K., Hunsmann, G., Schmidt, D.K., King, N.W. and Desrosiers, R.C. (1987). Long-term persistent infection of macaque monkeys with the simian immunodeficiency virus, *J. Gen. Virol.* **68**, 3183-3189.

Desrosiers, R.C., Wyand, M.S., Kodama, T., Ringler, D.J., Arthur, L.O., Sehgal, P.K., Letvin, N.L., King, N.W. and Daniel, M.D. (1989). Vaccine protection against simian immunodeficiency virus infection, *Proc. Natl. Acad. Sci. USA* **86**, 6353-6357.

Dimmock, N.J. (1993). Neutralization of animal viruses, *Curr. Top. Microbiol.* **183**, 1-149.

Dittmer, U., Brooks, D.M. and Hasenkrug, K.J. (1998). Characterization of a live-attenuated retroviral vaccine demonstrates protection via immune mechanisms, *J. Virol.* **72**, 6554-6558.

Dittmer, U., Brooks, D.M. and Hasenkrug, K.J. (1999). Requirement for multiple lymphocyte subsets in protection by a live attenuated vaccine against retroviral infection, *Nature Med.* **5**, 189-193.

Ditzel, H.J., Parren, P.W.H.I., Binley, J.M., Sodroski, J., Moore, J.P., Barbas, C.F. and Burton, D.R. (1997). Mapping the protein surface of human immunodeficiency virus type-1 gp120 using human monoclonal antibodies from phage display libraries, *J. Mol. Biol.* **267**, 684-695.

El-Amad, Z., Murthy, K.K., Higgins, K., Cobb, E.K., Haigwood, N.L., Levy, J.A. and Steimer, K.S. (1995). Resistance of chimpanzees immunized with recombinant gp120 SF2 to challenge by HIV-1 SF2, *AIDS* **9**, 1313-1322.

Emini, E.A., Schleif, W.A., Nunberg, J.H., Conley, A.J., Eda, Y., Tokiyoshi, S., Putney, S.D., Matsushita, S., Cobb, K.E., Jett, C.M., Eichberg, J.W. and Murthy, K.K. (1992). Prevention of HIV-1 infection in chimpanzees by gp120 V3 domain-specific monoclonal antibody, *Nature* **355**, 728-730.

Evans, D.J., McKeating, J., Meredith, J.M., Burke, K.L., Katrak, K., John, A., Ferguson, M., Minor, P.D., Weiss, R.A. and Almond, J.W. (1989). An engineered poliovirus chimaera elicits broadly reactive HIV-1-neutralising antibodies, *Nature* **339**, 385-388.

Foresman, L., Jia, F., Li, Z., Wang, C., Stephens, E.B., Sahni, M., Naryan, O. and Joag, S.V. (1998). Neutralizing antibodies administered before, but not after virulent SHIV prevent infection in macaques, *AIDS Res. Human Retrovir.* **12**, 1035-1043.

Fortin, J.-F., Cantin, R., Lamontagne, G. and Tremblay, M. (1997). Host-derived ICAM-1 glycoproteins incorporated on human immunodeficiency virus type 1 are biologically active and enhance viral infectivity, *J. Virol.* **71**, 3588-3596.

Fouchier, R.A.M., Groenink, M., Kootstra, N.A., Tersmette, M., Huisman, J.G., Miedema, F. and Schuitemaker, H. (1992). Phenotype-associated sequence variation in the third variable domain of the human immunodeficiency virus type-1 gp120 molecule, *J. Virol.* **66**, 3183-3187.

Fouts, T.R., Binley, J.M., Trkola, A., Robinson, J.E. and Moore, J.P. (1997). Neutralization of the human immunodeficiency virus type-1 primary isolate JR-FL by human monoclonal antibodies correlates with antibody binding to the oligometic form of the envelope glycoprotein complex, *J. Virol.* **71**, 2779-2785.

Frankel, S.S., Steinman, R.M., Michael, N.L., Kim, S.R., Bhardwaj, N., Pope, M., Louder, M.K., Ehrenberg, P.K., Parren, P.W.H.I., Burton, D.R., Katinger, H., VanCott, T.C., Robb, M.L., Birx, D.L. and Mascola, J.R. (1998). Neutralizing monoclonal antibodies block human immunodeficiency virus type-1 infection of dendritic cells and transmission to T cells, *J. Virol.* **72**, 9788-9794.

Fultz, P.N., Nara, P., Barre-Sinoussi, F., Chaput, A., Greenberg, M.L., Muchmore, E., Kieny, M.-P. and Girard, M. (1992). Vaccine protection of chimpanzees against challenge with HIV-1-infected peripheral blood mononuclear cells, *Science* **256**, 1687-1690.

Gallaher, W.R., Ball, J.M., Garry, R.F., Griffin, M.C. and Montelaro, R.C. (1989). A general model for the transmembrane proteins of HIV and other retrovirus, *AIDS Res. Human Retrovir.* **5**, 431-440.

Gardner, M., Rosenthal, A., Jennings, M., Yee, J., Antipa, L. and Robinson, W.E. (1995). Passive immunization of rhesus macaques against SIV infection and disease, *AIDS Res. Human Retrovir.* **11**, 843-854.

Gardner, M.B. and Luciw, P.A. (1989). Animal models of AIDS, *FASEB J.* **3**, 2593-2606.

Gauduin, M.C., Safrit, J.T., Weir, R., Fung, M.S. and Koup, R.A. (1995). Pre- and post-exposure protection against human immunodeficiency virus type-1 infection mediated by a monoclonal antibody, *J. Infect. Dis.* **171**, 1203-1209.

Gauduin, M.C., Allaway, G.P., Maddon, P.J., Barbas, C.F., Burton, D.R. and Koup, R.A. (1996). Effective *ex-vivo* neutralization of human immunodeficiency virus type 1 in plasma HIV-1 by recombinant immunoglobulin molecules, *J. Virol.* **70 (4)**, 2586-2592.

Gauduin, M.C., Parren, P.W.H.I., Weir, R., Barbas, C.F., Burton, D.R. and Koup, R.A. (1997). Passive immunization with a human monoclonal antibody protects hu-PBL-SCID mice against challenge by primary isolates of HIV-1, *Nature Med.* **3**, 1389-1393.

Gauduin, M.C., Allaway, G.P., Olson, W.C., Weir, R., Maddon, P.J. and Koup, R.A. (1998). CD4-immunoglobulin G2 protects hu-PBL-SCID mice against challenge by primary human immunodeficiency virus type-1 isolates, *J. Virol.* **72**, 3475-3478.

Girard, M., Kieny, M.P., Pinter, A., Barré-Sinoussi, F., Nara, P.L., Kolbe, H., Kusumi, K., Chaput, A., Reinhart, T., Muchmore, E., Ronco, J., Kaczorek, M., Gomard, E., Gluckman, J.-C. and Fultz, P.N. (1991). Immunization of chimpanzees confers protection against challenge with human immunodeficiency virus, *Proc. Natl. Acad. Sci. USA* **88**, 542-546.

Girard, M., Meignier, B., Barré-Sinoussi, F., Kieny, M.-P., Matthews, T., Muchmore, E., Nara, P.L., Wei, Q., Rimsky, L., Weinhold, K. and Fultz, P.N. (1995). Vaccine-induced protection of chimpanzees against infection by a heterologous human immunodeficiency virus type 1, *J. Virol.* **69**, 6239-6248.

Gomez, M.B. and Hildreth, J.E.K. (1995). Antibody to adhesion molecule LFA-1 enhances plasma neutralization of human immunodeficiency virus type 1, *J. Virol.* **69**, 4628-4632.

Gorny, M.K., Conley, A.J., Karwowska, S., Buchbinder, A., Xu, J.-Y., Emini, E.A., Koenig, S. and Zolla-Pazner, S. (1992). Neutralization of diverse human immunodeficiency virus type-1 variants by an anti-V3 human monoclonal antibody, *J. Virol.* **66**, 7538-7542.

Gorny, M.K., Xu, J.-Y., Karwowska, S., Buchbinder, A. and Zolla-Pazner, S. (1993). Repertoire of neutralizing human monoclonal antibodies specific for the V3 domain of HIV-1 gp120, *J. Immunol.* **150**, 635-643.

Gorny, M.K., Moore, J.P., Conley, A.J., Karwowska, S., Sodroski, J., Williams, C., Burda, S., Boots,

L.J. and Zolla-Pazner, S. (1994). Human anti-V2 monoclonal antibody that neutralizes primary but not laboratory isolates of human immunodeficiency virus type 1, *J. Virol.* **68**, 8312-8320.

Gorny, M.K., Mascola, J.R., Israel, Z.R., VanCott, T.C., Williams, C., Balfe, P., Hioe, C., Brodine, S., Burda, S. and Zolla-Pazner, S. (1998). A human monoclonal antibody specific for the V3 loop of HIV type-1 clade E cross-reacts with other HIV type-1 clades, *AIDS Res. Human Retrovir.* **14**, 213-221.

Graham, B.S., Keefer, M.C., McElrath, M.J., Gorse, G.J., Schwartz, D.H., Weinhold, K., Matthews, T.J., Esterlitz, J.R., Sinangil, F., Fast, P.E. and NIAID AIDS Vaccine Evaluation Group (1996). Safety and immunogenicity of a candidate HIV-1 vaccine in healthy adults: recombinant glycoprotein (rgp) 120 - a randomized, double-blind trial, *Ann. Intern. Med.* **125**, 270-279.

Groenink, M., Moore, J.P., Broersen, S. and Schuitemaker, H. (1995). Equal levels of gp120 retention and neutralization resistance of phenotypically distinct primary human immunodeficiency virus type-1 variants upon soluble CD4 treatment, *J. Virol.* **69**, 523-527.

Haigwood, N.L., Watson, A., Sutton, W.F., McClure, J., Lewis, A., Ranchalis, J., Travis, B., Voss, G., Letvin, N.L., Hu, S.L., Hirsch, V.M. and Johnson, P.R. (1996). Passive immune globulin therapy in the SIV/macaque model: early intervention can alter disease profile, *Immunol. Lett.* **51**, 107-114.

Harrer, T., Harrer, E., Kalams, S.A., Elbeik, T., Staprans, S.I., Feinberg, M.B., Cao, Y., Ho, D.D., Yilma, T., Caliendo, A.M., Johnson, R.P., Buchbinder, S.P. and Walker, B.D. (1996). Strong cytotoxic T-cell and weak neutralizing antibody responses in a subset of persons with stable nonprogressing HIV type-1 infection, *AIDS Res. Human Retrovir.* **12**, 585-592.

Hart, T.K., Kirsh, R., Ellens, H., Sweet, R.W., Lambert, D.M., Petteway, S.R., Leary, J. and Bugelski, P.J. (1991). Binding of soluble CD4 proteins to human immunodeficiency virus type 1 and infected cells induces release of envelope glycoprotein gp120, *Proc. Natl. Acad. Sci. USA* **88**, 2189-2193.

Hasenkrug, K.J. and Chesebro, B. (1997). Immunity to retroviral infection: the Friend virus model, *Proc. Natl. Acad. Sci. USA* **94**, 7811-7816.

Haynes, B.F. (1996). HIV vaccines: where we are and where we are going, *Lancet* **348**, 933-937.

Hesselgesser, J., Hlaks-Miller, M., DelVecchio, V., Peiper, S.C., Hoxie, J., Holson, D.L., Taub, D. and Horuk, R. (1997). CD4-independent association between HIV-1 gp120 and CXCR4: functional chemokine receptors are expressed in human neurons, *Curr. Biol.* **7**, 112-121.

Hewat, E.A., Verdaguer, N., Fita, N., Blakemore, W., Brookes, S., King, A., Newman, J., Domingo, E., Mateu, M.G. and Stuart, D.I. (1997). Structure of the complex of an Fab fragment of a neutralizing antibody with foot-and-mouth disease virus: positioning of a highly mobile antigenic loop, *EMBO J.* **16**, 1492-1500.

Ho, D.D., Neumann, A.U., Perelson, A.S., Chen, W., Leonard, J.M. and Markowitz, M. (1995). Rapid turnover of plasma virions and CD4 lymphocytes in HIV-1 infection, *Nature* **373**, 123-126.

Hogervorst, E., De Jong, J., Van Wijk, A., Bakker, M., Valk, M., Nara, P. and Goudsmit, J. (1995). Insertion of primary syncytium-inducing (SI) and non-SI envelope V3 loops in human immunodeficiency virus type 1 (HIV-1) LAI reduces neutralization sensitivity to autologous, but not heterologous, HIV-1 antibodies, *J. Virol.* **69**, 6342-6351.

Honnen, W.J., Wu, Z., Kayman, S.C. and Pinter, A. (1996). Potent neutralization of a macrophage-tropic HIV-1 isolate by antibodies against the V1/V2 domain of HIV-1 gp120. In: *Vaccins 96* (F. Brown, E. Norrby, D.R. Burton and J. Mekalanos, eds.), New York: Cold Spring Harbor Laboratory Press, pp. 289-297.

Hulskotte, E.G.J., Geretti, A.M. and Osterhaus, A.D.M.E. (1998). Towards an HIV-1 vaccine: lessons from studies in macaque models, *Vaccine* **16**, 904-915.

Javaherian, K., Langlois, A.J., LaRosa, G.J., Profy, A.T., Bolgnesi, D.P., Herlihy, W.C., Putney, S.D. and Matthews, T.J. (1990). Broadly neutralizing antibodies elicited by the hypervariable neutralizing determinant of HIV-1, *Science* **250**, 1590-1593.

Joag, S.V., Li, Z., Foresman, L., Stephens, E.B., Zhao, L.J., Adany, I., Pinson, D.M., McClure, H.M. and Narayan, O. (1996). Chimeric simian human immunodeficiency virus that causes progressive loss of $CD4^+$ T cells and AIDS in pig-tailed macaques, *J. Virol.* **70**, 3189-3197.

Joag, S.V., Li, Z., Foresman, L., Pinson, D.M., Raghavan, R., Zhu, G.W., Adany, I., Wang, C., Jia, F., Sheffer, D., Ranchalis, J., Watson, A. and Narayan, O. (1997). Characterization of the pathogenic KU-SHIV model of AIDS in macaques, *AIDS Res. Human Retrovir.* **13**, 635-645.

Jones, P.L., Korte, T. and Blumenthal, R. (1998). Conformational changes in cell surface HIV-1 envelope glycoproteins are triggered by cooperation between cell surface CD4 and co-receptors, *J. Biol. Chem.* **273**, 404-409.

Katinger, H., Purtscher, M., Muster, T., Steindl, F., Dopper, S., Vetter, N., Armbruster, C. and Gelbmann, H. (1995). A small phase I clinical trial with the human anti-HIV-1 mAb 2F5. In: *Retroviruses of Human AIDS and Related Animal Diseases*, Dixième Colloque des Cent Gardes (M. Girard and B. Dodet, eds.), Fondation Marcel Merieux, Lyon, France, pp. 291-297.

Kennedy, M.S., Orloff, S., Ibegbu, C.C., Odell, C.D., Maddon, P.J. and McDougal, J.S. (1991). Analysis of synergism/antagonism between HIV-1 antibody-positive human sera and soluble CD4 in blocking HIV-1 binding and infectivity, *AIDS Res. Human Retrovir.* **7**, 975-981.

Kent, K.A., Kitchin, P., Mills, K.H.G., Page, M., Taffs, F., Corcoran, T., Silvera, P., Flanagan, B., Powell, C., Rose, J., Ling, C., Aubertin, A.M. and Scott, E.J. (1994). Passive immunization of cynomolgus macaques with immune sera or a pool of neutralizing monoclonal antibodies failed to protect against challenge with SIVmac251, *AIDS Res. Huamn Retrovir.* **10**, 189-194.

Kestler, H., Kodama, T., Ringler, D., Marthas, M., Pedersen, N., Lackner, A., Regier, D., Sehgal, P., Daniel, M., King, N. and Desrosiers, R.C. (1990). Induction of AIDS in rhesus monkeys by molecularly cloned simian immunodeficiency virus, *Science* **248**, 1109-1112.

Klasse, P.J. and Moore, J.P. (1996). Quantitative model of antibody- and soluble CD4-mediated neutralization of primary isolates and T-cell line-adapted strains of human immunodeficiency virus type 1, *J. Virol.* **70**, 3668-3677.

Kostrikis, L.G., Cao, Y., Ngai, H., Moore, J.P. and Ho, D.D. (1996). Quantitative analysis of serum neutralization of human immunodeficiency virus type 1 from subtypes A, B, C, D, E, F and I: lack of direct correlation between neutralization serotypes and genetic subtypes and evidence for prevalent serum-dependent infectivity enhancement, *J. Virol.* **70**, 445-458.

Kozak, S.L., Platt, E.J., Madani, N., Ferro, F.E.J., Peden, K. and Kabat, D. (1997). CD4, CXCR4 and CCR5 dependencies for infections by primary patient and laboratory-adapted isolates of human immunodeficiency virus type 1, *J. Virol.* **71**, 873-882.

Kwong, P.D., Wyatt, R., Robinson, J., Sweet, R.W., Sodroski, J. and Hendrickson, W.A. (1998). Structure of an HIV gp120 envelope glycoprotein in complex with the CD4 receptor and a neutralizing human antibody, *Nature* **393**, 648-659.

Laal, S., Burda, S., Gorny, M.K., Karwowska, S., Buchbinder, A. and Zolla-Pazner, S. (1994). Synergistic neutralization of human immunodeficiency virus type 1 by combinations of human monoclonal antibodies, *J. Virol.* **68**, 4001-4008.

LaCasse, R.A., Follis, K.E., Moudgil, T., Trahey, M., Binley, J.M., Planelles, V., Zolla-Pazner, S. and Nunberg, J.H. (1998). Coreceptor utilization by human immunodeficiency virus type 1 is not a primary determinant of neutralization sensitivity, *J. Virol.* **72**, 2491-2495.

LaCasse, R.A., Follis, K.E., Trahey, M., Scarborough, J.D., Littman, D.R. and Nunberg, J.H. (1999). Fusion-competent vaccines: broad neutralization of primary isolates of HIV, *Science* **283**, 357-362.

Lambert, J.S., Mofenson, L.M., Fletcher, C.V., Moye, J., Stiehm, E.R., Meyer, W.A.I., Nemo, G.J., Mathieson, B.J., Hirsch, G., Sapan, C.V., Cummins, L.M., Jiminez, E., O'Neill, E., Kovacs, A., Stek, A. and the Pediatric AIDS Clinical Trials Group Protocol 185 Pharmacokinetic Study Group (1997). Safety and pharmacokinetics of hyperimmune anti-human immunodeficiency virus (HIV) immunoglobulin administered to HIV-infected pregnant women and their newborns, *J. Infect. Dis.* **175**, 283-291.

Lathey, J.L., Pratt, R.D. and Spector, S.A. (1997). Appearance of autologous neutralizing antibody correlates with reduction in virus load and phenotype switch during primary infection with human immunodeficiency virus type 1, *J. Infect. Dis.* **175**, 231-232.

Lee, W.-R., Syu, W.-J., Du, B., Matsuda, M., Tan, S., Wolf, A., Essex, M. and Lee, T.-H. (1992). Nonrandom distribution of gp120 N-linked glycosylation sites important for infectivity of human immunodeficiency virus type 1, *Proc. Natl. Acad. Sci. USA* **89**, 2213-2217.

Leonard, C.K., Spellman, M.W., Riddle, L., Harris, R.J., Thomas, J.N. and Gregory, T.J. (1990). Assignment of intrachain disulfide bonds and characterization of potential glycosylation sites of the type-1 recombinant human immunodeficiency virus envelope glycoprotein (gp120) expressed in Chinese hamster ovary cells, *J. Biol. Chem.* **265**, 10373-10382.

Letvin, N.L. (1998). Progress in the development of an HIV-1 vaccine, *Science* **280**, 1875-1880.

Levy, J., Youvan, T., Lee, M.L. and Passive Hyperimmune Therapy Study Group (1994). Passive hyperimmune plasma therapy in the treatment of acquired immunodeficiency syndrome: results of a 12-month multicenter double-blind controlled trial, *Blood* **84**, 2130-2135.

Lewis, M.G., Elkins, W.R., McCutchan, F., Benveniste, R.E., Lai, C.Y., Montefiori, D.C., Burke, D.S.,

Eddy, G.A. and Shafferman, A. (1993). Passively transferred antibodies directed against conserved regions of SIV envelope protect macaques from SIV infection, *Vaccine* **11 (13)**, 1347-1355.

Li, A., Katinger, H., Posner, M.R., Cavacini, L., Zolla-Pazner, S., Gorny, M.K., Sodroski, J., Chou, T.C., Baba, T.W. and Ruprecht, R.M. (1998). Synergistic neutralization of simian-human immunodeficiency virus SHIV-vpu$^+$ by triple and quadruple combinations of human monoclonal antibodies and high-titer anti-human immunodeficiency virus type-1 immunoglobulins, *J. Virol.* **72**, 3235-3240.

Li, J.T., Halloran, M., Lord, C.I., Watson, A., Ranchalis, J., Fung, M., Letvin, N.L. and Sodroski, J.G. (1995). Persistent infection of macaques with simian-human immunodeficiency viruses, *J. Virol.* **69**, 7061-7071.

Li, Y., Luo, L., Rasool, N. and Kang, C.Y. (1993). Glycosylation is necessary for the correct folding of human immunodeficiency virus gp120 in CD4 binding, *J. Virol.* **67**, 584-588.

Linsley, P.S., Ledbetter, J.A., Thomas, E.K. and Hu, S.-L. (1988). Effects of anti-gp120 monoclonal antibodies on CD4 receptor binding by the env protein of human immunodeficiency virus type 1, *J. Virol.* **62**, 3695-3702.

Louisirirotchanakul, S., Beddows, S., Cheingsong-Popov, R., Shaffer, N., Mastro, T.D., Auewarakul, P., Likanonsakul, S., Wasi, C. and Weber, J. (1998). Characterization of sera from subjects infected with HIV-1 subtypes B and E in Thailand by antibody binding and neutralization, *J. Acq. Immune Def. Syndr. Human Retrovir.* **19**, 315-320.

Lu, S., Putney, S.D. and Robinson, H.L. (1992). Human immunodeficiency virus type-1 entry into T cells: more rapid escape from an anti-V3 loop than from an anti-receptor antibody, *J. Virol.* **66**, 2547-2550.

Lubeck, M.D., Natur, R., Myagkikh, M., Kalyan, N., Aldrich, K., Sinangil, F., Alipanah, S., Murthy, S.C.S., Chanda, P.K., Nigida, S.M., Markham, P.D., Zolla-Pazner, S., Steimer, K., Wade, M., Reitz Jr., M.S., Arthur, L.O., Mizutani, S., Davis, A., Hung, P.P., Gallo, R.C., Eichberg, J. and Robert-Guroff, M. (1997). Long-term protection of chimpanzees against HIV-1 challenge induced by immunization, *Nature Med.* **3**, 651-658.

Luciw, P.A., Pratt-Lowe, E., Shaw, K.E., Levy, J.A. and Cheng-Mayer, C. (1995). Persistent infection of rhesus macaques with T-cell-line-tropic and macrophage-tropic clones of simian/human immunodeficiency virus (SHIV), *Proc. Natl. Acad. Sci. USA* **92**, 7490-7494.

Mascola, J.R., Louwagie, J., McCutchan, F.E., Fischer, C.L., Hegerich, P.A., Wagner, K.F., Fowler, A.K., McNeil, J.G. and Burke, D.S. (1994). Two antigenically distinct subtypes of human immunodeficiency virus type 1: viral genotype predicts neutralization serotype, *J. Infect. Dis.* **169**, 48-54.

Mascola, J.R., Louder, M.K., Surman, S.R., VanCott, T.C., Yu, X.F., Bradac, J., Porter, K.R., Nelson, K.E., Girard, M., McNeil,. J.G., McCutchan, F.E., Birx, D.L. and Burke, D.S. (1996a). Human immunodeficiency virus type-1 neutralizing antibody serotyping using serum pools and an infectivity reduction assay, *AIDS Res. Human Retrovir.* **12**, 1319-1328.

Mascola, J.R., Snyder, S.W., Weislow, O.S., Belay, S.M., Belshe, R.B., Schwartz, D.H., Clements, M.L., Dolin, R., Graham, B.S., Gorse, G.J., Keefer, M.C., McElrath, M.J., Walker, M.C., Wagner, K.F., McNeil, J.G., McCutchan, F.E., Burke, D.S. and The NIAID AIDS Vaccine Evaluation Group (1996b). Immunization with envelope subunit vaccine products elicits neutralizing antibodies against laboratory-adapted but not primary isolates of human immunodeficiency virus type 1, *J. Infect. Dis.* **173**, 340-348.

Mascola, J.R., Louder, M.K., VanCott, T.C., Sapan, C.V., Lambert, J.S., Muenz, L.R., Bunow, B., Birx, D.L. and Robb, M.L. (1997). Potent and synergistic neutralization of human immunodeficiency virus (HIV) type-1 primary isolates by hyperimmune anti-HIV immunoglobulin combined with monoclonal antibodies 2F5 and 2G12, *J. Virol.* **71**, 7198-7206

McDougal, J.S., Kennedy, M.S., Orloff, S.L., Nicholson, J.K.A. and Spira, T.J. (1996). Mechanisms of human immunodeficiency virus type-1 (HIV-1) neutralization: irreversible inactivation of infectivity by anti-HIV-1 antibody, *J. Virol.* **70**, 5236-5245.

McInerney, T.L., McLain, L., Armstrong, S.J. and Dimmock, N.J. (1997). A human IgG1 (b12) specific for the CD4 binding site of HIV-1 neutralizes by inhibiting the virus fusion entry process, but b12 Fab neutralizes by inhibiting a post-fusion event, *Virology* **233**, 313-326.

McKeating, J.A., Moore, J.P., Ferguson, M., Marsden, H.S., Graham, S., Almond, J.W., Evans, D.J. and Weiss, R.A. (1992a). Monoclonal antibodies to the C4 region of human immunodeficiency virus type-1 gp120: use in topological analysis of a CD4 binding site, *AIDS Res. Human Retrovir.* **8**, 451-459.

McKeating, J.A., Cordell, J., Dean, C.J. and Balfe, P. (1992b). Synergistic interaction between ligands binding to the CD4 binding site and V3 domain of human immunodeficiency virus type-1 gp120, *Virology* **191**, 732-742.

McKeating, J.A., Shotton, C., Cordell, J., Graham, S., Balfe, P., Sullivan, N., Charles, M., Page, M., Bolmstedt, A., Olofsson, S., Kayman, S.C., Wu, Z., Pinter, A., Dean, C., Sodroski, J. and Weiss, R.A. (1993). Characterization of neutralizing monoclonal antibodies to linear and conformation-dependent epitopes within the first and second variable domains of human immunodeficiency virus type-1 gp120, *J. Virol.* **67**, 4932-4944.

McKnight, A. and Clapham, P.R. (1995). Immune escape and tropism of HIV, *Trends Microbiol.* **3**, 356-361.

Means, R.E., Greenough, T. and Desrosiers, R.C. (1997). Neutralization sensitivity of cell culture-passaged simian immunodeficiency virus, *J. Virol.* **71**, 7895-7902.

Mondor, I., Ugolini, S. and Sattentau, Q.J. (1998a). HIV-1 attachment to HeLa CD4 cells is CD4-independent, gp120-dependent and requires cell surface heparans, *J. Virol.* **72**, 3623-3634.

Mondor, I., Moulard, M., Ugolini, S., Klasse, P.J., Hoxie, J., Amara, A., Delaunay, T., Wyatt, R., Sodroski, J. and Sattentau, Q.J. (1998b). Interactions among HIV gp120, CD4 and CXCR4: Dependence on CD4 expression level, gp120 viral origin, conservation of the gp120 COOH- and NH_2-termini and V1/V2 and V3 loops, and sensitivity to neutralizing antibodies, *Virology* **248**, 394-405.

Montefiori, D.C., Graham, B.S., Zhou, J., Schwartz, D.H., Cavacini, L.A., Posner, M.R. and NIH-NIAID AIDS Vaccine Clinical Trials Network (1993). V3-specific neutralizing antibodies in sera from HIV-1 gp160-immunized volunteers block virus fusion and act synergistically with human monoclonal antibody to the conformation-dependent CD4 binding site of gp120, *J. Clin. Invest.* **92**, 840-847.

Montefiori, D.C., Baba, T.W., Li, A., Bilska, M. and Ruprecht, R.M. (1996a). Neutralizing and infection-enhancing antibody responses do not correlate with the differential pathogenicity of SIVmac239D3 in adult and infant rhesus monkeys, *J. Immunol.* **157**, 5528-5535.

Montefiori, D.C., Pantaleo, G., Fink, L.M., Zhou, J.T., Zhou, J.Y., Bilska, M., Miralles, G.D. and Fauci, A.S. (1996b). Neutralizing and infection-enhancing antibody responses to human immunodeficiency virus type 1 in long-term non-progressors, *J. Infect. Dis.* **173**, 60-67.

Montefiori, D.C., Collman, R.G., Fouts, T.R., Zhou, J.Y., Bilska, M., Hoxie, J.A., Moore, J.P. and Bolognesi, D.P. (1998a). Evidence that antibody-mediated neutralization of human immunodeficiency virus type 1 by sera from infected individuals is independent of co-receptor usage, *J. Virol.* **72**, 1886-1893.

Montefiori, D.C., Reimann, K.A., Wyand, M.S., Manson, K., Lewis, M.G., Collman, R.G., Sodroski, J.G., Bolognesi, D.P. and Letvin, N.L. (1998b). Neutralizing antibodies in sera from macaques infected with chimeric simian-human immunodeficiency virus containing the envelope glycoproteins of either a laboratory-adapted variant or a primary isolate of human immunodeficiency virus type 1, *J. Virol.* **72**, 3427-3431.

Moog, C., Fleury, H.J.A., Pellegrin, I., Kirn, A. and Aubertin, A.M. (1997). Autologous and heterologous neutralizing antibody responses following initial seroconversion in human immunodeficiency virus type-1-infected individuals, *J. Virol.* **71**, 3734-3741.

Moore, J.P., McKeating, J.A., Weiss, R.A. and Sattentau, Q.J. (1990). Dissociation of gp120 from HIV-1 virions induced by soluble CD4, *Science* **250**, 1139-1142.

Moore, J.P., McKeating, J.A., Norton, W.A. and Sattentau, Q.J. (1991). Direct measurement of soluble CD4 binding to human immunodeficiency virus type-1 virions: gp120 dissociation and its implications for virus-cell binding and fusion reactions and their neutralisation by soluble CD4, *J. Virol.* **65**, 1133-1140.

Moore, J.P. and Nara, P.L. (1991). The role of the V3 loop in HIV infection, *AIDS* **5 (suppl. 2)**, S21-S33.

Moore, J.P., McKeating, J.A., Huang, Y., Ashkenazi, A. and Ho, D.D. (1992). Virions of primary human immunodeficiency virus type-1 isolates resistant to soluble CD4 (sCD4) neutralization differ in sCD4 binding and glycoprotein gp120 retention from sCD4-sensitive isolates, *J. Virol.* **66**, 235-243.

Moore, J.P., Thali, M., Jameson, B.A., Vignaux, F., Lewis, G.K., Poon, S.-W., Charles, M., Fung, M.S., Sun, B., Durda, P.J., Akerblom, L., Wahren, B., Ho, D.D., Sattentau, Q.J. and Sodroski, J. (1993a). Immunochemical analysis of the gp120 surface glycoprotein of human immunodeficiency virus type 1: probing the structure of the C4 and V4 domains and the interaction of the C4 domain with the V3 loop, *J. Virol.* **67**, 4785-4796.

Moore, J.P., Sattentau, Q.J., Yoshiyama, H., Thali, M., Charles, M., Sullivan, N., Poon, S.W., Fung, M.S., Traincard, F., Pinkus, M., Robey, G., Robinson, J.E., Ho, D.D. and Sodroski, J. (1993b). Probing the structure of the V2 domain of human immunodeficiency virus type-1 surface glycoprotein gp120 with a panel of eight monoclonal antibodies: human immune response to the V1 and V2 domains, *J. Virol.* **67**, 6136-6151.

Moore, J.P., McCutchan, F.E., Poon, S.-W., Mascola, J., Liu, J., Cao, Y. and Ho, D.D. (1994a). Exploration of antigenic variation in gp120 from clades A through F of human immunodeficiency virus type 1 by using monoclonal antibodies, *J. Virol.* **68**, 8350-8364.

Moore, J.P., Sattentau, Q.J., Wyatt, R. and Sodroski, J. (1994b). Probing the structure of the human immunodeficiency virus surface glycoprotein gp120 with a panel of monoclonal antibodies, *J. Virol.* **68**, 469-484.

Moore, J.P. and Ho, D.D. (1995). HIV-1 neutralization: the consequences of viral adaptation to growth on transformed T cells, *AIDS* **9 (suppl. A)**, S117-S136.

Moore, J.P. (1995). HIV vaccines. Back to primary school, *Nature* **376**, 115.

Moore, J.P., Trkola, A., Korber, B., Boots, L.J., Kessler, J.A.I., McCutchan, F.E., Mascola, J., Ho, D.D., Robinson, J. and Conley, A.J. (1995a). A human monoclonal antibody to a complex epitope in the V3 region of gp120 of human immunodeficiency virus type 1 has broad reactivity within and outside clade B, *J. Virol.* **69**, 122-130.

Moore, J.P., Cao, Y., Qing, L., Sattentau, Q.J., Pyati, J., Koduri, R., Robinson, J., Barbas, C.F., Burton, D.R. and Ho, D.D. (1995b). Primary isolates of human immunodeficiency virus type 1 are relatively resistant to neutralization by monoclonal antibodies to gp120, and their neutralization is not predicted by studies with monomeric gp120, *J. Virol.* **69 (1)**, 101-109.

Moore, J.P. and Sodroski, J. (1996). Antibody cross-competition analysis of the human immunodeficiency virus type-1 gp120 exterior envelope glycoprotein, *J. Virol.* **70**, 1863-1872.

Moore, J.P., Cao, Y., Leu, J., Qin, L., Korber, B. and Ho, D.D. (1996). Inter- and intraclade neutralization of human immunodeficiency virus type 1: genetic clades do not correspond to neutralization serotypes but partially correspond to gp120 antigenic serotypes, *J. Virol.* **70**, 427-444.

Moore, J.P., Trkola, A. and Dragic, T. (1997). Co-receptors for HIV-1 entry, *Curr. Opin. Immunol.* **9**, 551-562.

Moore, J.P. and Montefiori, D.C. (1997). Neutralization assays using the BZ167 strain of human immunodeficiency virus type 1, *J. Infect. Dis.* **175**, 764-774.

Moore, J.P. (1998). Clinical trials of HIVIG, *J. Infect. Dis.* **178**, 597-598.

Moore, J.P. and Binley, J. (1998). Envelope's letters boxed into shape, *Nature* **393**, 630-631.

Murphey-Corb, M., Martin, L.N., Davison-Fairburn, N., Montelaro, R.C., Miller, M., West, M., Ohkawa, S., Baskin, G.B., Zhang, J.Y., Putney, S.D., Allison, A.C. and Eppstein, D.A. (1989). A formalin-inactivated whole SIV vaccine confers protection in macaques, *Science* **246**, 1293-1297.

Muster, T., Steindl, F., Purtscher, M., Trkola, A., Klima, A., Himmler, G., Rüker, F. and Katinger, H. (1993). A conserved neutralizing epitope on gp41 of human immunodeficiency virus type 1, *J. Virol.* **67**, 6642-6647.

Muster, T., Guinea, R., Trkola, A., Purtscher, M., Klima, A., Steindl, F. and Palese, P. (1994). Cross-neutralizing activity against divergent human immunodeficiency virus type-1 isolates induced by the gp41 sequence ELDKWAS, *J. Virol.* **68**, 4031-4034.

Muster, T., Ferko, B., Klima, A., Purtscher, M., Trkola, A., Schulz, P., Grassauer, O., Engelhardt, G., Garcia-Sastre, A., Palese, P. and Katinger, H. (1995). Mucosal model of immunization against human immunodeficiency virus type 1 with a chimeric influenza virus, *J. Virol.* **69**, 6678-6686.

Nara, P.L. and Garrity, R. (1998). Deceptive imprinting: a cosmopolitan strategy for complicating vaccination, *Vaccine* **16**, 1780-1787.

Norrby, E. and Matthews, T. (1993). B-cell antigenic sites in the envelope proteins of primate lentiviruses and their role in vaccine development, *AIDS* **7 (suppl. 1)**, S127-S133.

Nyambi, P.N., Nkengasong, J., Lewi, P., Andries, K., Janssens, W., Fransen, K., Heyndrickx, L., Piot, P. and Van der Groen, G. (1996). Multivariate analysis of human immunodeficiency virus type-1 neutralization data, *J. Virol.* **70**, 6235-6243.

Nyambi, P.N., Lewi, P., Peeters, M., Janssens, W., Heyndrickx, L., Fransen, K., Andries, K., Vanden Haesevelde, M., Heeney, J., Piot, P. and Van der Groen, G. (1997). Study of the dynamics of neutralization escape mutants in a chimpanzee naturally infected with the simian immunodeficiency virus SIVcpz-ant, *J. Virol.* **71**, 2320-2330.

Ohgimoto, S., Shioda, T., Mori, K., Nakayama, E.E., Hu, H. and Nagai, Y. (1998). Location-specific unequal contribution of the N glycans in simian immunodeficiency virus gp120 to viral infectivity and removal of multiple glycans without disturbing infectivity, *J. Virol.* **72**, 8365-8370.

Overbaugh, J., Rudensey, L.M., Papenhausen, M.D., Benveniste, R.E. and Morton, W.R. (1991). Variation in simian immunodeficiency virus env is confined to V1 and V4 during progression to AIDS, *J. Virol.* **65**, 7025-7031.

Pal, R., Hoke, G.M. and Sarngadharan, M.G. (1989). Role of oligosaccharides in the processing and maturation of envelope glycoproteins of human immunodeficiency virus type 1, *Proc. Natl. Acad. Sci. USA* **86**, 3384-3388.

Pantaleo, G., Demarest, J.F., Vaccarezza, M., Graziosi, C., Bansal, G.P., Koenig, S. and Fauci, A.S. (1995). Effect of anti-V3 antibodies on cell-free and cell-to-cell human immunodeficiency virus transmission, *Eur. J. Immunol.* **25**, 226-231.

Parren, P.W.H.I., Ditzel, H.J., Gulizia, R.J., Binley, J.M., Barbas, C.F., Burton, D.R. and Mosier, D.E. (1995). Protection against HIV-1 infection in hu-PBL-SCID mice by passive immunization with a neutralizing human monoclonal antibody against the gp120 CD4-binding site, *AIDS* **9**, F1-F6.

Parren, P.W.H.I., Fisicaro, P., Labrijn, A.F., Binley, J.M., Yang, W.-P., Ditzel, H.J., Barbas, C.F. and Burton, D.R. (1996). *In-vitro* antigen challenge of human antibody libraries for vaccine evaluation: the human immunodeficiency virus type-1 envelope, *J. Virol.* **70 (12)**, 9046-9050.

Parren, P.W.H.I. and Burton, D.R. (1997). Antibodies against HIV-1 from phage display libraries: mapping of an immune response and progress towards anti-viral immunotherapy, *Chem. Immunol.* **65**, 18-56.

Parren, P.W.H.I., Sattentau, Q.J. and Burton, D.R. (1997a). HIV-1 antibody - debris or virion?, *Nature Med.* **3**, 367-368.

Parren, P.W.H.I., Sattentau, Q.J. and Burton, D.R. (1997b). Importance of anti-HIV-1 antibodies, *Nature Med.* **3**, 591.

Parren, P.W.H.I., Gauduin, M.-C., Koup, R.A., Poignard, P., Sattentau, Q.J., Fisicaro, P. and Burton, D.R. (1997c). Relevance of the antibody response against human immunodeficiency virus type-1 envelope to vaccine design, *Immunol. Lett.* **58**, 125-132.

Parren, P.W.H.I., Naniche, D., Mondor, I., Ditzel, H.J., Klasse, P.J., Burton, D.R. and Sattentau, Q.J. (1998). Neutralization of HIV-1 by antibody to gp120 is determined primarily by occupancy of sites on the virion irrespective of epitope specificity, *J. Virol.* **72**, 3512-3519.

Pelchen-Matthews, A., Clapham, P. and Marsh, M. (1995). Role of CD4 endocytosis in human immunodeficiency virus infection, *J. Virol.* **69**, 8164-8168.

Pelletier, E. and Wain-Hobson, S. (1996). AIDS is not caused by the extreme genetic variability of HIV, *J. NIH Res.* **8**, 45-49.

Perelson, A.S., Neumann, A.U., Markowitz, M., Leonard, J.M. and Ho, D.D. (1996). HIV-1 dynamics *in vivo*: virion clearance rate, infected cell life-span, and viral generation time, *Science* **271**, 1582-1586.

Pilgrim, A.K., Pantaleo, G., Cohen, O.J., Fink, L.M., Zhou, J.Y., Zhou, J.T., Bolognesi, D.P., Fauci, A.S. and Montefiori, D.C. (1997). Neutralizing antibody responses to human immunodeficiency virus type 1 in primary infection and long-term non-progressive infection, *J. Infect. Dis.* **176**, 924-932.

Pinter, A., Honnen, W.J., Kayman, S.C., Trochev, O. and Wu, Z. (1998). Potent neutralization of primary HIV-1 isolates by antibodies directed against epitopes present in the V1/V2 domain of HIV-1 gp120, *Vaccine* **16**, 1803-1811.

Planz, O., Ehl, S., Furrer, E., Horvath, E., Brundler, M.-A., Hengartner, H. and Zinkernagel, R.M. (1997). A critical role for neutralizing antibody-producing B cell, $CD4^+$ T cells, and interferons in persistent and acute infections of mice with lymphocyte choriomeningitis virus: implications for adoptive immunotherapy of virus carriers, *Proc. Natl. Acad. Sci. USA* **94**, 6874-6879.

Platt, E.J., Madani, N., Kozak, S.L. and Kabat, D. (1997). Infectious properties of human immunodeficiency virus type-1 mutants with distinct affinities for the CD4 receptor, *J. Virol.* **71**, 883-890.

Poignard, P., Klasse, P.J. and Sattentau, Q.J. (1996a). Antibody neutralization of HIV-1, *Immunol. Today* **17**, 239-246.

Poignard, P., Fouts, T., Naniche, D., Moore, J.P. and Sattentau, Q.J. (1996b). Neutralizing antibodies to human immunodeficiency virus type-1 gp120 induce envelope glycoprotein subunit dissociation, *J. Exp. Med.* **183**, 473-484.

Poignard, P., Sabbe, S., Picchio, G.R., Wang, M., Gulizia, R.J., Katinger, H., Parren, P.W.H.I., Mosier, D.E. and Burton, D.R. (1999). Neutralizing antibodies have limited effects on the control of established HIV-1 infection (unpublished).

Potts, B.J., Field, K.G., Wu, Y., Posner, M., Cavacini, L. and White-Scharf, M. (1993). Synergistic inhibition of HIV-1 by CD4-binding domain reagents and V3-directed monoclonal antibodies, *Virology* **197**, 415-419.

Pratt, R.D., Shapiro, J.F., McKinney, N., Kwok, S. and Spector, S.A. (1995). Virologic characterization of primary human immunodeficiency virus type-1 infection in a health care worker following needlestick injury, *J. Infect. Dis.* **172**, 851-854.

Prince, A.M., Horowitz, B., Baker, L., Shulman, R.W., Ralph, H., Valinsky, J., Cundell, A., Brotman, B., Boehle, W., Rey, F., Piet, M., Reesink, H.W., Lelie, P.N., Tersmette, M., Miedema, F., Barbosa, L., Nemo, G., Nastala, C.L., Allan, J.S., Lee, D.R. and Eichberg, J.W. (1988). Failure of a human immunodeficiency virus (HIV) immune globulin to protect chimpanzees against experi-mental challenge with HIV, *Proc. Natl. Acad. Sci. USA* **85**, 6944-6948.

Prince, A.M., Reesink, H.W., Pascual, D., Horowitz, B., Hewlett, I., Murthy, K.K., Cobb, K.E. and Eichberg, J.W. (1991). Prevention of HIV infection by passive immunization with HIV immuno-globulin, *AIDS Res. Human Retrovir.* **7**, 971-973.

Pu, R., Okada, S., Little, E.R., Xu, B., Stoffs, W.V. and Yamamoto, J.K. (1995). Protection of neonatal kittens against feline immunodeficiency virus infection with passive maternal antiviral antibodies, *AIDS* **9**, 235-242.

Putkonen, P., Thorstensson, R., Ghavamzadeh, L., Albert, J., Hild, K., Biberfeld, G. and Norrby, E. (1991). Prevention of HIV-2 and SIVsm infection by passive immunization in cynomolgus monkeys, *Nature* **352**, 436-438.

Reimann, K.A., Li, J.T., Veazey, R., Halloran, M., Park, I.-W., Karlsson, G.B., Sodroski, J. and Letvin, N.L. (1996a). A chimeric simian/human immunodeficiency virus expressing a primary patient human immunodeficiency virus type-1 isolate env causes an AIDS-like disease after *in-vivo* passage in rhesus monkeys, *J. Virol.* **70**, 6922-6928.

Reimann, K.A., Li, J.T., Voss, G., Lekutis, C., Tenner-Racz, K., Racz, P., Lin, W., Montefiori, D.C., Lee-Parritz, D.E., Lu, Y.C., Collman, R.G., Sodroski, J. and Letvin, N.L. (1996b). An env gene derived from a primary HIV-1 isolate confers high *in-vivo* replicative capacity to a chimeric simian/human immunodeficiency virus in rhesus monkeys, *J. Virol.* **70**, 3198-3206.

Reitter, J.N. and Desrosiers, R.C. (1998). Identification of replication-competent strains of simian immunodeficiency virus lacking multiple attachment sites for N-linked carbohydrates in variable regions 1 and 2 of the surface envelope protein, *J. Virol.* **72**, 5399-5407.

Reitter, J.N., Means, R.E. and Desrosiers, R.C. (1998). A role for carbohydrates in immune evasion in AIDS, *Nature Med.* **4**, 679-684.

Rizzuto, C.D. and Sodroski, J.G. (1997). Contribution of virion ICAM-1 to human immunodeficiency virus infectivity and sensitivity to neutralization, *J. Virol.* **71**, 4847-4851.

Rizzuto, C.D., Wyatt, R., Hernandez-Ramos, N., Sun, Y., Kwong, P.D., Hendrickson, W.A. and Sodroski, J. (1998). A conserved HIV gp120 glycoprotein structure involved in chemokine receptor binding, *Science* **280**, 1949-1953.

Roderiquez, G., Oravecz, T., Yanagishita, M., Bou-Habib, D.C., Mostowski, H. and Norcross, M.A. (1995). Mediation of human immunodeficiency virus type-1 binding by interaction of cell surface heparan sulfate proteoglycans with the V3 region of envelope gp120-gp41, *J. Virol.* **69**, 2233-2239.

Rossio, J.L., Esser, M.T., Suryanarayana, K., Schneider, D.K., Bess Jr., J.W., Vasquez, G.M., Wiltrout, T.A., Chertova, E., Grimes, M.K., Sattentau, Q.J., Arthur, L.O., Henderson, L.E. and Lifson, J.D. (1998). Inactivation of human immunodeficiency virus type-1 infectivity with preservation of conformational and functional integrity of viron surface proteins, *J. Virol.* **72**, 7992-8001.

Rudensey, L.M., Kimata, J.T., Long, E.M., Chackerian, B. and Overbaugh, J. (1998). Changes in the extracellular envelope glycoprotein of variants that evolve during the course of simian immunodeficiency virus SIVMne infection affect neutralizing antibody recognition, syncytium formation, and macrophage tropism but not replication, cytopathicity, or CCR-5 coreceptor recognition, *J. Virol.* **72**, 209-217.

Safrit, J.T., Fung, M.S.C., Andrews, C.A., Braun, D.G., Sun, W.N.C., Chang, T.W. and Koup, R.A. (1993). Hu-PBL-SCID mice can be protected from HIV-1 infection by passive transfer of monoclonal antibody to the principal neutralizing determinant of envelope gp120, *AIDS* **7**, 15-21.

Sattentau, Q.J. and Moore, J.P. (1991). Conformational changes induced in the human immunodeficiency virus envelope glycoprotein by soluble CD4 binding, *J. Exp. Med.* **174**, 407-415.

Sattentau, Q.J., Moore, J.P., Vignaux, F., Traincard, F. and Poignard, P. (1993). Conformational changes induced in the envelope glycoproteins of the human and simian immunodeficiency viruses by soluble receptor binding, *J. Virol.* **67**, 7383-7393.

Sattentau, Q.J. (1995). Conservation of HIV-1 gp120 neutralization epitopes after formalin inactivation, *AIDS* **9**, 1383-1385.

Sattentau, Q.J., Zolla-Pazner, S. and Poignard, P. (1995). Epitope exposure on functional, oligomeric HIV-1 gp41 molecules, *Virology* **206**, 713-717.

Sattentau, Q.J. and Moore, J.P. (1995). Human immunodeficiency virus type-1 neutralization is determined by epitope exposure on the gp120 oligomer, *J. Exp. Med.* **182**, 185-196.

Sattentau, Q.J. (1996). Neutralization of HIV-1 by antibody, *Curr. Opin. Immunol.* **8**, 540-545.

Sattentau, Q.J. (1998). HIV gp120: double lock strategy foils host defences, *Structure* **6**, 945-949.

Schønning, K., Jansson, B., Olofsson, S. and Hansen, J.-E.S. (1996a). Rapid selection for an N-linked oligosaccharide by monoclonal antibodies directed against the V3 loop of human immunodeficiency virus type 1, *J. Gen. Virol.* **77**, 753-758.

Schønning, K., Jansson, B., Olofsson, S., Nielsen, J.O. and Hansen, J.-E.S. (1996b). Resistance to V3-directed neutralization caused by an N-linked oligosaccharide depends on the quaternary structure of HIV-1 envelope oligomer, *Virology* **218**, 134-140.

Schønning, K., Bolmstedt, A., Novotny, J., Søgaard Lund, O., Olofsson, S. and Hansen, J.-E.S. (1998). Induction of antibodies against epitopes inaccessible on the HIV type-1 envelope oligomer by immunization with recombinant monomeric glycoprotein 120, *AIDS Res. Human Retrovir.* **16**, 1451-1456.

Schultz, A.M. and Hu, S.L. (1993). Primate models for HIV vaccines, *AIDS* **7**, S161-S170.

Schutten, M., Andeweg, A.C., Bosch, M.L. and Osterhaus, A.D.M.E. (1995). Enhancement of infectivity of a non-syncytium-inducing HIV-1 by sCD4 and by human antibodies that neutralize syncytium-inducing HIV-1, *Scand. J. Immunol.* **41**, 18-22.

Schutten, M., Tenner-Racz, K., Racz, P., Van Bekkum, D.W. and Osterhaus, A.D.M.E. (1996). Human antibodies that neutralize primary human immunodeficiency virus type 1 *in vitro* do not provide protection in an *in-vivo* model, *J. Gen. Virol.* **77**, 1667-1675.

Schutten, M., Andeweg, A.C., Rimmelzwaan, G.F. and Osterhaus, A.D.M.E. (1997). Modulation of primary human immunodeficiency virus type-1 envelope glycoprotein-mediated entry by human antibodies, *J. Gen Virol.* **78**, 999-1006.

Scott Jr., C.F., Silver, S., Profy, A.T., Putney, S.D., Langlois, A., Weinhold, K. and Robinson, J.E. (1990). Human monoclonal antibody that recognizes the V3 region of human immunodeficiency virus gp120 and neutralizes the human T-lymphotropic virus type IIIMN strain, *Proc. Natl. Acad. Sci. USA* **87**, 8597-8601.

Shibata, R., Igarashi, T., Haigwood, N., Buckler-White, A., Ogert, R., Ross, W., Willey, R., Cho, M.W. and Martin, M.A. (1999). Neutralizing antibodies directed against the HIV-1 envelope glycoprotein can completely block HIV-1/SIV chimeric virus infections of macaque monkeys, *Nature Med.* **5**, 205-210.

Shotton, C., Arnold, C., Sattentau, Q.J., Sodroski, J. and McKeating, J.A. (1995). Identification and characterization of monoclonal antibodies specific for polymorphic antigenic determinants within the V2 region of the human immunodeficiency virus type-1 envelope glycoprotein, *J. Virol.* **69**, 222-230.

Skinner, M.A., Langlois, A.J., McDanal, C.B., McDougal, J.S., Bolognesi, D.P. and Matthews, T.J. (1988). Neutralising antibodies to an immunodominant envelope sequence do not prevent gp120 binding to CD4, *J. Virol.* **62**, 4195-4200.

Smith, T.J., Chase, E.S., Schmidt, T.J., Olson, N.H. and Baker, T.S. (1996). Neutralising antibody to human rhinovirus 14 penetrates the receptor-binding canyon, *Nature* **383**, 350-354.

Spear, G.T., Takefman, D.M., Sharpe, S., Ghassemi, M. and Zolla-Pazner, S. (1994). Antibodies to the HIV-1 V3 loop in serum from infected persons contribute a major proportion of immune effector functions including complement activation, antibody binding and neutralization, *Virology* **204**, 609-615.

Spenlehauser, C., Saragosti, S., Fleury, H.J., Kirn, A., Aubertin, A.-M. and Moog, C. (1998). Study of the V3 loop as a target epitope for antibodies involved in the neutralization of primary isolates *versus* T-cell-line-adapted strains of human immunodeficiency virus type 1, *J. Virol.* **72**, 9855-9864.

Stamatatos, L. and Cheng-Mayer, C. (1995). Structural modulations of the envelope gp120 glycoprotein of human immunodeficiency virus type 1 upon oligomerization and differential V3-loop epitope exposure of isolates displaying distinct tropism upon virion-soluble receptor binding, *J. Virol.* **69**, 6191-6198.

Stamatatos, L., Zolla-Pazner, S., Gorny, M.K. and Cheng-Mayer, C. (1997). Binding of antibodies to virion-associated gp120 molecules of primary-like human immunodeficiency virus type-1 (HIV-1) isolates: effect on HIV-1 infection of macrophages and peripheral blood mononuclear cells, *Virology* **229**, 360-369.

Stamatos, N.M., Mascola, J.R., Kalyanaraman, V.S., Louder, M.K., Frampton, L.M., Birx, D.L. and VanCott, T.C. (1998). Neutralizing antibodies from the sera of human immunodeficiency virus type-1-infected individuals bind to monomeric gp120 and oligomeric gp140, *J. Virol.* **72**, 9656-9667.

Sullivan, N., Sun, Y., Li, J., Hofmann, W. and Sodroski, J. (1995). Replicative function and neutralization sensitivity of envelope glycoproteins from primary and T-cell line-passaged human immunodeficiency virus type-1 isolates, *J. Virol.* **69**, 4413-4422.

Sullivan, N., Sun, Y., Binley, J., Lee, J., Barbas III, C.F., Parren, P.W.H.I., Burton, D.R. and Sodroski, J. (1998). Determinants of human immunodeficiency virus type-1 envelope glycoprotein activation by soluble CD4 and monoclonal antibodies, *J. Virol.* **72**, 6332-6338.

Sun, N.-C., Ho, D.D., Sun, C.R.Y., Liou, R.-S., Gordon, W., Fung, M.S.C., Li, X.-L., Ting, R.C., Lee, T.H., Chang, N.T. and Chang, T.-W. (1989). Generation and characterisation of monoclonal antibodies to the putative CD4-binding domain of human immunodeficiency virus type-1 gp120, *J. Virol.* **63**, 3579-3585.

Thali, M., Olshevshy, U., Furman, C., Gabuzda, D., Posner, M. and Sodroski, J. (1991). Characterization of a discontinuous human immunodeficiency virus type-1 gp120 epitope recognised by a broadly reactive neutralizing human monoclonal antibody, *J. Virol.* **65**, 6188-6193.

Thali, M., Moore, J.P., Furman, C., Charles, M., Ho, D.D., Robinson, J. and Sodroski, J. (1993). Characterization of conserved human immunodeficiency virus type-1 gp120 neutralization epitopes exposed upon gp120-CD4 binding, *J. Virol.* **67**, 3978-3988.

Tilley, S.A., Honnen, W.J., Racho, M.E., Hilgartner, M. and Pinter, A. (1991). A human monoclonal antibody against the CD4-binding site of HIV-1 gp120 exhibits potent, broadly neutralizing activity, *Res. Virol.* **142**, 247-259.

Tilley, S.A., Honnen, W.J., Racho, M.E., Chou, T.C. and Pinter, A. (1992). Synergistic neutralization of HIV-1 by human monoclonal antibodies against the V3 loop and the CD4-binding site of gp120, *AIDS Res. Human Retrovir.* **8**, 461-467.

Trkola, A., Dragic, T., Arthos, J., Binley, J.M., Olson, W.C., Allaway, G.P., Cheng-Mayer, C., Robinson, J., Maddon, P.J. and Moore, J.P. (1996a). CD4-dependent, antibody-sensitive interactions between HIV-1 and its co-receptor CCR-5, *Nature* **384**, 184-187.

Trkola, A., Purtscher, M., Muster, T., Ballaun, C., Buchacher, A., Sullivan, N., Srinivasan, K., Sodroski, J., Moore, J.P. and Katinger, H. (1996b). Human monoclonal antibody 2G12 defines a distinctive neutralization epitope on the gp120 glycoprotein of human immunodeficiency virus type 1, *J. Virol.* **70**, 1100-1108.

Trkola, A., Ketas, T., Kewalramani, V.N., Endorf, F., Binley, J.M., Katinger, H., Robinson, J., Littman, D.R. and Moore, J.P. (1998). Neutralization sensitivity of human immunodeficiency virus type-1 primary isolates to antibodies and CD4-based reagents is independent of coreceptor usage, *J. Virol.* **72**, 1876-1885.

Ugolini, S., Mondor, I., Parren, P.W.H.I., Burton, D.R., Tilley, S.A., Klasse, P.J. and Sattentau, Q.J. (1997). Inhibition of attachment to $CD4^+$ target cells is a major mechanism of T-cell line-adapted HIV-1 neutralization, *J. Exp. Med.* **186**, 1287-1298.

Ugolini, S., Mondor, I. and Sattentau, Q.J. (1999). HIV-1 attachment: another look, *Trends Microbiol.* (in press).

Valenzuela, A., Blanco, J., Krust, B., Franco, R. and Hovanessian, A.G. (1997). Neutralizing antibodies against the V3 loop of human immunodeficiency virus type-1 gp120 block the CD4-dependent and -independent binding of virus to cells, *J. Virol.* **71**, 8289-8298.

VanCott, T.C., Polonis, V.R., Loomis, L.D., Michael, N.L., Nara, P.L. and Birx, D.L. (1995a). Differential role of V3-specific antibodies in neutralization assays involving primary and laboratory-adapted isolates of HIV type 1, *AIDS Res. Human Retrovir.* **11**, 1379-1391.

VanCott, T.C., Bethke, F.R., Burke, D.S., Refield, R.R. and Birx, D.L. (1995b). Lack of induction of antibodies specific for conserved, discontinuous epitopes of HIV-1 envelope glycoprotein by candidate AIDS vaccines, *J. Immunol.* **155**, 4100-4110.

VanCott, T.C., Mascola, J.R., Kaminski, R.W., Kalyanaraman, V., Hallberg, P.L., Burnett, P.R., Ulrich, J.T., Rechtman, D.J. and Birx, D.L. (1997). Antibodies with specificity to native gp120 and neutralization activity against primary human immunodeficiency virus type-1 isolates elicited by immunization with oligomeric gp160, *J. Virol.* **71**, 4319-4330.

Vijh-Warrier, S., Pinter, A., Honnen, W.J. and Tilley, S.A. (1996). Synergistic neutralization of human immunodeficiency virus type 1 by a chimpanzee monoclonal antibody against the V2 domain of gp120 in combination with monoclonal antibodies against the V3 loop and the CD4-binding site, *J. Virol.* **70**, 4466-4473.

Vittecoq, D., Chevret, S., Morand-Joubert, L., Heshmati, F., Audat, F., Bary, M., Dusautoir, T., Bismuth, A., Viard, J.P., Barré-Sinoussi, F., Bach, J.F. and Lefrère, J.J. (1995). Passive immunotherapy in AIDS: a double-blind randomized study based on transfusions of plasma rich in anti-human immunodeficiency virus-1 antibodies *versus* transfusions of seronegative plasma, *Proc. Natl. Acad. Sci. USA* **92**, 1195-1199.

Vogel, T., Kurth, R. and Norley, S. (1994). The majority of neutralizing abs in HIV-1-infected patients recognize linear V3-loop sequences. Studies using HIV-1MN multiple antigenic peptides, *J. Immunol.* **153**, 1895-1904.

Weber, J., Fenyö, E.-M., Beddows, S., Kaleebu, P., Bjorndal, A. and the WHO Network for HIV Isolation and Characterization (1996). Neutralization serotypes of human immunodeficiency virus type-1 field isolates are not predicted by genetic subtype, *J. Virol.* **70**, 7827-7832.

Wei, X., Ghosh, S.K., Taylor, M.E., Johnson, V.A., Emini, E.A., Deutsch, P., Lifson, J.D., Bonhoeffer, S., Nowak, M.A., Hahn, B.H., Saag, M.S. and Shaw, G.M. (1995). Viral dynamics in human immunodeficiency virus type-1 infection, *Nature* **373**, 117-122.

Weissenhorn, W., Dessen, A., Harrison, S.C., Skehel, J.J. and Wiley, D.C. (1997). Atomic structure of the ectodomain from HIV-1 gp41, *Nature* **387**, 426-430.

White-Scharf, M.E., Potts, B.J., Smith, L.M., Sokolowski, K.A., Rusche, J.R. and Silver, S. (1993). Broadly neutralizing monoclonal antibodies to the V3 region of HIV-1 can be elicited by peptide immunization, *Virology* **192**, 197-206.

Willey, R.L., Shibata, R., Freed, E.O., Cho, M.W. and Martin, M.A. (1996). Differential glycosylation, virion incorporation, and sensitivity to neutralizing antibodies of human immunodeficiency virus type-1 envelope produced from infected primary T-lymphocyte and macrophage cultures, *J. Virol.* **70**, 6431-6436.

Wu, H., Kwong, P.D. and Hendrickson, W.A. (1997). Dimeric association and segmental variability in the structure of human CD4, *Nature* **387**, 527-530.

Wu, L., Gerard, N., Wyatt, R., Choe, H., Parolin, C., Ruffing, N., Borsetti, A., Cordoso, A., Desjardin, E., Newman, W., Gerard, C. and Sodroski, J. (1996). CD4-induced interaction of primary HIV-1 gp120 glycoproteins with the chemokine receptor CCR-5, *Nature* **384**, 179-183.

Wyatt, R., Moore, J., Accola, M., Desjardin, E., Robinson, J. and Sodroski, J. (1995). Involvement of the V1/V2 variable loop structure in the exposure of human immunodeficiency virus type-1 gp120 epitopes induced by receptor binding, *J. Virol.* **69**, 5723-5733.

Wyatt, R., Kwong, P.D., Desjardin, E., Sweet, R.W., Robinson, J., Hendrickson, W.A. and Sodroski, J.G. (1998). The antigenic structure of the HIV gp120 envelope glycoprotein, *Nature* **393**, 705-711.

Wyatt, R. and Sodroski, J. (1998). The HIV-1 envelope glycoproteins: fusogens, antigens and immunogens, *Science* **280**, 1884-1888.

Xu, J.-Y., Gorny, M.K., Palker, T., Karwowska, S. and Zolla-Pazner, S. (1991). Epitope mapping of two immunodominant domains of gp41, the transmembrane protein of human immunodeficiency virus type 1, using ten human monoclonal antibodies, *J. Virol.* **65**, 4832-4838.

Zhou, J.Y. and Montefiori, D.C. (1997). Antibody-mediated neutralization of primary isolates of human immunodeficiency virus type 1 in peripheral blood mononuclear cells is not affected by the initial activation state of the cells, *J. Virol.* **71**, 2512-2517.

Zolla-Pazner, S. and Sharpe, S. (1995). A resting cell assay for improved detection of antibody-mediated neutralization of HIV type-1 primary isolates, *AIDS Res. Human Retrovir.* **11**, 1449-1457.

CHAPTER 8

SUPPRESSION OF PRIMATE IMMUNODEFICIENCY LENTI-RETROVIRUSES BY CD8$^+$ T-CELL-DERIVED SOLUBLE FACTORS

PAOLO LUSSO

Unit of Human Virology, DIBIT, San Raffaele Scientific Institute, Milan, Italy

1. Foreword

Three lentiretroviruses of primates have been hitherto recognised, namely, human immunodeficiency virus type 1 (HIV-1) and type 2 (HIV-2), and simian immunodeficiency virus (SIV), all of which cause immunodeficiency and establish a chronic active infection persisting for the entire lifetime of the host [1]. Since the early days of AIDS research, studies by *in-situ* hybridisation and immunohistochemistry have demonstrated that virological latency never occurs during the natural history of HIV infection [2-4]; this observation was later confirmed using more sophisticated molecular techniques [5,6]. The introduction of sensitive methods for the quantitation of viral RNA in plasma has permitted to document a continuous viral replication throughout the course of the infection, even during the prolonged phase of clinical latency that precedes the development of the overt immunodeficiency [7-10]. Indeed, the level of viral RNA in plasma or serum represents one of the best predictive markers of disease progression. These observations imply that the host's immune system is never able to fully control, nor to clear the infection, in spite of the diverse antiviral mechanisms operating *in vivo*, including both humoral and cell-mediated antigen-specific immune responses. Such mechanisms are nevertheless believed to help restrain the replication and spread of the virus, thereby prolonging the phase of clinical latency.

The dissection of the complex interplay between virus and host, particularly during the asymptomatic stage of infection by HIV or SIV, has always attracted considerable attention. Understanding the mechanisms of immune containment of primate lentiretroviruses may indeed permit to devise novel therapeutic approaches. However, no clearcut correlation has been established between the classic markers of protective immunity, such as neutralising antibodies and cytotoxic T lymphocytes (CTL), and the natural course of HIV infection. It is in this context that the interest in an unconventional type of antiviral immune response, apparently unique to infection by immunodeficiency lentiretroviruses, has originated and thrived [11-15]. Since the original phenomenological observation by Walker and colleagues in 1986, who described the antiviral activity of CD8$^+$ T cells derived

H. Schuitemaker and F. Miedema (ed.), Aids Pathogenesis, 133–153.

from HIV-infected individuals [11], a great deal of information on the biology and clinical relevance of soluble factors with HIV- or SIV-inhibitory activity has been accumulated. Most intriguing, unlike the traditional correlates of protective immunity, the production of virus-suppressive factors by CD8$^+$ T cells was found to correlate with the clinical stage of the infection [16].

Since the discovery of HIV, a large number of cytokines with *in-vitro* antiviral activity has been described [17]. However, none of them, by itself, seems to display biological features compatible with those predicted from experiments with crude CD8$^+$ T-lymphocyte culture supernatants. Thus, the identity of the CD8$^+$ T-cell antiviral factor(s) has remained elusive for almost a decade. Great ferment in the field was stirred at the end of 1995, when three members of the chemokine (chemotactic cytokine) superfamily, namely, RANTES, MIP-1α and MIP-1β, were identified as potent natural HIV-suppressive factors produced by CD8$^+$ T cells [18]. These proteins were subsequently found to act by directly engaging and blocking critical viral coreceptors expressed on the cellular surface membrane [19]. Although chemokines undoubtedly play an important role in the CD8$^+$ T-cell-mediated anti-viral activity, it is increasingly evident that they are not the only component of this complex phenomenon, which involves an intricate network of positive and negative regulatory mechanisms acting at different stages of the retroviral life-cycle. Such complexity is also reflected in the variety of *in-vitro* assays that have been developed for testing the HIV-suppressive activity of different cell types and cell-free culture fluids, raising important methodological issues related to the transferability of results from one system to the other. This chapter summarises the body of knowledge accumulated during the past decade on soluble factors with HIV- or SIV-suppressive activity, discussing the major technical and biological questions that still remain unsolved in this rapidly evolving field.

2. Suppression of Primate Immunodeficiency Lentiretroviruses by CD8$^+$ T Cells: The Early Descriptions

In 1986, Walker and colleagues at the University of California, San Francisco, noticed that their success in isolating HIV-1 from cultured peripheral blood mononuclear cells of infected individuals was dramatically enhanced by removal of CD8$^+$ T cells before *in-vitro* activation. When added back to the cultures, CD8$^+$ T cells were found to suppress endogenous HIV-1 replication in autologous CD4$^+$ T cells in a number-dependent fashion [11]. Both autologous and heterologous CD8$^+$ T cells exerted an inhibitory effect, although suppression was apparently more pronounced in the autologous system. The authors suggested that the anti-HIV activity of CD8$^+$ T cells could be relevant to the control of HIV-1 replication *in vivo* and thus could be exploited for devising new therapeutic approaches to AIDS [11]. This study opened an entirely new area of investigation: although CD8$^+$ T cells were traditionally viewed as antiviral effector cells, by virtue of their CTL effect on virally infected cells, the anti-HIV activity described by Walker and colleagues was unconventional, because it was not major histocompatibility complex (MHC)-restricted and was not associated with target-cell lysis.

In the subsequent years, new insights into the nature of the non-cytotoxic CD8⁺ T-cell-mediated anti-HIV activity came from several laboratories. The original results of the San Francisco group were reproduced by others [12,20-22] and analogous observations were made in non-human primate models [14,23-28] and in a feline model [15], suggesting the existence of common CD8⁺ T-cell-mediated antiviral mechanisms against all immunodeficiency lentiretroviruses. By the use of two-chamber culture systems or by addition of cell-free culture supernatants, several investigators confirmed the initial hypothesis that the HIV-suppressive activity was to be ascribed, at least in part, to soluble factor(s) [12,13,26,29], although one group later reported no evidence of a soluble viral inhibitor [30]. A virus-suppressive activity was detected by using not only naturally infected CD4⁺ T cells, but also primary CD4⁺ T cells acutely infected *in vitro* with exogenous HIV-1 strains [29,31]. Instead, continuous CD4⁺ T-cell lines, either acutely or chronically infected with HIV, appeared to be insensitive to the suppressive activity of CD8⁺ T cells, although this matter has never been rigorously investigated: whether this phenomenon is related to the peculiar biological properties of the viral strains adapted to infection of cell lines or, alternatively, to an inherent lack of sensitivity of such cells to the anti-viral factors (*e.g.*, failure to express specific membrane receptors) remains to be elucidated. The concept of a "soluble antiviral factor" raised immediate interest, because its identification and cloning would have had major implications for the prevention and treatment of HIV infection. The working hypothesis originally formulated by Levy and co-workers was that a single soluble factor (subsequently designated CD8⁺ T-cell Antiviral Factor, or "CAF") is responsible for the antiviral activity of CD8⁺ T cells [11]. This theory, which has been detailed in several review articles [32-35], has represented the reference paradigm in the field during the decade between 1986 and 1995.

3. The "CAF" Decalogue

The "CAF" paradigm is founded on the "single-factor" hypothesis, according to which CD8⁺ T cells derived from HIV-infected patients (or SIV-infected monkeys) produce a single dominant antiviral factor. The following is a summary of the major postulates of the "CAF" theory: 1) suppression of HIV by CD8⁺ T cells is due to one soluble factor ("CAF") which is exclusively produced by CD8⁺ T cells and not by other blood cells, such as B, natural killer (NK) or mononuclear phagocytic cells [32-36]; 2) "CAF" is elaborated by a specific subset of activated CD8⁺ T cells characterized by the expression of the HLA-DR and CD28 antigens [37]; 3) the ability to produce "CAF" is a selective property of CD8⁺ T cells derived from HIV-infected patients [31]; 4) "CAF" is responsible for all the antiviral activities detectable in different *in-vitro* systems (*i.e.*, endogenous-infection, acute-infection, long terminal repeat [LTR]-activation tests, etc.) [32-35]; 5) "CAF" acts in a non-MHC-restricted fashion [11,12,31]; 6) the suppressive effect of "CAF" is non-cytolytic and does not affect cellular activation or proliferation [11,12,29,31,37]; 7) "CAF"-mediated blockade of viral replication is independent of the biological subtype of HIV or SIV [31-35,38]; 8) the mechanism of "CAF" acts downstream of the viral entry step [36,39] and more specifically at the level of HIV-LTR-directed viral gene transcription [39]; 9) "CAF" is not identical to any known cytokine or chemokine [35,40]; 10) "CAF" is filterable, less than 30,000 dalton in molecular weight, and both trypsin- and acid-resistant [34,35]. It is important to emphasise that, in several instances, no primary experimental data have been reported in support of these postulates, although they have been repeatedly asserted in review articles [32-35]. Indeed, as discussed in detail below, a critical analysis of the experimental evidence hitherto accumulated reveals important inconsistencies with some of the above principles.

4. Methodological Issues: Multiple Tests for a Single Factor?

A variety of test systems has been developed for the measurement of $CD8^+$ T-cell-mediated viral suppression *in vitro*. This multiplication of assays, together with the subtle technical modifications introduced in different laboratories and the inherent difficulties in standardising biological systems based on short-term culture of primary cells, has contributed to generate confusion in the field, as data obtained in one test system were often inferred to be valid also in other systems, in the absence of direct scientific evidence. In some instances, such extrapolations have resulted in simplistic interpretations of a complex phenomenon.

The major discrepancies in the methodologies used for testing $CD8^+$ T-cell-mediated viral suppression are summarised in Table 1.

TABLE 1. Different Methodological Approaches to the Study of $CD8^+$ T-Cell-Mediated Lentiretroviral Suppression

1) Effector ($CD8^+$)-target ($CD4^+$) Cell Interaction:
- Direct co-cultivation (at variable ratio)
- Two-chamber system (no cell-to-cell contact)
- Treatment of target cells with cell-free $CD8^+$ culture supernatant

2) Target-cell Infection Procedure:
- Endogenous (*ex-vivo* activated)
- Acute exogenous (uninfected donor)
- Cell-to-cell transmission

3) *In-Vitro* Manipulation of Effector Cells:
- No *ex vivo* activation
- Activation by lectins, anti-T-cell receptor antibodies, cytokines
- *In-vitro* immortalization with human oncoretroviruses or simian herpesviruses

4) Assay Read-out
- Productive infection (reverse transcriptase, p24 antigen release, viral RNA expression)
- Viral long-terminal-repeat [LTR]-directed gene expression

5) Retroviral Phenotype
- MT2-negative (CXCR4-negative)
- MT2-tropic (CXCR4-positive)

One of the most important methodological issues is related to the mode of interaction between effector ($CD8^+$) and target ($CD4^+$) cells. Since the early descriptions of the phenomenon of $CD8^+$ T-cell-mediated viral suppression, emphasis has been posed on soluble factors [11-13]; this notwithstanding, most of the *in-vitro* data have in fact been collected using co-cultivation systems, without the barrier of a semi-permeable membrane separating effector and target cells. A large body of evidence indicates that the HIV-suppressive effect of $CD8^+$ T cells placed in direct contact with infected targets is markedly stronger than that exerted by soluble factors [12,13,21,29,35], suggesting that a wider range of suppressive mechanisms is operational in co-culture experiments. Furthermore, even the two-chamber system cannot be automatically equated to the treatment with cell-free culture supernatants, as the cross-talk between two live cell cultures involves subtle modulatory mechanisms which cannot be reproduced using a static conditioned medium derived from a single time-point. The modality of target cell infection may also greatly affect the outcome of the suppression test, as there are inherent difficulties in standardizing the conditions of the endogenous assay. First, because the viral strains harboured by *in-vivo*-infected $CD4^+$ T cells may have a different phenotype. Second, because the frequency of *in-vivo*-infected $CD4^+$ T cells can be dramatically different among patients. In contrast, the conditions of the acute infection assay can be better standardised, at least with regard to the viral phenotype and the size of the viral inoculum [31]. However, reproducibility may be problematic even in this case since $CD4^+$ T cells from different individuals display a variable susceptibility to HIV infection under the influence of both constitutive and acquired factors, including chemokine and chemokine-receptor genotypes (see below). Another important issue is the read-out employed in different tests being invariably delayed from the onset of the cultures, implying that several cycles of intra-cultural viral replication and thereby of cell-to-cell spread occur before the collection of the results. Thus, irrespective of the step targeted by the soluble factor(s) in the viral life-cycle, all types of assay read-outs will be affected alike. Standardisation problems are also encountered with effector $CD8^+$ T cells, as different modes of *in-vitro* manipulation can influence the outcome of the tests. In most studies, $CD8^+$ T cells were activated *ex vivo* with polyclonal T-cell activators, usually with the addition of exogenous IL-2 [11-13,29-35]. Only in fewer instances were $CD8^+$ T cells directly tested for their spontaneous ability to suppress HIV infection, in the absence of exogenous stimuli [37,41]. The latter setting reflects more reliably the status of $CD8^+$ T-cell activation and functional competence *in vivo*. Such discrepancies could help explain the conflicting results obtained by different groups on the *in-vivo* correlates of the $CD8^+$ T-cell-mediated suppressive activity (see below).

5. A Copernican Revolution: From One "CAF" to Multiple "CAFs"

Our understanding of the phenomenon of CD8$^+$ T-cell-mediated HIV suppression has greatly advanced during the past decade. Nevertheless, it must be emphasised that the rigid application of the "CAF" postulates may have sometimes hindered the progress in dissecting the complex nature of this phenomenon. An important bias was undoubtedly introduced by the devout acceptance of the "single-factor" hypothesis, with two direct consequences on the experimental strategies: first, a reductionist approach to the identification of the HIV-suppressive factors produced by CD8$^+$ T cells, as cytokine-neutralization experiments were invariably conducted using a single antiserum at a time [40,42], thus limiting the appreciation of the combined effect of multiple factors; second, a continuous extrapolation of data from one experimental system to the others, based on the assumption that all the different assays would measure the activity of the same suppressive factor. This has resulted in the systematic attribution to a single putative entity of all the biological and biochemical properties observed in different experimental systems. However, a novel viewpoint has recently started to emerge: there is a growing perception that the antiviral activity of CD8$^+$ T cells is complex and multifarious, encompassing a cocktail of suppressive factors, that may vary according to the lineage origin, the differentiation level and the activation state of the producer-cell system. Moreover, it is increasingly appreciated that the choice of the assay system, particularly with regard to the biological phenotype of the viral strains used, may dictate the array of factors that can be measured.

A critical re-evaluation of one of the earliest experiments of cytokine neutralisation [40] may offer a new interpretative perspective. Using a single cytokine-neutralising antiserum at a time, Mackewicz *et al.* concluded that none of several known cytokines is identical to "CAF". Nevertheless, a limited, though reproducible, effect was documented using different neutralising antisera: anti-IFN-α reversed HIV suppression by an average of 18%, anti-IFN-β by 10%, anti-TNFα and TNFβ each by 8%, anti-IL-4 by 3%, anti-IL-6 by 20% and anti-IL-8 by 4% [40]. If we sum all these "partial" inhibitory activities, we obtain a remarkable 71% reversion of HIV suppression, without considering that no primary data on IL-10 neutralisation were reported. What can be inferred from these data? That the combined effect of several cytokines, each showing a partial anti-HIV activity, may result in a significant suppressive effect which cannot be abrogated using any individual neutralising antiserum. Indeed, as conclusively demonstrated with the recent identification of the first HIV-inhibitory chemokines [18], only a mixture of neutralising antisera can block the activity of crude CD8$^+$ T-cell supernatants containing multiple suppressive factors, whereas single antisera produce only a partial, if any, reduction of the suppressive effect. Another point emerging from the study of chemokines is that the suppressive effect may be critically dependent on the biological phenotype of the viral strain used: indeed, CCR5-binding chemokines potently suppress viral strains that fail to infect continuous CD4$^+$ cell lines (MT2-negative), but are totally ineffective against T-cell line-tropic ones (MT2-positive) [18]. Similar data were recently reported using crude CD8$^+$ T-cell culture supernatants [43]. As all the patients studied by Mackewicz *et al.* using the two-chamber transwell system were clinically aymptomatic [40], we can assume with reasonable certainty that their blood cells were infected by MT2-negative viral strains. Conversely, only a few experiments were performed using an acute-infection assay based on the MT2-tropic strain HIV-1_{SF33}, but the role of several cytokines, including the IFNs, IL-6, TNFβ and IL-10, was not evaluated in this system.

There is still considerable confusion in the literature regarding the HIV-suppressive activities measured in tests allowing cell-to-cell contact. Such activities have often been considered to be different manifestations of an identical suppressive mechanism, even though several reports have documented a markedly lower efficiency of the soluble HIV-suppressive activity [12,13,21,29,30,34], suggesting that the two effects may be at least partly due to unrelated mechanisms or factors. Of course, in co-cultivation experiments, one can assume all the diverse components to be present at the same time. In this respect, it is important to recall that $CD8^+$ T cells co-cultured with $CD4^+$ T cells exert classical CTL-like effects on virus-expressing target cells (when MHC-matched), as well as non-MHC-restricted cytotoxic effects on both infected and uninfected $CD4^+$ T cells [44,45]. As expected, $CD8^+$ T-cell populations and clones with dual-function (both classical CTL and soluble factor-mediated effects) have been characterised [30,46]. Nevertheless, the cytolytic component appears to be insufficient to account for the superior antiviral activity documented when cell-to-cell contact is allowed, suggesting the presence of additional mechanisms, possibly mediated by membrane-anchored suppressive molecules.

A further indication that different factors (or sets of factors) are involved in different test systems comes from the observation of the variable spectrum of $CD8^+$ T-cell activity documented in HIV-infected *versus* -uninfected subjects. Some investigators have reported that while endogenous infection is suppressed by $CD8^+$ T cells from both infected and uninfected individuals, only the former are effective in the acute-infection assay [12,35,38,47]. A major difference between the two assays resides in the viral phenotype, as the former is generally performed with MT2-tropic HIV-1 strains, whereas any strain subtype may be present in the latter, although this test was shown to function predominantly in asymptomatic patients [16,35] who harbour almost invariably MT2-negative strains. Indeed, Mackewicz and Levy state that "slower replicating strains appear more sensitive in most cases" [32], although primary data supporting this statement have been published only recently by other investigators [43]. Additional evidence in this direction, albeit only anecdotal, is the apparent inability of $CD8^+$ T cells to suppress HIV replication in continuous $CD4^+$ T-cell lines, which is by definition sustained by T-cell line-adapted (MT2-tropic) viral strains. Levy *et al.* state that in people not previously exposed to HIV $CD8^+$ T- cell-mediated HIV-inhibitory activity is a rare occurrence, which they interpret as the result of "cross-reactivity with non-HIV antigens" [35]. Only in the chimpanzee model did the same group document $CD8^+$ T-cell-mediated viral suppression in the absence of previous exposure to HIV [27]: this type of "natural" protection was suggested as a possible mechanism for the lack of disease progression in chimpanzees chronically infected with HIV-1. In evident contrast with these findings, different groups reported an efficient suppression of biologically diverse HIV strains by $CD8^+$ T cells derived from uninfected individuals, including CD8$^+$ T cells from neonatal cord blood [43,48,49]. These observations suggest that previous exposure to HIV is not an absolute requirement for the development of CD8$^+$ T-cell-mediated virus-suppressive activity.

Several reports have documented a CD8$^+$ T-cell-mediated HIV suppressive activity mediated by transcriptional inhibition of the retroviral LTR activation [23,39,50-53], although such activity does not appear to be identical to that observed in the classic endogenous or exogenous infection tests. A decreased activation of a reporter gene linked to the viral LTR was seen upon co-cultivation of transfected target cells with activated CD8$^+$ T cells, as well as upon treatment with cell-free culture supernatants. In two studies [52,53], CD8$^+$ T cells derived from HIV-infected individuals were immortalised *in vitro* with a monkey virus, herpesvirus saimiri (HVS), a strategy subsequently exploited by others in an attempt to increase the yield of soluble suppressive factor [54]. However, these cell lines had peculiar properties, suggesting a different mechanism of suppression with respect to other systems: first, their range of antiviral activity was unusually broad, with inhibition not only of lentiviruses, but also of human T-cell leukemia virus (HTLV)-I and Rous sarcoma virus (RSV) [51]; second, opposite effects were documented in two different experimental systems, with suppression of LTR activation in the CD4$^+$ T-cell line Jurkat, but enhancement in U937 [53], a cell line exhibiting pro-monocytic characteristics (although infectable only by T-cell-line-adapted HIV-1 strains). Another experimental system was based on a resident LTR-reporter gene construct in Jurkat cells, which can be activated following acute infection with MT2-tropic HIV-1 [39]. However, this model system is not specific for suppression at the LTR level, as intracultural spread of live virus does occur, being required for trespassing the threshold of signal detection. Thus in this test, suppression at any stage in the viral life-cycle will result in a decreased LTR activation, which represents only the final read-out of the test. It is unquestionable that the suppression of the HIV-LTR activation may, in specific experimental systems, represent an important component of the CD8$^+$ T cell-mediated antiviral activity. Nevertheless, only few attempts were made until now to identify the factors responsible for this particular form of suppression. In this respect, it should be emphasized that several known cytokines may cause suppression of the HIV-LTR activation.

6. The Role of Known Antiviral Cytokines: Specific Versus Non-Specific Inhibitory Mechanisms

The antiviral activity of natural soluble factors, such as cytokines, has been the object of intensive investigation since the outbreak of the AIDS pandemic, particularly in the perspective of identifying new tools for potential therapeutic use *in vivo*. As with any complex biological system, variability has been a keyword in this area of investigation, with striking discrepancies among results obtained with the same factor in different experimental systems or at different doses. Members of the prototypic class of antiviral cytokines, the IFNs, were initially considered as suitable candidates for therapeutic interventions. All three major types, IFN-α, -β and -γ, suppress HIV infection *in vitro*, although for the latter the evidence is not univocal

[17]. An early observation that has been instrumental for the first identification of the AIDS virus in 1984 was that the addition of neutralising antibodies to IFN-α markedly increased the efficiency of viral isolation and propagation *in vitro* [55]. However, the results of clinical trials with IFN-α or -β have been somewhat disappointing [56.57], possibly as a consequence of the inherent anti-proliferative effect of these cytokines on lymphocytes, as well as of the ability of HIV-1 to escape from IFN control *in vivo*.

Several other cytokines are known to block HIV infection at different levels in the viral life-cycle *in vitro*. Despite claims that "CAF" is not identical to any known cytokine [35,40,42], the role of some of these factors needs to be carefully re-evaluated. In particular, in light of the arguments presented above, the potential combined effect of multiple cytokines should be investigated. Besides the IFNs, cytokines exhibiting anti-HIV activity *in vitro*, at least in specific experimental systems, include IL-4 [40,58], IL-8 [40], IL-10 [40,58], IL-13 [59], IL-16 [60,61], TGFβ [17,40]. and TNFα [17,40,62]. Moreover, as described in detail in the next section, both C-C and C-X-C chemokines have been identified as potent and specific HIV-suppressive factors. Some of these molecules are currently under consideration for *in-vivo* use in HIV-infected patients.

One of the major questions related to the suppressive effect of cytokines is the specificity of their action, as it is well established that any factor exerting non-specific immunosuppressive or anti-proliferative effects may determine a dramatic reduction of HIV replication *in vitro*. Indeed, immunosuppressive agents, such as cyclosporin A, have even been proposed as potential anti-HIV drugs [63], but their use in a disease characterized by a progressive and ultimately fatal immunodeficiency has encountered strong opposition. Among the antiviral cytokines, IFN-α, TGF-β, IL-10 and IL-16 are known to exert different types of immunosuppressive, anti-proliferative or anti-differentiative effects [17,64,65]. However, while in the case of IFN-α the anti-proliferative effect is coupled with a powerful and specific antiviral mechanism (*e.g.*, synthesis of 2'-5' oligoadenylate synthetase), in other instances (*e.g.*, IL-16, which blocks $CD4^+$ T-cell activation in mixed lymphocyte reactions, possibly due to direct engagement of the CD4 glycoprotein [65]), it is still unclear whether any specific mechanism exists, besides the immunosuppressive effect.

Considerable attention has recently been given to cytokines that can influence the HIV-suppressive capability of $CD8^+$ T lymphocytes, in the perspective of exploiting such factors to modulate HIV replication *in vivo*. The somewhat "manichaean" classification of the immune responses into T-helper type 1 (Th_1) and type 2 (Th_2) [66] has provided the background paradigm for investigation in this area. Using the acute infection assay with the MT2-tropic strain HIV-1_{SF33}, Barker *et al.* found that addition of the Th_1 cytokine IL-2 enhanced the antiviral activity of $CD8^+$ T cells from 21 subjects with non-progressive HIV infection, as well as from some of the patients with progressive infection [67]. By contrast, the Th_2 cytokines IL-4 and IL-10 caused a decrease in the HIV-suppressive ability of $CD8^+$ T cells, although treatment with IL-2 could reverse their effect. Kinter *et al.* confirmed the enhancing power of IL-2, while observing no effect with another Th_1 cytokine, IL-12 [68]. However, no clearcut correlation was found at the clonal level between the subset of $CD8^+$ T cells responsible for HIV suppression and the Th_1/Th_2 classification [69,70].

In conclusion, several known cytokines may be involved, possibly in an additive or synergistic fashion, in the phenomenon of $CD8^+$ T-cell-mediated HIV suppression. Conversely, other cytokines dramatically enhance HIV replication [17]. The combined role of these factors may vary according to the test system employed, suggesting that the results obtained in one experimental model should not be auto-matically extrapolated to other models. Indeed, different cell types and viral variants often exhibit a different sensitivity to cytokines, which may explain the remarkable heterogeneity of the results reported in the literature. The most critical question, however, remains the specificity of the antiviral mechanism, particularly for cytokines that are being considered for potential therapeutic use, as non-specific immunosuppressive agents could have serious detrimental effects on an already damaged immune system.

7. HIV-Suppressive Chemokines: Natural Coreceptor Antagonists of HIV and SIV

The discovery of the three prototypic HIV-suppressive chemokines (RANTES, MIP-1α and MIP-1β) by Cocchi *et al.* in 1995 [18] was a direct result of the long quest for the identification of $CD8^+$ T-cell-derived HIV-suppressive factors. Most likely, the failure of previous attempts to purify and clone any such factor had been due to the inefficient secretion of inhibitory molecules by primary $CD8^+$ T cells, as well as to the poor reproducibility of test systems which utilise primary $CD4^+$ T cells [35]. The strategy used by Cocchi and colleagues was based on the establishment of high-producer $CD8^+$ T-cell lines and clones, after *in-vitro* immortalisation with HTLV-I, as well as of highly standardised and reproducible test systems based on the $CD4^+$ T-cell clone PM1: cell-free culture supernatants from a $CD8^+$ T-cell clone (FC36.22) were subjected to extensive biochemical fractionation and the purified fractions were tested against acute infection of PM1 by either MT2-tropic (IIIB) or MT2-negative (Ba-L) HIV-1. Partial amino-acid sequencing of active fractions revealed identity with the three related C-C chemokines RANTES, MIP-1α and MIP-1β. These chemokines were found to be effective not only on HIV-1, but also on HIV-2 and SIV, whereas unrelated viruses (such as human herpesvirus 6 and 7, and HTLV-I) were insensitive, suggesting a lentiretrovirus-specific suppressive mechanism. An important observation was that not all the biological variants of HIV were equally sensitive to inhibition by RANTES, MIP-1α and MIP-1β: while MT2-negative R5 M-tropic viral strains, passaged exclusively in primary lymphocytes or macrophages, were potently blocked, T-cell line-adapted laboratory isolates were resistant to C-C chemokine-mediated suppression [18].

A few months after the first "close encounter" between HIV and chemokines, another major breakthrough provided a key to understanding the mechanism of antiviral action of chemokines in 1996: Feng *et al.* reported that a previously cloned "orphan" chemokine receptor-like protein serves as a critical cofactor (or co-receptor) for the entry of T-cell line-adapted HIV-1 into cells [19]. Such molecule, initially named "fusin" for its fusogenic function in combination with CD4, was subsequently designated CXCR4 after the identification of its ligand, the C-X-C chemokine stromal cell-derived factor 1 (SDF-1). The combination of these landmark discoveries triggered an extraordinary chain-reaction of events, leading in a few months to the elucidation of some fundamental aspects of the physiology of HIV infection, which had remained elusive for more than a decade. Of major importance, CCR5 (a specific receptor for RANTES, MIP-1α and MIP-1β) was identified as the major coreceptor for MT2-negative strains, those most commonly transmitted *in vivo* and predominating during the early stages of the infection [reviewed in 19,71,72]. The essential role of CCR5 in the biology of HIV-1 was confirmed by epidemiological studies showing that a homozygous defect in the CCR5 gene (CCR5/Δ32) confers a near-complete protection from HIV-1 infection, even in people with multi-ple high-risk sexual contacts [19,71,72].

The role of chemokines in the phenomenon of $CD8^+$ T-cell-derived HIV suppression is currently under intensive investigation. RANTES, MIP-1α and MIP-1β were originally identified as major components of the suppressive activity produced by primary $CD8^+$ T cells of asymptomatic HIV-infected subjects, as conclusively demonstrated with the use of neutralising polyclonal antisera: however, antibodies to a single chemokine used separately caused little, if any, inhibition of the suppressive activity, whereas the combined use of antibodies to all three factors permitted to re-establish the level of HIV replication observed in the controls [18]. Subsequent studies have confirmed the critical role of this chemokine trio in $CD8^+$ T-cell-mediated HIV or SIV suppression [73-76]. However, it is important to emphasise that the same limitations discussed in the previous sections (*i.e.*, multiplicity of factors released by $CD8^+$ T cells, variable factor-specificities of the different *in-vitro* assay systems, *etc.*) do apply also in the case of chemokines, as RANTES, MIP-1α and MIP-1β are active exclusively on MT2-negative, CCR5-dependent viral strains, whereas SDF-1 suppresses specifically MT2-positive, CXCR4-using viral strains. Likewise, as expectable on the basis of the predicted mechanism of action of RANTES, MIP-1α and MIP-1β, the suppressive activity of $CD8^+$ T cells on the viral LTR activation is not mediated by these chemokines [77,78]. Using different experimental systems, additional evidence that these chemokines are not the only factors responsible for the $CD8^+$ T-cell antiviral activity was reported [79-81].

A question of substantial importance is whether chemokines have a *bona fide* physiologic relevance in the control of HIV infection *in vivo*. In this respect, evidence has been provided by Lehner *et al.* [74] and by Heeney *et al.* [82] who showed a correlation between *ex-vivo* chemokine production by lymph-node $CD8^+$ T cells and vaccine-induced protection in macaques (see below). Other observations supporting a role of C-C chemokines *in vivo* include the increased RANTES, MIP-1α and MIP-1β secretion by $CD4^+$ T lymphocytes derived from HIV-exposed uninfected individuals [76], the high efficiency of virus isolation documented *in vitro* following treatment with neutralising antibodies to these three chemokines [83], the efficient release of RANTES and MIP-1α upon *ex-vivo* stimulation of $CD8^+$ T cells derived from infected lymph nodes with HIV-1 envelope-expressing autologous B cells [84], the augmented production of inhibitory chemokines by T-cell clones derived from HIV-infected long-term non-progressors [70], the adaptive evolution of HIV-1 *in vivo* in the course of the disease progression, leading to escape from C-C chemokine-mediated control [85] and the dramatic enhancement of RANTES plasma levels in patients with acute primary HIV-1 infection [86].

An important change of perspective introduced by the study of chemokines has been the recognition that the ability to produce soluble HIV-suppressive factors is not an exclusive property of $CD8^+$ T cells. Indeed, other primary immune cells, such as differentiated macrophages [87], NK cells [88] and $CD4^+$ T cells [76], were found to release antiviral chemokines upon stimulation. The ability of macrophage-culture supernatants to suppress infection by HIV-1_{BaL} was completely abro-

TABLE 2. Chemokine Receptors Serving as HIV Coreceptors and Specific Chemokine Ligands Blocking the Coreceptor Function.

Chemokine Receptor	Tissue Expression	High-Affinity Ligands	HIV Variant Specificity	Frequency of Use
CCR5	T_{act},Mc,MΦ DC,Th_1 clones	RANTES MIP-1α,MIP-1β	Primary isolates (all phenotypes) + SIV	Very high
CXCR4	T_{rest},T_{act},Mc, MΦ, DC,myeloid progenitors	SDF-1	Primary MT2-tropic (+ T-cell line-adapted)	High
CCR2b	Mc,MΦ,DC	MCP-1 to -4	Dual-tropic	Rare
CCR3	Eosinophils, DC,Th_2 clones	Eotaxin-1,-2 MCP-3,RANTES	Primary isolates (all phenotypes)	Moderate (High?)
CCR8	Thymocytes, T_{act},MΦ	I-309	Primary isolates (all phenotypes)	Low
STRL33 /Bonzo	T_{rest},T_{act},Mc	Unknown	MT2-negative + SIV	Rare
GPR15 /BOB	T_{act}	Unknown	MT2-negative + SIV	Rare
GPR1	Mc,MΦ	Unknown	? + SIV	Rare
APJ	?	Unknown ?		Low
V28	T_{act},Mc, neural tissue	Fractalkine	MT2-tropic?	Rare

Legend: T_{rest} = resting T cells; T_{act} = activated T cells; Mc = monocytes; MΦ = differentiated macrophages; DC= dendritic cells.

gated by pre-absorption on a cocktail of immobilised antibodies to RANTES, MIP-1α and MIP-1β [87].

In recent months, the number of chemokines and chemokine receptors with potential relevance for HIV infection has gradually increased (Table 2).

The C-X-C chemokine SDF-1 is able to block CXCR4-using HIV strains [19], although its efficiency on primary isolates is generaly low [85] and it is not expressed at significant levels by primary $CD8^+$ T cells. Other chemokines which bind to minor HIV coreceptors, such as eotaxin and MCP-3 (CCR3 ligands), I-309 (CCR8 ligand) and fractalkine (CX_3CR1 ligand), might block HIV in selected anatomical sites, but are ineffective on $CD4^+$ T cells or macrophages. Moreover, they do not seem to be produced by activated $CD8^+$ T cells. A C-C chemokine that has recently attracted considerable attention as a broad-spectrum HIV inhibitor is macrophage-derived chemokine (MDC) [89], a CCR4 ligand, although its effective ability to block HIV is controversial.

In conclusion, the discovery of the HIV-inhibitory chemokines has undoubtedly brought a remarkable change of perspective in the field. Although they do not account for the entire spectrum of soluble $CD8^+$ T-cell-derived HIV-suppressive factors, chemokines represent important players in specific experimental systems (*i.e.*, against CXCR4-negative viral strains) and, may play a critical role in the control of HIV infection *in vivo*. The rapid progress in this field may ultimately lead to the development of novel therapeutic approaches.

8. $CD8^+$ T-Cell-mediated Viral Suppression in the Natural History of HIV Infection

One of the major reasons behind the interest raised by the phenomenon of $CD8^+$ T-cell-mediated viral suppression has been its putative correlation with the course of HIV disease, which suggested an involvement of this unique type of immune response in the control of primate lentiretrovirus infection *in vivo*. However, it is still difficult to draw a reliable picture of the correlations between the prevalence and potency of such antiviral mechanisms and the natural history of HIV disease. The main limitations of the studies conducted thus far include the small numbers of patients analysed, the lack of well-defined inclusion criteria and the substantial differences in the methodological approaches, which hamper a comparative analysis of the results. Furthermore, it is increasingly evident that the multiplicity of suppressive factors produced by $CD8^+$ T lymphocytes, some of which have now been identified, demands a critical re-evaluation of previous studies.

A first question is whether the observations made *in vitro*, particularly after polyclonal activation or other types of *ex-vivo* manipulation of the effector cells, can be directly extrapolated to the *in-vivo* situation. In most studies, $CD8^+$ T lymphocytes were maximally activated *in vitro* with plant lectins, which is more likely to reflect the potential functional reservoir of $CD8^+$ T cells, rather than their actual *in-vivo* activity. Conversely, testing the spontaneous activity of isolated $CD8^+$ T lymphocytes, without exogenous stimuli, should provide a more accurate representation of their *in-vivo* function. The question of the correlation between $CD8^+$ T-cell pre-activation *in vivo*, production of soluble HIV-suppressive factors and risk of disease progression is still controversial. Irrespective of the test systems used, the level of CD8$^+$ T-cell-mediated HIV suppression has been associated with an *in-vivo*-activated phenotype [36,37,41]. In HIV-infected patients, a variable proportion of circulating CD8$^+$ T cells indeed expresses activation markers, such as HLA-DR, CD38, or both [90-92]: the level of activation is generally associated with an increased risk of disease progression, although the expression of HLA-DR in the absence of CD38 has been correlated with a higher level of CD8$^+$ T-cell-mediated viral suppression and ultimately with a better prognosis [37,41]. One study, however, performed without an *ex-vivo* activation step, documented a correlation between the production of HIV-suppressive factor(s) and high levels of CD38 expression, increased viral load and abnormal CD4$^+$ T-cell counts, as well as with CTL activity against specific viral proteins [41].

A striking variability in the prevalence of HIV-suppressive activity by CD8$^+$ T-cells was reported in different studies, ranging from 25-30% to virtually 100% [11-13,30,41]. These discrepancies may be related to methodological issues, including a poor standardisation of the patient populations and/or of the test systems. Cross-sectional studies have shown a correlation between high levels of CD8$^+$ T-cell-mediated viral suppression and early disease stage [12,16,30,37]. Consistent with this concept, an amelioration of the CD8$^+$ T-cell-mediated antiviral activity was observed after starting antiretroviral treatment with AZT [93]. However, others have documented a strong antiviral activity even in patients who have progressed to AIDS [43,47]. Some authors observe an association between high levels of CD8$^+$ T-cell antiviral activity and elevated numbers of peripheral blood CD4$^+$ T cells, asymptomatic status and high proportions of CD28-expressing blood cells [30,37], whereas others failed to document a correlation with either CD4$^+$ T cells or viral load in blood cells [43]. Consistent with this concept, a loss of CD28 expression in peripheral blood lymphocytes was associated with a poor prognosis [43,94]. By contrast, another study reported a lack of suppressive activity in patients with long-standing HIV infection and a number of circulating CD4$^+$ T lymphocytes greater than 1000/ mm^3 [41]. It is important to emphasise that, at least in some cases, the loss of suppressive function detected in the endogenous test may be only apparent, as the viral strains emerged with the progression of the disease may have lost sensitivity to the predominant suppressive factors, such as C-C chemokines. This kind of viral escape from CD8$^+$ T-cell-mediated viral suppression was recently demonstrated in a limited number of patients [43]. The *in-vivo* evolution of HIV-1, leading to resistance to C-C chemokines [85], provides additional ground to this concept. Interestingly, CD8$^+$ T-cell-mediated viral suppression was suggested as a major mechanism for the resistance of cultured mononuclear cells from asymptomatic HIV seropositive individuals to *in-vitro* superinfection with exogenous HIV-1 strains [22,95].

It has been speculated that CD8$^+$ T-cell-mediated HIV suppression may play a major role in the course of primary HIV infection, contributing to the initial containment of viral replication before the appearance of antigen-specific antiviral mechanisms, such as neutralising antibodies or CTL. Surprisingly, however, in spite of the obvious interest raised by this question, only a single report has analysed the level of CD8$^+$ T-cell antiviral activity during this early stage of the infection [96]. An acute infection assay *in vitro* was used, with autologous virus, not characterised biologically, isolated from the same patient at a previous visit and CD8$^+$ T cells most likely activated *ex vivo* (although not expressedly stated in the paper). At the time of antibody seroconversion, all the patients studied displayed CD8$^+$ T-cell-mediated HIV suppression, while antibodies capable of neutralising the autologous virus appeared only in 2 of 7 patients studied, and only at a later time. An inverse relationship was observed between the suppressive activity and the level of plasma viraemia.

9. Animal Models

Soon after the first phenomenological recognition of the HIV-suppressive capability of CD8$^+$ T lymphocytes, several groups reported the identification of a similar activity in non-human primates infected with HIV-1 or its simian counterpart, SIV. Suppression of SIV infection was documented with CD8$^+$ T lymphocytes derived from sooty mangabeys [23,24,97], macaques [14,21,25], and African green monkeys [26], whereas an HIV-inhibitory activity was recognised in chimpanzees infected with HIV-1 [27], and more recently in baboons infected with HIV-2 [28], suggesting the existence of common CD8$^+$ T-cell-mediated antiviral mechanisms against all types of primate lentiretroviruses. A similar activity has recently been detected in cats infected with FIV [15]. Using SCID mice repopulated with adult human leukocytes, cloned CD8$^+$ T cells were shown to protect the animals from HIV-1 infection *in vivo*: the possible role of the CD8$^+$ T-cell-mediated antiviral activity was suggested by the observation that protection occurred in a non-MHC-restricted fashion [98].

The case of SIV-infected African green monkeys [26] is particularly intriguing. These animals naturally possess remarkably high proportions of circulating CD8$^+$ T cells and display a vigorous CD8$^+$ T-cell-mediated antiviral activity. However, a seemingly homeostatic relationship is reached in these animals between virus and host, because moderate to high levels of SIV viraemia occur for prolonged periods of time, without evidence of immune system deterioration or other pathological sequelae. This phenomenon may reflect an immunologic tolerance developed by African green monkeys, possibly as a consequence of a long-lasting persistence of the SIV in this species.

Evidence for a protective role of CD8$^+$ T-cell-mediated virus suppression *in vivo*, associated with the production of antiviral C-C chemokines, was reported in vaccinated monkeys [74,82]. In one study, macaques were immunized with recombinant Gag and Env proteins of SIV using a novel "targeted" iliac lymph-node immunisation protocol, and subsequently evaluated for several immunological parameters, including the functional capability of purified CD8$^+$ T cells extracted from immunised and unimmunised lymph nodes of the same animal. Strikingly, the *ex-vivo* production of crude SIV-suppressive factors and the release of RANTES, MIP-1α and MIP-1β by activated CD8$^+$ T cells from immunised lymph nodes were found to be the best correlate of protection. By contrast, CD8$^+$ T cells from unimmunised nodes yielded levels comparable to control animals. When the crude culture supernatants were treated with a cocktail of neutralising antibodies to the three chemokines, their suppressive effect was totally abrogated (Lehner T., personal communication). Of note, the only unimmunised monkey that was naturally protected from SIV infection also had a dramatically enhanced chemokine release, compared to monkeys that became infected [74]. In a second study comparing the efficacy of different immunisation strategies based on the same vaccine strain, the production of the three CCR5-binding chemokines by $CD8^+$ T cells was markedly increased in the group of protected monkeys, but in none of the vaccinated or control animals which became infected [82].

10. Concluding Remarks

The recent identification of factors involved in $CD8^+$ T-cell-mediated virus suppression has greatly boosted research in this area of investigation and important new developments can be expected in the upcoming years. Besides the elucidation of the entire spectrum of physiologically relevant antiviral soluble factors, whether derived from $CD8^+$ T cells or other immune cells, other questions remain unsolved. Is the non-lytic control enacted by $CD8^+$ T cells in fact more beneficial to the virus or to the host? What is the relative contribution of the cytolytic and non-cytolytic functions of $CD8^+$ T cells to the suppression of HIV *in vivo*? At present, these Dr. Jekyll and Mr. Hyde facets of the $CD8^+$ T-cell "personality" seem difficult to reconcile. Unlike the CTL activity, which induces destruction of the infected cells with potentially severe inflammatory side-effects, the non-lytic mechanisms, albeit unable to eradicate the infection, could exert a partial control for prolonged periods of time with only a limited proportion of cells containing actively replicating virus at any given moment. While non-lytic virus control may be an important survival mechanism for the host, it might also represent a calculated strategy of the virus for reducing the aggressiveness of CTL effectors and thereby escaping both clearance and rapid host deterioration. We could guess that the dynamic relation that we observe today represents an equilibrium reached in the course of a long and carefully played chess game. As such, it is likely to provide both the host and the virus a reasonable share of the benefits.

11. References

1. Paul, We.: Can the immune response control HIV infection?, *Cell* **82** (1995), 177-182.
2. Racz, P., Tenner-Racz, K., Kahl, C., Feller, A.C., Kern, P. and Dietrich, M.: Spectrum of morphologic changes of lymph nodes from patients with AIDS or AIDS-related complexes, *Progr. Allergy* **37** (1986), 81-181.
3. Tenner-Racz, K., Racz, P., Bofill, M., Schulz-Meyer, A., Dietrich, M., Kern, P., Weber, J., Pinching, A.J., Veronese-Dimarzo, F. and Popovic, M.: HTLV-III/LAV viral antigens in lymph nodes of homosexual men with persistent generalized lymphadenopathy and AIDS, *Am. J. Pathol.* **123** (1986), 9-15.

4. Harper, M.E., Marselle, L.M. and Gallo, R.C.: Detection of lymphocytes expressing human T-lymphotropic virus type III in lymph nodes and peripheral blood from infected individuals by *in-situ* hybridization, *Proc. Natl. Acad. Sci. USA* **83** (1985), 772-776.
5. Embretson, J., Zupancic, M., Ribas, J.L., Burke, A., Racz, P., Tenner-Racz, K. and Haase, A.T.: Massive covert infection of helper T lymphocytes and macrophages by HIV during the incubation period of AIDS, *Nature* **362** (1993), 359.
6. Pantaleo, G., Graziosi, C., Demarest, J.F., Butini, L., Montroni, M., Fox, C.H., Orenstein, J.M., Kotler, D.P. and Fauci, A.S.: HIV infection is active and progressive in lymphoid tissue during the clinically latent stage of disease, *Nature* **362** (1993), 355-258.
7. Coombs, R.W., Collier, A.C. and Allain, J.P.: Plasma viremia in human immunodeficiency virus infection, *New Engl. J. Med.* **321** (1989), 1626-1631.
8. Holodniy, M., Katzenstein, D.A., Sengupta, S., Wang, A.M., Casipit, C., Schwartz, D.H., Konrad, M., Groves, E. and Merigan, T.C.: Detection and quantification of human immunodeficiency virus RNA in patient serum by use of polymerase chain reaction, *J. Infect. Dis.* **163** (1991), 862-866.
9. Piatak Jr., M., Saag, M.S., Yang, L.C., Clark, S.J., Kappes, J.C., Luk, K.-C., Hahn, B.H., Shaw, G.M. and Lifson, J.D.: High levels of HIV in plasma during all stages of infection determined by competitive PCR, *Science* **259** (1993), 1749-1754.
10. Bagnarelli, P., Valenza, A., Menzo, S., Manzin, A., Scalise, G., Varaldo, P.E. and Clementi, M.: Dynamics of molecular parameters of human immunodeficiency virus type-1 activity *in vivo*, *J. Virol.* **66** (1994), 7328-7335.
11. Walker, C.M., Moody, D.J., Stites, D.P. and Levy, J.A.: $CD8^+$ lymphocytes can control HIV infection *in vitro* by suppressing virus replication, *Science* **234** (1986), 1563-1566.
12. Brinchmann, J.E., Gaudernack, G. and Vardtal, F.: $CD8^+$ T cells inhibit HIV replication in naturally infected $CD4^+$ cells. Evidence for a soluble inhibitor, *J. Immunol.* **144** (1990), 2961.
13. Walker, C.M. and Levy, J.A.: A diffusible lymphokine produced by $CD8^+$ T lymphocytes suppresses HIV replication, *Immunology* **66** (1989), 628-630.
14. Kannagi, M., Chalifoux, L.V., Lord, C.I. and Letvin, N.L.: Suppression of simian immunodeficiency virus replication *in vitro* by $CD8^+$ lymphocytes, *J. Immunol.* **140** (1988), 2237-2242.
15. Jeng, C.R., English, R.V., Childers, T., Tompkins, M.B. and Tompkins, W.A.: Evidence of $CD8^+$ antiviral activity in cats infected with feline immunodeficiency virus, *J. Virol.* **70** (1996), 2474-2480.
16. Mackewicz, C.E., Ortega, H.W. and Levy, J.A.: CD8 cell anti-HIV activity correlates with the clinical state of the infected individual, *J. Clin. Invest.* **87** (1991), 1462.
17. Poli, G. and Fauci, A.S.: Cytokine modulation of HIV expression, *Semin. Immunol.* **5** (1993), 304.1.
18. Cocchi, F., De Vico, A.L., Garzino-Demo, A., Arya, S.K., Gallo, R.C. and Lusso, P.: Identification of RANTES, MIP-1α and MIP-1ß as the major HIV-suppressive factors produced by $CD8^+$ T cells, *Science* **270** (1995), 1811-1815.
19. Berger, E.A.: HIV entry and tropism: the chemokine receptor connection, *AIDS* **11** (1997), S3-S16.
20. Brinchmann, J.E., Gaudernack, G., Thorsby, E., Jonassen, T.O. and Vardtal, F.: Reliable isolation of human immunodeficiency virus from cultures of naturally infected $CD4^+$ cells, *J. Virol. Methods* **25** (1989), 293.
21. Tsubota, H., Lord, C.I., Watkins, D.I., Morimoto, C. and Letvin, N.L.: A cytotoxic T lymphocyte inhibits acquired immunodeficiency syndrome virus replication in peripheral blood lymphocytes, *J. Exp. Med.* **169** (1989), 1421-1434.
22. Kannagi, M., Masuda, T., Hattori, T., Kanoh, T., Nasu, K., Yamamoto, N. and Harada, S.: Interference with human immunodeficiency virus (HIV) replication by $CD8^+$ T cells in peripheral blood leukocytes of asymptomatic HIV carriers *in vitro*, *J. Virol.* **64** (1990), 3399.
23. Powell, J.D., Yehuda-Cohen, T., Villinger, F., McClure, H.M., Sell, K.W. and Ahmed-Ansari, A.: Inhibition of SIV/SMM replication *in vitro* by $CD8^+$ cells from SIV/SMM-infected seropositive clinically asymptomatic sooty mangabeys, *J. Med. Primatol.* **19** (1990), 239-249.
24. Kunchel, M., Bednarik, D.P. and Chikkala, N.: Biphasic *in-vitro* regulation of retroviral replication by $CD8^+$ cells from nonhuman primates, *J. AIDS* **7** (1994), 438-446.

25. Blackbourn, D.J., Chuang, L.F., Killam Jr., K.F. and Chuang, R.Y.: Inhibition of simian immunodeficiency virus (SIV) replication by CD8$^+$ cells of SIV-infected rhesus macaques: implications for immunopathogenesis, *J. Med. Primatol.* **23** (1994), 343-354.
26. Ennen, J., Findeklee, H., Dittmar, M.T., Norley, S., Ernst, M. and Kurth, R.: CD8$^+$ lymphocytes of African green monkeys secrete an immunodeficiency virus-suppressing lymphokine, *Proc. Natl. Acad. Sci. USA* **91** (1994), 7207-7211.
27. Castro, B.A., Walker, C.M., Eichberg, J.W. and Levy, J.A.: Suppression of human immunodeficiency virus replication by CD8$^+$ cells from infected and uninfected chimpanzees, *Cell Immunol.* **132** (1991), 246-255.
28. Blackbourn, K.D., Brasky, K.M. and Levy, J.A.: CD8$^+$ cells from HIV-2-infected baboons control HIV replication, *AIDS* **11** (1997), 737-746.
29. Walker, C.M., Erickson, A.L., Hsueh, F.C. and Levy, J.A.: Inhibition of human immunodeficiency virus replication in acutely infected CD4$^+$ cells by CD8$^+$ cells involves a noncytotoxic mechanism, *J. Virol.* **65** (1991), 5921-5927.
30. Gomez, A.M., Smail, F.M. and Rosenthal, K.L.: Inhibition of HIV replication by CD8$^+$ T cells correlates with CD4 counts and clinical stage of disease, *Clin. Exp. Immunol.* **97** (1994), 68-75.
31. Walker, C.M., Thomson-Honnebier, G.A., Hsueh, F.C., Erickson, A.L., Pan, L.Z. and Levy, J.A.: CD8$^+$ cells from HIV-1-infected individuals inhibit acute infection by human and primate immunodeficiency viruses, *Cell Immunol.* **137** (1991), 420-428.
32. Mackewicz, C.E. and Levy, J.A.: CD8$^+$ cell anti-HIV activity: nonlytic suppression of virus replication, *AIDS Res. Human Retrovir.* **8** (1992), 1039-1050.
33. Levy, J.A.: Pathogenesis of human immunodeficiency virus infection, *Microbiol. Rev.* **57** (1993), 183-289.
34. Blackbourn, D.J., Mackewicz, E., Barker, E. and Levy, J.A.: Human CD8$^+$ cell non-cytolytic anti-HIV activity mediated by a novel cytokine, in: *60th Forum in Immunology*, 1994, pp. 653-658.
35. Levy, J.A., Mackewicz, C.E. and Barker, E.: Controlling HIV pathogenesis: the role of the noncytotoxic anti-HIV response of CD8$^+$ T cells, *Immunol. Today* **17** (1996), 217-224.
36. Wiviott, L.D., Walker, C.M. and Eichberg, J.W.: CD8$^+$ lymphocytes suppress HIV production by autologous CD4$^+$ cells without eliminating the infected cells from culture, *Cell Immunol.* **128** (1990), 628-634.
37. Landay, A.L., Mackewicz, C. and Levy, J.A.: An activated CD8$^+$ T-cell phenotype correlates with anti-HIV activity and asymptomatic clinical status, *Clin. Immunol. Immunopathol.* **69** (1993), 106-116.
38. Hsueh, F.W., Walker, C.M., Blackbourn, D.J. and Levy, J.A.: Suppression of HIV replication by CD8$^+$ cell clones derived from HIV-infected and uninfected individuals, *Cell Immunol.* **159** (1994), 271-279.
39. Mackewicz, C.E., Blackbourn, D.J. and Levy, J.A.: CD8$^+$ T cells suppress human immunodeficiency virus replication by inhibiting viral transcription, *Proc. Natl. Acad. Sci. USA* **92** (1995), 2308-2312.
40. Mackewicz, C.E., Ortega, H. and Levy, J.A.: Effect of cytokines on HIV replication in CD4$^+$ lymphocytes: lack of identity with CD8$^+$ cell antiviral factor, *Cell Immunol.* **153** (1994), 329-343.
41. Ferbas, J., Kaplan, A.H., Hausner, A., Hultin, L.E., Matud, J.L., Liu, Z., Panicali, D.L., Nerng-Ho, H., Detels, R. and Giorgi, J.V.: Virus burden in long-term survivors of human immunodeficiency virus (HIV) infection is determinant of anti-HIV CD8$^+$ lymphocyte activity, *J. Infect. Dis.* **172** (1995), 329-339.
42. Brinchmann, J.E., Gaudernack, G. and Vardtal, F.: *In-vitro* replication of HIV-1 in naturally infected CD4$^+$ T cells is inhibited by rIFN-α_2 and by a soluble factor secreted by activated CD8$^+$ T cells but not by rIFN-ß, rIFN-γ, or by recombinant tumor necrosis factor-α, *J. Acq. Immune Def. Syndr. Human Retrovir.* **4** (1991), 480.
43. Kootstra, N.A., Miedema, F. and Schuitemaker, H.: Analysis of CD8$^+$ T-lymphocyte-mediated nonlytic suppression of autologous and heterologous primary human immunodeficiency virus type-1 isolates, *AIDS Res. Human Retrovir.* **13** (1997), 685-693.
44. Zarling, J.M., Ledbetter, J.A., Sias, J., Fultz, P., Eichberg, J., Gjerset, G. and Moran, P.A.: HIV-infected humans, but not chimpanzees, have circulating cytotoxic T lymphocytes that lyse uninfected CD4$^+$ cells, *J. Immunol.* **144** (1990), 2992-2998.

45. Grant, G.E., Smail, F.M. and Rosenthal, K.L.: Lysis of CD4[+] lymphocytes by non-HLA-restricted cytotoxic T lymphocytes from HIV-infected individuals, *Clin. Exp. Immunol.* **93** (1993), 356-362.

46. Bolliger, R.C., Quinn, T.C., Liu, A.Y., Stanhope, P.E., Hammond, S.A., Viveen, R., Clements, M.L. and Siliciano, R.F.: Cytokines from vaccine-induced HIV-1-specific cytotoxic T lymphocytes: effects on viral replication, *AIDS Res. Human Retrovir.* **11** (1993), 1067-1077.

47. Barker, T.D., Weissman, D., Daucher, J.A., Roche, K.M. and Fauci, A.S.: Identification of multiple and distinct CD8[+] T-cell suppressor activities, *J. Immunol.* **156** (1996), 4476-4483.

48. Brühl, P., Kerschbaum, A., Zimmermann, K., Eibl, M.M. and Mannhalter, J.W.: Allostimulated lymphocytes inhibit replication of HIV type 1, *AIDS Res. Human Retrovir.* **12** (1996), 31-37.

49. Rosok, B., Voltersvik, P., Larsson, B., Albert, J., Brinchmann, J.E. and Asjo, B.: CD8[+] T cells from HIV type-1-seronegative individuals suppress virus replication in acutely infected cells, *AIDS Res. Human Retrovir.* **13** (1997), 79-85.

50. Chen, C.H., Weinhold, K.J., Barlett, J.A., Bolognesi, D.P. and Greenberg, M.L.: CD8[+] T-lymphocyte-mediated inhibition of HIV-1 long-terminal repeat transcription: a novel antiviral mechanism, *AIDS Res. Human Retrovir.* **9** (1993), 1079-1086.

51. Copeland, K.F.T., McKay, P.J. and Rosenthal, K.L.: Suppression of activation of the human immunodeficiency virus long-terminal repeat by CD8[+] cells is not lentivirus-specific, *AIDS Res. Human Retrovir.* **11** (1995), 1321-1326.

52. Copeland, K.F.T., Leith, J.G., McKay, P.J., Kelleher, L., Smaill, F.M. and Rosenthal, K.L.: CD8[+] T-cell supernatants of HIV type-1-infected individuals have opposite effects on long-terminal repeat-mediated transcription in T cells and monocytes, *AIDS Res. Human Retrovir.* **13** (1997), 71-77.

53. Copeland, K.F., Leith, J.G., McKay, P.J., Kelleher, L., Smaill, F.M. and Rosenthal, K.L.: CD8[+] T-cell-mediated suppression of HIV long-terminal repeat-driven gene expression is not associated with improved clinical status, *AIDS* **11 (5)** (1997), 581-586.

54. Mackewicz, C.E., Orque, R., Jung, J. and Levy, J.A.: Derivation of Herpesvirus saimiri-transformed CD8[+] T-cell lines with noncytotoxic anti-HIV activity, *Clin. Immunol. Immuniopathol.* **82** (1997), 274-281.

55. Barré-Sinoussi, F., Chemann, J.C., Rey, F., Nugeyre, M.T., Chamaret, S., Gruest, J., Dauguet, C., Axler-Blin, C., Vézinet-Brun, F., Rouzioux, C., Rozenbaum, W. and Montagnier, L.: Isolation of T-lymphotropic retrovirus from a patient at risk for acquired immune deficiency syn-drome (AIDS), *Science* **220** (1983), 868-871.

56. Francis, M.L., Meltzer, M.S. and Gendelman, H.E.: Interferons in the persistence, patho-genesis and treatment of HIV infection, *AIDS Res. Human Retrovir.* **8** (1992), 199-207.

57. Lane, H.C., Davey, V.,Kovacs, J.A., Feinberg, J., Metcalf, J.A., Herpin, B., Walker, R., Deyton, L., Davey Jr., R.T. and Falloon, J.: Interferon-α in patients with asymptomatic human immunodeficiency virus (HIV) infection, *Ann. Intern. Med.* **112** (1990), 805-811.

58. Schuitemaker, H.: IL-4 and IL-10 as potent inhibitors of HIV-1 replication in macrophages *in vitro*: a role for cytokines in the *in-vivo* virus host range?, *Res. Immunol.* **145** (1994), 588-592.

59. Montaner, L.J., Doyle, A.G., Collin, M., Herbein, G., Illei, P., James, W., Minty, A., Caput, D., Ferrara, P. and Gordon, S.: Interleukin 13 inhibits human immunodeficiency virus type-1 production in primary blood-derived human macrophages *in vitro*, *J. Exp. Med.* **178** (1993), 743-747.

60. Baier, M., Werner, A., Bannert, N., Bannert, N., Metzner, K. and Kurth, R.: HIV suppression by interleukin-16, *Nature* **378** (1995), 563.

61. Maciaszek, J., Parada, N.A., Cruikshank, K., Center, D.M., Kornfeld, H. and Viglianti, G.A.: IL-16 represses HIV-1 promoter activity, *J. Immunol.* **158** (1997), 5-8.

62. Lokensgard, J.R., Gekker, G., Ehrlich, L.C., Hu, S., Chao, C.C. and Peterson, P.K.: Proinflammatory cytokines inhibit HIV-1_{SF162} expression in acutely infected human brain cell cultures, *J. Immunol.* **158** (1997), 2449-2455.

63. Thali, M.: Cyclosporins: immunosuppressive drugs with anti-HIV-1 activity, *Mol. Med. Today* **1** (1995), 287-291.

64. Chang, J., Naif, H.M., Li, S., Jozwiak, R., Ho-Shon, M. and Cunningham, A.L.: The inhibition of HIV replication in monocytes by interleukin 10 is linked to inhibition of cell differentiation, *AIDS Res. Human Retrovir.* **12** (1996), 1227-1235.
65. Theodore, A.C., Center, D.M., Nicoll, J., Fine, G., Kornfeld, H. and Cruikshank, W.W.: CD4 ligand IL-16 inhibits the mixed lymphocyte reaction, *J. Immunol.* **157** (1996), 1958-1964.
66. Clerici, M. and Shearer, G.M.: The Th1-Th2 hypothesis of HIV infection: new insights, *Immunol. Today* **15** (1994), 575-581.
67. Barker, E., Mackewicz, C.E. and Levy, J.A.: Effects of TH1 and TH2 cytokines on CD8$^+$ cell response against human immunodeficiency virus: implications for long-term survival, *Proc. Natl. Acad. Sci. USA* **92** (1995), 11135-11139.
68. Kinter, A.L., Ostrowski, M., Goletti, D., Oliva, A., Weissman, D., Gantt, K., Hardy, E., Jackson, R., Ehler, L. and Fauci, A.S.: HIV replication of CD4$^+$ T cells of HIV-infected individuals is regulated by a balance between the viral suppressive effects of endogenous ß-chemokines and the viral inductive effects of other endogenous cytokines, *Proc. Natl. Acad. Sci. USA* **93** (1996), 14076-14081.
69. Toso, J.F., Chen, C.-H., Mohr, J.R., Piglia, L., Oei, C., Ferrari, G., Greenberg, M.L. and Weinhold, K.L.: Oligoclonal CD8$^+$ lymphocytes from asymptomatic HIV-1infected individuals inhibit HIV-1 replication, *J. Infect. Dis.* **172** (1995), 964-973.
70. Scala, E., D'Offizi, G., Rosso, R. and Turriziani, O.: C-C chemokines, IL-16, and soluble antiviral factor activity are increased in cloned T cells from subjects with long-term non-progressive HIV infection, *J. Immunol.* **158** (1997), 4485-4492.
71. Lusso, P. and Gallo, R.C.: Chemokines and HIV infection, *Curr. Opin. Infect. Dis.* **10** (1997), 12-17.
72. Moore, J.P., Trkola, A. and Dragic, T.: Co-receptor for HIV entry, *Curr. Opin. Immunol.* **9** (1997), 450-468.
73. Greenberg, M.L., Lacey, S.L. and Chen, C.H.: Noncytolytic CD8 T-cell-mediated suppression of HIV replication, *Springer Semin. Immunopathol.* **18** (1997), 355-369.
74. Lehner, T., Wang, Y. and Cranage, M.: Protective mucosal immunity elicited by targeted iliac lymph-node immunization with a subunit SIV envelope and core vaccine in macaques, *Nature Med.* **7** (1996), 767-775.
75. Riley, J., Carroll, R.G., Levine, B.L., Bernstein, W., St. Louis, D.C., Weislow, O.S. and June, C.H.: Intrinsic resistance to T-cell infection with HIV type-1 induced by CD28 costimulation, *J. Immunol.* **158** (1997), 5545-5553.
76. Paxton, W., Martin, S.R., Tse, D., O'Brien, T.R., Skurnick, J., VanDevanter, N.L., Padian, N., Braun, J.F., Kotler, D.P., Wolinsky, S.M. and Koup, R.A.: Relative resistance to HIV-1 infection of CD4 lymphocytes from persons who remain uninfected despite multiple high-risk sexual exposures, *Nature Med.* **2** (1996), 412-417.
77. Leith, J.G., Copeland, K.F. and McKay, P.J.: CD8$^+$ T-cell-mediated suppression of HIV-1 long terminal repeat-driven gene expression is not modulated by the CC chemokines RANTES, macrophage inflammatory protein (MIP)-1α and MIP-1ß, *AIDS* **5** (1997), 575-580.
78. Garzino Demo, A., Arya, S.K., Devico, A., Cocchi, F., Lusso, P. and Gallo, R.C.: C-C chemokine RANTES and HIV long terminal repeat-driven gene expression, *AIDS Res. Human Retrovir.* **13** (1997), 1367-1371.
79. Rubbert, A., Weissman, D., Combadiere, C., Pettrone, K.A., Daucher, J.A., Murphy, P.M., Fauci, A.S. and Fauci, A.S.: Multifactoral nature of noncytolytic CD8$^+$ T-cell-mediated suppression of HIV replication: ß-chemokine-dependent and -independent effects, *AIDS Res. Human Retrovir.* **13** (1997), 63-69.
80. Moriuchi, H., Moriuki, M., Combadiere, C., Murphy, P.M. and Fauci, A.S.: CD8$^+$ T-cell-derived soluble factor(s), but not ß-chemokines RANTES, MIP-1α, and MIP-1ß, suppress HIV-1 replication in monocytes/macrophages, *Proc. Natl. Acad. Sci. USA* **43** (1996), 15341-15345.
81. Yang, O., Kalams, S.A., Trocha, A., Cao, H., Luster, A., Johnson, R.P. and Walker, B.D.: Suppression of human immunodeficiency virus type-1 replication by CD8$^+$ cells: evidence for HLA class-I-restricted triggering of cytolytic and noncytolytic mechanisms, *J. Virol.* **71** (1997), 3120-3128.

82. Heeney, J.L., Teeuwsen, V.J., Van Gils, M., Bogers, W.M., De Giuli Morghen, C., Radaelli, A., Barnett, S., Morein, B., Akerblom, L., Wang, Y., Lehner, T. and Davis, D.: Beta-chemokines and neutralizing antibody titers correlate with sterilizing immunity generated in HIV-1-vaccinated macaques, *Proc. Natl. Acad. Sci. USA* **95** (1998), 10803-10808.

83. Mackewicz, C.E., Barker, E. and Greco, G.: Do beta-chemokines have clinical relevance in HIV infection?, *J. Clin. Invest.* **100** (1997), 921-930.

84. Triozzi, P.L., Bresler, H.S. and Aldrich, W.A.: HIV type-1-reactive chemokine-producing $CD8^+$ and $CD4^+$ cells expanded from infected lymph nodes, *AIDS Res. Human Retrovir.* **14** (1998), 643-649.

85. Scarlatti, G., Tresoldi, E., Bjorndal, A., Fredriksson, R., Colognesi, C., Deng, H.K., Malnati, M.S., Plebani, A., Siccardi, A.G., Littman, D.R., Fenyö, E.M. and Lusso, P.: *In-vivo* evolution of HIV-1 coreceptor usage and sensitivity to chemokine-mediated suppression, *Nature Med.* **3** (1997), 1259-1265.

86. Malnati, M., Tambussi, G., Clerici, E., Polo, S., Algeri, M., Nardese, V., Furci, L., Lazzarin, A. and Lusso, P.: Increased plasma levels of the C-C chemokine RANTES in patients with primary HIV-1 infection, *J. Biol. Homeost. Ag.* **11** (1997), 40-42.

87. Verani, A., Scarlatti, G., Comar, M., Tresoldi, E., Polo, S., Giacca, M., Lusso, P., Siccardi, A.G. and Vercelli, D.: C-C chemokines released by lipopolysaccharide (LPS)-stimulated human macrophages suppress HIV-1 infection in both macrophages and T cells, *J. Exp. Med.* **185** (1997), 805-816.

88. Oliva, A., Kinter, A.L., Vaccarezza, M., Rubbert, A., Catanzaro, A., Moir, S., Monaco, J., Ehler, L., Mizell, S., Jackson, R., Li, Y., Romano, J.W. and Fauci, A.S.: Natural killer cells from human immunodeficiency virus (HIV)-infected individuals are an important source of CC-chemokines and suppress HIV-1 entry and replication *in vitro*, *J. Clin. Invest.* **102** (1998), 223-231.

89. Pal, R., Garzino-Demo, A., Markham, P.D., Burns, J., Brown, M., Gallo, R.C. and DeVico, A.L.: Inhibition of HIV-1 infection by the ß-chemokine MDC, *Science* **278** (1997), 695-698.

90. Lang, W., Perkins, H., Anderson, R.E., Rouyce, R., Jewell, N. and Winkelstein, W.: Patterns of T-lymphocyte changes with human immunodeficiency virus infection: from seroconversion to the development of AIDS, *J. Acq. Immune Def. Syndr. Human Retrovir.* **2** (1989), 63-69.

91. Kestens, L., Vanham, G., Gigase, P., Young, G., Hannet, I., Vanlangendonck, F., Hulstaert, F. and Bach, B.A.: Expression of activation antigens, HLA-DR and CD38, on CD8 lymphocytes during HIV-1 infection, *AIDS* **6** (1992), 793.

92. Giorgi, J.V., Liu, Z., Hultin, L.E., Cumberland, W.G., Hennessey, K. and Detels, R.: Elevated level of $CD38^+CD8^+$ cells in HIV infection add to the prognostic value of low $CD4^+$ T-cell levels: results of 6 years follow-up, *J. Acq. Immune Def. Syndr. Human Retrovir.* **6** (1993), 904-912.

93. Mackewicz, C.E., Landay, A., Hollander, H. and Levy, J.A.: Effect of zidovudine therapy on $CD8^+$ T-cell anti-HIV activity, *Clin. Immunol. Immunopathol.* **73** (1994), 80-87.

94. Choremi-Papadopoulos, H., Viglis, V., Gargalianos, P., Gargalianos, P., Kordossis, T., Iniotaki-Theodoraki, A. and Kosmidis, J.: Downregulation of CD28 surface antigen on $CD4^+$ and $CD8^+$ T lymphocytes during HIV-1 infection, *J. AIDS* **7** (1994), 245-253.

95. Barker, E., Bossart, K.N., Locher, C.P., Patterson, B.K. and Levy, J.A.: $CD8^+$ cells from asymptomatic human immunodeficiency virus-infected individuals suppress superinfection of their peripheral blood mononuclear cells, *J. Gen. Virol.* **77** (1996), 2953-2962.

96. Mackewicz, C.E., Yang, L.C., Lifson, J.D. and Levy, J.A.: Noncytolytic CD8 T-cell anti-HIV-1 infection, *Lancet* **344** (1994), 1671-1673.

97. Powell, J.D., Bednarik, D.P., Folks, T.M., Jehuda-Cohen, T., Villinger, F., Sell, K.W. and Ansari, A.A.: Inhibition of cellular activation of retroviral replication by $CD8^+$ T cells derived from non-human primates, *Clin. Exp. Immunol.* **91** (1993), 473-481.

98. Van Kuyk, R., Torbett, B.E., Gulizia, R.J., Leath, S., Mosier, D.E. and Koenig, S.: Cloned human $CD8^+$ cytotoxic lymphocytes protect human peripheral blood leukocyte-severe combined immunodeficient mice from HIV-1 infection by an HLA-unrestricted mechanism, *J. Immunol.* **153** (1994), 4826-4833.

CHAPTER 9

OPPORTUNISTIC INFECTIONS

JANNIK HELWEG-LARSEN, M.D., THOMAS L. BENFIELD, M.D. AND JENS D. LUNDGREN, M.D., D.MSC.

Jens D. Lundgren, M.D., D.MSc, Project leader for EuroSIDA, Department of Infectious Diseases, Hvidovre Hospital, Hvidovre, Denmark

1. Introduction

From the beginning of the Eighties with the first observations of *Pneumocystis carinii* pneumonia among homosexuals in San Francisco and New York to the late Nineties HIV pandemic, drastic changes have occurred in the presentation, epidemiology and understanding of opportunistic infections.

After a brief general discussion of risk factors, epidemiology and a discussion of the interaction of opportunistic infection and HIV replication, we shall focus on recent advances in the understanding of the natural history and immunology in some of the most frequently detected opportunistic infections seen in patients with chronic HIV-infection.

2. Risk Factors for Opportunistic Infections in HIV-infected Persons

The risk of developing opportunistic disease is primarily influenced by exposure, host immunocompetence and access to and quality of patient care.

Exposure to a potential pathogen is required before disease results. Hence ubiquitous opportunistic pathogens cause infection in a large proportion of HIV-infected individuals (*e.g. Mycobacterium avium* complex (MAC), *Cytomegalovirus, Pneumocystis carinii, Candida albicans*), whereas the risk of acquiring non-ubiquitous infectious diseases (*e.g. Tuberculosis, Histoplasmosis, Cryptosporidiosis, Cryptococcosis, Leishmaniasis*) depends on whether the pathogen is endemic in the background population or not.

Host immunocompetence is obviously critical for the development of specific opportunistic diseases (Table I). The risk of acquiring any opportunistic infections increases as the CD4 count approaches 0. However, the association

H. Schuitemaker and F. Miedema (ed.), Aids Pathogenesis, 155–176.

between risk of infection and the CD4 count varies for different pathogens (Table I). Certain common or high-grade pathogens, such as *Mycobacterium tuberculosis*, *Streptococcus pneumonia* and *Varicella zoster* virus are able to cause disease in the early stage of infection (CD4 cell count >200 cells/µl), while *P. carinii*, and *Toxoplasma gondii* become important in the later stage of disease (CD4 count <100 cells/µl). *Cytomegalovirus* and MAC require a virtual destruction of the immune system to develop (*i.e.* CD4 <50 cells/µl. [1].

TABLE I. CD4 counts in relation to opportunistic pathogens. Infections are listed at the highest range of CD4 count where they typically occur. With decreasing CD4 counts the risk of all infections increase and the risk of dissemination increases.

	500>CD4>200	**200>CD4>100**	**100>CD4**
Bacteria	Mycobacterium tuberculosis Bacterial gastroenteritis: Salmonella, Shigella. Bacterial pneumonia: Pneumococcus, Hemophilus, Staphylococcus aureus.	Syfilis	Mycobacterium avium
Parasites	Giardia lamblia	Leishmania Cryptosporidiosis Isosporiasis	Toxoplasmosis Microsporidiosis
Fungi	Candida spp.	Pneumocystis carinii	Cryptococcosis Histoplasmosis Coccidiodomycosis Aspergillosis
Virus	HSV VZV	HHV-8 (Kaposi's sarcoma)	CMV JC-virus (PML) EBV (immunoblast lymphoma)

Abbrevations : CMV : *Cytomegalovirus*, EBV : *Epstein-Barr* Virus, HHV-8 : Human Herpesvirus 8, HSV:*Herpes Simplex* Virus, PML : Progressive multifocal leukoencephalopathy, VZV: *Varicella Zoster* Virus.

Access to care is the perhaps most important risk factor. Although major advances have been made in the therapy of HIV infection and the prophylaxis and therapy of opportunistic infections, these treatments are too costly and unavailable to the great majority of HIV-infected living in the poor part of the world.

3. Epidemiology

The occurrence and risk of opportunistic infections in HIV-infected patients show large geographical variations and even variations within different patient groups in a region. Although the natural history of HIV-infection in developing countries is relatively poorly described, it is clear that tuberculosis (TB) and bacterial infections prevail here [2]. In fact, until recently, it was discussed whether the most common

AIDS-defining disease in industrialised countries, *Pneumocystis carinii* pneumonia (PCP), were seen at all in African HIV-infected patients [3]. Even within industrialised countries, regional differences in the epidemiology of opportunistic infections are present. In a European study of patients diagnosed with AIDS in the period 1979-1989, 30% of AIDS patients in Southern Europe had PCP and 25% had TB compared to 51% with PCP and 6% with TB of patients in Northern Europe [4].

In the last 10 years the incidence of opportunistic infections in the western world has changed. Now many opportunistic infections tend to occur at more advanced stages of immunosuppression compared to the beginning of the 1980s [5]. One explanation is the introduction of prophylactic regimens against opportunistic infections, in particular *P.carinii* prophylaxis. Before the widespread use of anti-pneumocystis prophylaxis, PCP was the major cause of AIDS in the developed countries. In 1996, PCP still ranked as one of the most common AIDS-indicator opportunistic infection in the US and Europe, but is now followed more closely in incidence by *Oesophageal candidiasis*, *Tuberculosis*, *Toxoplasmosis*, *Cytomegalovirus* disease, MAC disease, *Herpes simplex* and *Cryptococcosis* (Fig. 1). As certain types of prophylaxis against PCP also protect against toxoplasmosis, the risk of contracting this disease has also diminished considerably.

FIGURE 1 AIDS-defining diagnosis in Europe 1994-1996 by percentage.

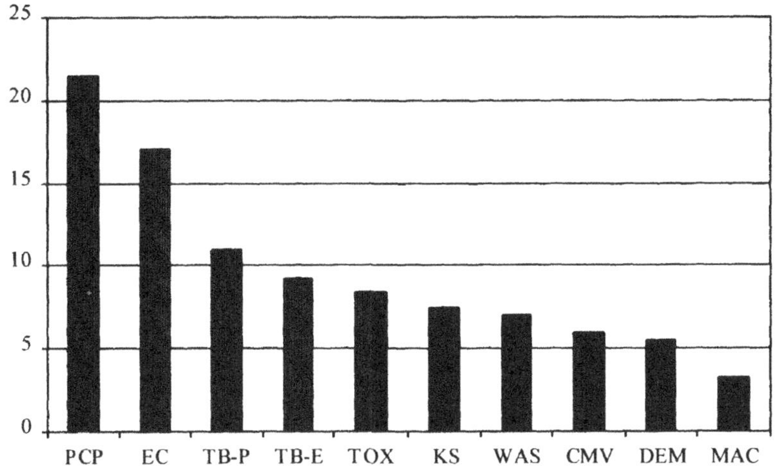

PCP:*Pneumocystis carinii* pneumonia, EC: Esophageal candidiasis, TB-P: Pulmonary Tuberculosis, TB-E: Extrapulmonary Tuberculosis, TOX: Cerebral toxoplasmosis, KS: Kaposi sarcoma, CMV: *Cytomegalovirus*-disease, WAS: Wasting, DEM: AIDS-dementia, MAC: *Mycobacterium avium* complex disease

4. Opportunistic Infections in the Area of HAART

Recent introduction of highly active anti-retroviral therapy (HAART) has led to dramatic changes in the natural history of HIV infection and opportunistic infections. Since 1996, the use of increasingly intensive anti-retroviral therapies has directly resulted in declines in morbidity and mortality among HIV-infected patients with advanced immune depletion. As highly active anti-retroviral therapy has become the standard of care in developed countries a change in the epidemiology of HIV-related opportunistic infections is being observed. In one study the incidence of any of the three major opportunistic infections, *Pneumocystis carinii* pneumonia, MAC disease and *Cytomegalovirus* retinitis declined from 21.9 per 100 person-years in 1994 to 3.7 by mid-1997 - in another study the risks of death were reduced by more than 80% in patients under active follow-up at clinical sites [6]. Furthermore, admission rates to in-patient wards has in some centres been reduced more than 5-fold. The most marked reduction has occurred at a time when the use of protease inhibitors became widespread [8].

In addition to this major impact on incidence of opportunistic infections, HAART has also effect on the clinical presentation of specific opportunistic infections. The reports of resolution or improvement of opportunistic conditions now include *Cytomegalovirus* retinitis, Kaposis sarcoma, progressive multifocal leucoencephalopathy, molluscum contagioum, cryptosporidiosis, microsporidosis, MAC and hepatitis B [9]. Furthermore, a new clinical syndrome "immune restoration syndrome" has been described among patients with advanced AIDS and clinical latent, but ongoing opportunistic infections. Within weeks of starting HAART, some patients mount a partial host response to pathogens which induces inflammation around established infection. The syndrome has been described in patients infected with *Cytomegalovirus*, TB, MAC-disease, hepatitis B and hepatitis C with the observation of vitreitis, MAC adenitis and aggravation of hepatitis following the initiation of HAART [11]. Patients with a low CD4 count (<100 cells/mm^3) when starting HAART are at particular risk of developing the immune reconstitution syndrome (due to the general increase risk of having opportunistic infections), which will develop within the first couple of months after the anti-retroviral therapy has begun. Precise prevalence data are lacking, but the syndrome is a rare event.

The observation of HAART-associated improvement in immune function has raised the question of whether prophylactic therapy, primary or secondary (maintenance), can be stopped. At present, a few studies with relatively short follow up and a limited number of patients have investigated withdrawal of prophylaxis with HAART. Previously, active life-long therapy was recommended for all patients diagnosed with *Cytomegalovirus*-disease, however recent reports observed that anti-*Cytomegalovirus* maintenance therapy could be safely discontinued in patients whose CD4 counts increases to a stable level above 200 cells/mm. Similarly, stopping primary (and less frequently secondary) prophylaxis against pneumocystosis in patients with a CD4 count stable above 200 cells/mm^3 appears to be safe. For both *Cytomegalovirus* and PCP, the current incidence of active disease appears very small in patients on HAART, which is a substantial reduction, compared with historical data in patients without immune recovery and with a comparable CD4 nadir. In patients with clinical side effects due to the prophylaxis and whom have responded to HAART, discontinuation appears to be a safe and reasonable strategy.

4.1 INTERACTION

Besides causing morbidity by themselves, opportunistic infections may also accelerate the underlying HIV infection. *In-vivo* studies have shown that several opportunistic infections increase HIV replication and demonstrated an increased HIV viral load during infection [13]. Although the occurrence of these infections is closely linked to the decrease in CD4 count, it has been shown that most opportunistic diseases increase the risk of death independently of CD4 cell count [14].

It has been suggested that the higher rate of HIV disease progression in Sub-Saharan Africa in part may be explained by immune activation associated with ongoing responses to parasites and other infections, causing a shift in the cytokine balance towards a T-helper 2-type response (with increased Interleukin-4, Interleukin-5 and Interleukin-10 production), making the host more susceptible to infection with HIV and its progression [15]. In particular, *Mycobacterium tuberculosis* seems to cause a more rapid progression of HIV in Africa. Observational studies have found that patients with TB compared to those without, are at increased risk of dying from the HIV-infection [16]. Studies have reported that active (but not latent) pulmonary tuberculosis is associated with a 5- to 160- fold increase in plasma HIV-1 viral load [17].

TABLE II. Mechanisms by which opportunistic infections could interact with HIV replication. Modified from [71].

Mechanism	Pathogen	reference
Cytokine production	Mycobacterium tuberculosis	[17]
	Mycobacterium avium	[13]
	Pneumocystis carinii	[72;73]
	Toxoplasma gondii	[74]
	CMV, EBV	[49;75]
Transactivation	Mycobacterium tuberculosis	[76] [77]
	CMV	[78]
	HSV	[79]
	HHV-6	[80]
	EBV	[81]

Abbrevations : CMV: Cytomegalovirus; EBV: Epstein-Barr Virus; HHV: Human Herpes Virus; HSV: Herpes Simplex Virus

A number of mechanisms (Table II) may explain the increase in HIV replication during co-infection with opportunistic infections. Recently, tissue macrophages have been shown to be a highly productive source of HIV in lymph nodes co-infected by HIV and common opportunistic infections, such as MAC and *P. carinii*. In patients with opportunistic infections, *in-situ* hybridisation revealed large amounts of virus associated with macrophages in opportunistic infections-infected lymph nodes, whereas non-infected nodes contained far less virus [18].

HIV replicates more efficiently in activated cells. Transcription of viral DNA can be triggered by host-cell activation by cytokines, mitogens, antigens or gene products of viruses, such as *Epstein-Barr* virus, *Cytomegalovirus*, *Herpes simplex* virus and *Human Herpes* virus-6. In both T cells and macrophages the inducible nuclear transcription factor, NF-κB is important. NF-κB upregulates viral transcription by binding to regulatory sequences in the long 5`terminal repeat (LTR) of the viral DNA and upregulates viral transcription [19]. Co-infection studies have shown that *Herpes simplex* virus-1 can reactivate transcription of latent HIV-1 and that *Cytomegalovirus* can transactivate the HIV-1 LTR promotor *in vitro*, resulting in increased transcription [20]. In addition, a number of cytokines are able to upregulate HIV replication. Upregulation of NF-κB can be induced by cytokines as interleukin-1β, interleukin-6, tumour necrosis factor-α. In addition, interleukin-2, interleukin-3, interleukin-4, tumour necrosis factor-β, macrophage-colony-stimulating factor, granulocyte-macrophage colony-stimulating factor are able to upregulate the replication of HIV. As activation of the immune system with increased cytokine production is a general feature of infection, opportunistic infections may thus result in activation of HIV-1 production. The effect of immune activation is further supported by the observation of transient increases in viraemia following vaccination of HIV-infected persons [21]. Thus, opportunistic infections may cause more rapid progression of HIV by increasing HIV production.

5. Pneumocystis Carinii

Pneumocystis carinii remains the most common cause of life-threatening pulmonary opportunistic infection in HIV-1-infected persons. Before the introduction of prophylaxis against PCP, PCP was seen in 80% of all patients with AIDS and represented the AIDS-defining illness for 60% of people with AIDS. Although PCP prophylaxis is now used extensively, approximately 20% of patients with HIV develop PCP. In 1996, PCP occurred as the AIDS-defining condition in 26% of cases in the U.S.

P. carinii was long mis-identified as a protozoan based on its morphological appearance, its failure to grow in culture, and the effect of anti-protozoan but not anti-fungal therapy. However, in 1988 a rRNA sequence homology study of *P. carinii* established a close phylogenetic link with fungi, especially *Saccharomyces cerevisiae* [22]. Since then, sequencing of several other genes has documented the fungal nature of the organism with rRNA sequencing placing *P.carinii* in an evolutionary tree between ascomycetes and basidiomycetes.

5.1 NATURAL HISTORY

In animal studies, *P.carinii* can be transmitted by air, and human *P.carinii*-DNA has been detected in air [23]. However, the inability to culture the organism *in vitro* has hampered studies on basic biology, and the environmental reservoir, lifecycle of the organism and the exact way of infection are still not definitely known. Based on the high prevalence of antibodies to *P.carinii* in early childhood, PCP was thought to arise by reactivation of a latent infection, which remained latent in the lung until immunosuppression of the host . This hypothesis is now being questioned. Studies using sensitive molecular probes have failed to find *P.carinii* in the lungs of immunocompetent persons. A PCR search in post-mortem samples of lung tissue from immunocompetent persons did not reveal *P.carinii* DNA [24] . Animal models have shown limited persistence of *P.carinii* after a primary episode of pneumonia [25]. In addition, hospital clusters of PCP have been reported, suggesting infection immediately prior to the time of disease. Taken together, these findings could imply that re-infection is the predominant mode of infection. However, the case is not clear-cut in immunocompromised patients. In some studies *P.carinii* DNA has been detected by PCR in bronchoalveolar fluids or sputum from asymptomatic HIV patients, and although some of these patients subsequently developed PCP, not all patients did [26]. Not all studies have confirmed these findings, and to what extent asymptomatic colonisation with *P.carinii* occurs in immunocompromised patients is not definitely known.

P.carinii causes infection in a number of mammals, with organisms morphologically indistinguishable from *P.carinii* in humans. However, host specificity is strict with phenotypic and genetic divergence between *P.carinii* infecting different animals and inability of transferring organisms from one species to another [27]. Genetic divergence occurs among isolates of human *P.carinii*. Several loci have shown genetic polymorphism, the most studied being the gene encoding mitochondrial large subunit rRNA (mtrRNA) and the internal transcribed spacer regions (ITS1 and ITS2) of rRNA [28]. A number of studies have examined *P.carinii* types in a small number of recurrent episodes of PCP (Table III).

TABLE III. Comparison of *Pneumocystis carinii* genotypes from the first and second episode in patients with recurrent *P. carinii* pneumonia. Results from three separate studies, using two different ways of genotyping. Number of patients with different or same type. Sixty percent had a different genotype, suggesting re-infec-tion. See text for details.

Study	Loci	N	Same type	Different type
Keely[82]	mtrRNA	12	6	6
Tsolaki [83]	ITS	7	3	4
Latouche[84]	ITS	6	1	5
Total (%)		25	10 (40%)	15 (60%)

Abbrevations : mtrRNA : Mitochondrial large subunit rRNA, ITS : Internal transcribed spacer regions of the nuclear rRNA operon.

Taken together, these studies show different types of *P.carinii* in first and second disease episode during more than half of cases. These findings suggest the occurrence of both reactivation of a previously acquired infection and re-infection from an exogenous source. However, one caveat to the interpretation of these results is the observation of mixed infection with several different ITS genotypes of *P. carinii* present in one episode of pneumonia. In approximately 10-30% of episodes mixed infection can be detected, depending on the sensitivity of the PCR and cloning technique. In addition, several new types of ITS have recently been described with over 50 different genotypes in a large study comparing *P.carinii* isolates from different parts of the world [30]. Thus, a genotype switch could have been the result of reactivation of organisms present but undetected during the first episode of pneumonia.

After inhalation, *P.carinii* takes up residence in the alveoli and attaches tightly to type-I pneumocytes. Adherence of *P.carinii* occurs via several different pathways and stimulates an oxidative burst, cytokine production, phagocytosis and killing of the organism. Adherence is primarily mediated by the 95-140 kDA major surface glycoprotein (MSG). MSG show high level antigenic variation. Genes encoding MSGs are highly polymorphic, repeated and distributed among all chromosomes of *P.carinii*. MSG expression is varied on the surface at high frequency by DNA re-arrangement of these genes with only one MSG isoform expressed per organism. In this way, the organism may evade host immune surveillance [31].

The pathology of PCP is typically a severe interstitial pneumonia. Inflammation is seen with an infiltrate of plasma cells, lymphocytes and neutrophils and erosion of type-I pneumocytes. Alveoli are filled with a protein-rich exudate consisting of *P.carinii* trophozoites and cellular debris. Alveolar membranes become thickened and parenchymal inflammation, oedema and fibrosis occur. In addition, alterations in the surfactant system have been shown in rodent models of PCP with a fall in phospholipid production. Hypoxaemia and ventilation-perfusion abnormalities follow, in severe PCP resulting in diffuse alveolar damage.

The immune response to *P.carinii* is essential both to the defence against infection and to the lung injury that follows infection. Cell-mediated immunity appears to be essential in the prevention and control of *P.carinii* infection. Studies in SCID mouse have shown that $CD4^+$ cells are essential for protection against infection and that CD8 cells contribute to host defence [32].

Containment of *P.carinii* within alveoli requires intact humoral immunity, in this way patients with pure B-lymphocyte dysfunction are susceptible to PCP and $CD4^+$ depleted rats can be protected against PCP by immunisation with *P.carinii* antigen [33]. In addition to the perturbations seen in cellular and humoral immunity, HIV may alter innate immunity. A recent study found impaired alveolar macrophage mannose receptor-mediated binding of *P.carinii* in HIV-positive individuals [34].

Evidence suggests that a superimposed inflammatory reaction contributes to the lung injury seen during PCP. Adjuvant corticosteroid therapy in addition to antimicrobial therapy decreases the risk of acute respiratory failure of moderate to severe (*i.e.* pO_2<9kPa) PCP by 40%. Secondly, an increase in bronchoalveolar neutrophilia confers a poor prognosis and correlates to the degree of lung impairment. Recent studies implicate cytokines in the pathogenesis. Cytokines contribute to defense against *P. carinii* but also to the lung injury by recruitment and induction of inflammatory cells with resulting alterations of local homeostasis and increased vas-cular permeability. Production of tumour necrosis factor α and the neutrophil chemoattactant interleukin-8 from monocytes and alveolar macrophages is increased, with levels of interleukin-8 in bronchoalveolar fluid correlated to the severity of pneumonia [35]. The release of cytokines may be aggravated by the initiation of antimicrobial therapy due to increased release of immunogenic material from *P. carinii*, similar to the situation seen after initiation of therapy against several helminths (the so-called Jarich-Herxheimer reaction).

6. Tuberculosis

Tuberculosis (TB) is the most common life-threatening HIV-related infection in the world. WHO estimates that of nearly 31 million people world-wide who were HIV-positive in 1997, around one-third were infected with TB. TB is responsible for almost one-third of AIDS deaths world-wide. In Africa, HIV is the single most important factor determining the increased incidence of TB in the last ten years. In developing countries a third of HIV-infected patients have TB as initial clinical presentation and a half have evidence of disease at *post-mortem* examinations. In the United States, TB cases increased 19% between 1985 and 1992, with an estimated excess cases of 52.000. This resurgence was mainly caused by HIV. European data for 1994 show that among 19.978 AIDS cases, 3805 (19%) were diagnosed with TB [36]. The highest incidence was seen in Spain and Portugal, where TB accounted for an astounding 51% and 42%, respectively, of AIDS cases.

6.1 NATURAL HISTORY

TB often occurs at a relatively early stage of immunodeficiency, reflecting the virulence of the organism. Active disease may be a re-activation of a latent infection or may directly follow the primary infection. In most immunocompetent persons the primary infection is contained within 2-10 weeks with development of a delayed-type hypersensitivity (DTH) response. However, about 5-10% will develop either progressive primary disease in the 2 years immediately following the primary infection, or re-activation of TB after a latent period. In HIV-infected persons with previous TB infection (as determined by skin testing to detect the DTH response), 5-12% per year will develop active TB (Table IV) [37].

TABLE IV. Risk of developing infection with Mycobacterium tuberculosis in persons with and without HIV infection.

	General Population	HIV co-infection
Primary disease	5% in 2 years	37% in 6 months
Reactivated disease	5%	5-12% per year
Cumulative life-long risk	< 10%	>50%

Nearly all latently infected HIV-infected patients will develop active infection within 10 years after primary infection. However, the annual risk of developing active tuberculosis is markedly higher for patients infected when their immune system is very impaired. Therefore, the ratio between primary infection and re-activation of latent infection is expected to increase as the CD4 count approaches

0 cells/μl. As a reflection of this, increased immunodeficiency is associated with an increased likelihood of dissemination of TB. Thus, patients with relatively high CD4⁺ counts (250-500 cells/μl) typically present with pulmonary TB, whereas patients with low CD4 counts present with disseminated disease.

Although incidence rates of HIV-associated TB is highest in countries with high TB rates in the general population, re-activation of latent TB cannot explain all cases. In Southern Europe HIV-associated TB is more common in young age and among injecting drug users. This suggests that contact transmission with *Mycobacterium tuberculosis* is common. This has been confirmed by studies using restriction-fragment length polymorphism typing, where about one-third of HIV-associated TB cases in cities in the U.S.A. were caused by recent infection [38].

After inhalation of aerosolised mycobacteria, bacteria multiply in the alveolar spaces or within mononuclear phagocytes, either as tissue macrophages or circulating monocytes. Mononuclear phagocytes are both the preferred habitat of mycobacteria and the chief effector cells for killing the bacteria. After phagocytosis of mycobacteria, the initial response of the macrophage determines whether productive infection ensues. Virulent strains of *Mycobacterium tuberculosis* fail to stimulate the early activation of macrophages, enabling the organisms to proliferate in macrophage vesicles [39].

Infected monocytes/macrophages act as antigen-presenting cells and as releasers of cytokines stimulating a variety of T-cell subsets [40]. The T-cell-mediated response is complex, involving interactions between different subsets of T cells, monocyte-macrophages, cytokines and other mediators. CD4 cell subsets are the principal effector cells by activating macrophages and are also able to directly lyse infected monocytes. Persistence of mycobacteria within macrophages provides chronic stimulation to T cells inducing cytokines that transform macrophages into epi-theloid cells which may fuse to form multinucleate giant cells and granulomas. Aggregation of epitheloid cells to form granulomas is dependent of tumour necrosis factor-α released by antigen-reactive T cells and activated macrophages. *In-vitro* concomitant exposure of HIV and tuberculosis antigens increases the release of tumour necrosis factor-α and interleukin-1β from monocytes compared with non-HIV-exposed cells. Thus, the inability of patients with severe immunodeficiency to form granulomas appears not to be due to a direct effect on cytokine production, but is rather due to impairment of cellular immunity.

Prevention of TB is important, but difficult to achieve. The effect of chemoprophylaxis has been evaluated in several studies. A recent meta-analysis reviewed 4 trials of preventive treatment for tuberculosis in HIV-infected individuals conducted in Uganda, Kenya, Haiti and the U.S. Preventive isoniasid treatment given for 3-12 months protected against tuberculosis in adults infected with HIV, at least in the short to medium term. Evidence of protection was observed exclusively in subjects positive for purified protein derivative (PPD), while the risk was not reduced in HIV-infected patients not responding to PPD. Overall, mortality was significantly reduced in PPD-positive individuals [41].

7. Mycobacterium Avium Complex

Infection with non-tuberculous mycobacteria is commonly seen in late-stage HIV infection, usually in patients with $CD4^+$ counts less than 50/μl. Disease in AIDS patients is caused principally by the common environmental saprophyte *Mycobacterium avium* complex, which encompasses *M.avium* and *M.intracellulare*.

Disseminated MAC infection is common in the developed world, with an annual incidence of approximately 20% after the occurrence of a first AIDS-defining event. MAC disease is associated with increased morbidity and shortened survival. Infection occurs in adults with advanced HIV infection in Sub-Saharan Africa, but is much less common than disseminated TB.

Both newer macrolides and rifabutine has been shown to decrease the risk of developing MAC and improve survival of susceptible patients (*i.e.* those with CD4 count <50-75 cells/μl) [42]. Chemoprophylaxis with either claritromycin or azitromycin is now recommended for patients with CD4 count <50 cells/μl. In the pre-HAART era prognosis was poor when disease developed, with median survival 8-9 months even with multidrug regimens.

7.1 NATURAL HISTORY

In AIDS patients, it is not clear whether MAC is a primary infection or a reactivated latent infection. Infection is acquired orally or inhaled. As MAC infection is distributed equally among patients in all HIV-transmission categories, a possible environmental source has to be universal. The source(s) of infection remains to be deter-mined, but potential candidates include water, soil, vegetables, unpasteurised milk and possible aerosols. MAC organisms have been isolated from drinking water and cases with demonstration of the same serotypes in both patients and hospital hot water supply has been described [44]. However, exposure to water is not associated with an increased risk of MAC and despite several attempts, it has not been possible to clearly delineate genotypes of MAC detected in AIDS patients and the genotypes of MAC found in their environment [45].

After the initial infection MAC survive within macrophages, granulomas are absent, but acid-fast bacilli are numerous within macrophages. MAC can be isolated from the respiratory or gastrointestinal tract in the absence of disseminated infection. Detection of MAC in stool or sputum is associated with an increased risk of contracting clinical disease. In one study patients with MAC, in either the respiratory or gastrointestinal tract and CD4 counts <50, had a ~60% risk of developing MAC bacteraemia within 1 year. But, in the same study two-thirds of the patients who eventually developed bacteraemia had negative cultures from the respiratory or gastrointestinal tract. The pathogenesis of disseminated MAC is thus not completely understood. One hypothesis is that MAC disease starts with colonisation of the respiratory or gastrointestinal tract, establishing a localised infection with intermittent MAC bacteraemia followed by seeding of other organs. After an unknown period of proliferation in the infected organs, the increasing organism burden eventually results in "spillover" sustained bacteraemia and clinical symptoms. This model is supported by studies from autopsies of HIV-infected patients with MAC bacteraemia, which report MAC concentrations in multiple organs to be many times higher than concentrations in the blood.

The most common clinical manifestations of MAC disease are anaemia, pancytopenia, fever, night sweats, wasting, diarrhoea, hepatic dysfunction and lymphadenopathy. Most patients exhibit a continuous bacteraemia with large number of organisms per ml of blood. The intensity of clinical symptoms are directly correlated with the bacterial burden as determined by the number of colony-forming units in the blood. Polyclonal infection is common, with more than one strain of MAC being observed in up to 38% of patients. This may provide an explanation for the inability to correlate the outcome of antibiotic treatment with susceptibility patterns.

As MAC is seen in connection with severe CD4 depletion, the most promising therapy is now anti-microbial therapy combined with HAART. Recently, cure of disseminated MAC with anti-microbial therapy combined with HAART was reported in 4 patients. At diagnosis of MAC all 4 patients had CD4 counts <10, but after 12 months of HAART and macrolide therapy the 4 patients had absolute CD4 counts >100 cells/μl and <10.000 copies/ml of HIV RNA and all mycobacterial cultures were sterile [10].

8. Cytomegalovirus

Like all herpes viruses, *Cytomegalovirus* remains latent in the host after primary infection. The replication of *Cytomegalovirus* is seen in a variety of cells, including epithelia, endothelia and fibroblasts and is able to cause intracellular inclusions, cell swelling and cell death. After a usually asymptomatic primary infection, *Cytomegalovirus* latently infects several different blood subsets, including lymphocytes, monocyte/macrophages, neutrophils and haematopoietic progenitor cells.

Transmission can occur by contact with blood, genital secretions, breast milk, saliva, urine or faeces. Excretion of virus is seen intermittently in asymptomatic immunocompetent individuals. *Cytomegalovirus* cannot ordinarily be detected in immunocompetent adults, but a high carriage frequency has been found in healthy homosexual men, where 30-40% shed *Cytomegalovirus* in semen or urine. As *Cytomegalovirus* and HIV are transmitted by the same routes (sexually), most AIDS patients (75-98%) are co-infected. With advancing HIV infection and impairment of cell-mediated immunity, latent *Cytomegalovirus* re-activation may occur. This is most likely with CD4 counts below 100 cells/μL. *Cytomegalovirus* disease occurs in 25-40% of patients with a CD4 count less than 50 cells/μL, with autopsy studies demonstrating that the virus actively replicate in up to 90% of dying AIDS patients (although the proportion of patients with clinical disease due to viral replication remains unanswered). Excretion of *Cytomegalovirus* increases with increasing immune dysfunction, which may account for the high rate of transmission among homosexual men.

The most common manifestations of *Cytomegalovirus* disease in HIV patients are retinitis, which are seen in up to 15% of AIDS patients, and gastrointestinal disease affecting up to 10% of patients. Retinitis accounts for 85% of AIDS-related *Cytomegalovirus* disease. Other clinical syndromes are encephalitis, polyradiculitis, hepatitis and pneumonitis. When *Cytomegalovirus* disease develops anti-*Cytomegalovirus* is usually able to halt progression but does not eradicate *Cytomegalovirus*. Life-long maintenance therapy is required but, in spite of this, the disease eventually progresses as a consequence of the emergence of drug-resistant *Cytomegalovirus*.

Why *Cytomegalovirus* predominantly affects the eye in HIV-infected persons is unknown, as it is rare in other patients with *Cytomegalovirus* infection and immunosuppresion. Suggestions are that HIV may induce microvascular damage to the eye with haematogenous spread through breaks in the capillary endothelium and subsequent infection of the retina [46]. Another unsolved question is how come *Cytomegalovirus* is an important pulmonary pathogen in transplant patients, but rarely causes pneumonia in HIV-infected patients. An immunological reaction causing *Cytomegalovirus pneumonitis* has been suggested, which should take place only in non-HIV-infected immunosuppressed patients.

Diagnosis of *Cytomegalovirus* disease may be difficult. Not all patients with blood, urine or tissue cultures positive for *Cytomegalovirus* have an illness related to the infection. Serology is not helpful as many have persistent IgM antibodies and culture of the virus may be positive without implication of active *Cytomegalovirus* disease. *Cytomegalovirus*-retinitis is diagnosed by the typical ophthalmoscopic appearance, but in other *Cytomegalovirus* diseases the verification of *Cytomegalovirus* disease requires histologic evidence of or demonstration of the virus or viral activity in the affected tissue. Newer methods for the detection of *Cytomegalovirus* disease has been developed. PCR test can detect virus in plasma and serum and is at least 10-fold more sensitive than viral cultures. More than 80% of patients with *Cytomegalovirus* retinitis are *Cytomegalovirus* PCR-positive at the time of diagnosis and almost two-thirds are *Cytomegalovirus*-positive up to one year before diagnosis. Recent studies have shown that the risk of organ-related *Cytomegalovirus* infections in severely immunosuppressed patients can be predicted by the detection of *Cytomegalovirus*-DNA in plasma as assessed by the PCR-method (relative risk 3.4 compared with patients without a positive reaction) [47]. Among patients with a positive PCR reaction, the level of *Cytomegalovirus*-DNA in plasma predicted a progressively increased risk of *Cytomegalovirus* disease (3.1-fold increased risk for each $\log_{10}$ increase in viral load (copies of DNA/mL)). Thus, organ-specific *Cytomegalovirus* disease is a result of a disseminated infection.

Cytomegalovirus strains can be compared by sequence analysis of glycoprotein genes after amplification with PCR. The gB and gH *Cytomegalovirus* glycoproteins are important targets for the formation of neutralising antibodies and essential for viral entry into host cells and cell-to-cell spread of *Cytomegalovirus*. Analysis of polymorphism within the gB genes has led to the definition of four gB and 2 gH genotypes [48]. An increased incidence of retinitis in HIV patients with the gB2 genotype has been reported. Class-I-restricted $CD8^+$ cytotoxic T cells and NK cells are necessary for recovery from *Cytomegalovirus* infection, with characteristic expansion and activation of $CD8^+$ and NK subsets during infection.

Cytomegalovirus infection has been proposed as a factor in HIV pathogenesis, but the evidence supporting a co-factor role for *Cytomegalovirus* is controversial. An *in-vitro* interaction with HIV has been documented, but direct *in-vivo* evidence for its role as a co-factor in the development of AIDS is missing [49]. *Cytomegalovirus* can enhance transcription of HIV`s genome in some cell types, but less than 1% of HIV-infected CD4 cells harbour *Cytomegalovirus*. Recent research has shown that *Cytomegalovirus* encodes a β-chemokine receptor US28 that is related to the chemokine receptors CCR5 and CXCR4. US28-bearing cells allow infection by HIV strains that normally use the CCCR5 and CXCR4 β-chemokines as co-factors for the entry in CD4 cells [50]. However, epidemiological studies have come to conflicting results concerning an association between initial *Cytomegalovirus* serostatus and progression to AIDS [51].

9. Kaposi's Sarcoma and Human Herpesvirus 8

Kaposi's sarcoma (KS) is the most frequent neoplasm among individuals infected with HIV in the Western hemisphere [52]. Although there has been a decline in the proportion of individuals presenting with KS at the time of AIDS, 30% will develop KS during the course of their HIV infection [53]. Similarly, the development of KS is associated with more increased immunosuppression in recent years compared to before 1985.

The epidemiology of KS indicates that an infectious agent other than HIV may be the etiologic cause of KS [54]. Homosexual men have a 10-fold increased risk of developing KS compared to all other transmission groups. Further, women infected through contact with bisexual men have an increased risk of developing KS compared with women infected via intravenous drug use or blood transfusion.

A novel herpesvirus belonging to the Rhadinovirus genus, and known as Kaposi's sarcoma-associated virus or Human Herpesvirus 8 (HHV-8), has been detected in KS lesions from patients with AIDS-associated KS, classic KS, and endemic African KS. These studies demonstrated the presence of HHV-8 DNA by PCR. Recently, HHV-8 was propagated from skin lesions of patients with AIDS-associated KS [55]. KS lesions are characterised histologically by a mixture of spindle cells and endothelial-lined vascular spaces. By *in-situ* PCR HHV-8 is found in both spindle cells and endothelial cells [56].

The pathogenesis behind the development of KS is unknown. Several growth factors, including interleukin-1β, interleukin-6, tumour necrosis factor-α, platelet-derived growth factor, basic fibroblast growth factor, and endothelial growth factor, regulate tumour growth in cultures of spindle cells from patients with AIDS. HIV-1 tat protein in combination with basic fibroblast growth factor have a synergistic effect on the induction of tumour growth [57]. Sequencing of the HHV-8 genome has identified a number of virally encoded proteins with close homology to human proteins. So far, genes encoding a constitutively activated G-protein-coupled receptor, and bcl-2, vIL-6 and cyclin-D homologs have been identified [58]. Common for these molecules is the *in-vitro* ability to act as oncogenes through transformation, proliferation and prevention of apoptosis. In this respect, it is of interest that intralesional administration of human chorionic gonadotropin induces regression of KS tumours, presumably through induction of apoptosis [59].

Several serologic assays have been applied to various populations and risk groups [60]. Generally, HHV-8 antibody seroprevalence is high among individuals with KS, persons living in endemic areas, and HIV-infected homosexual men with and without KS. Seroprevalence is low among blood donors, intravenous drug users and HIV-infected haemophiliacs (Table V).

TABLE V. Seroprevalence of HHV-8 among selected risk groups and populations.

HIV				
	Homo-/bisexual men n=198	AIDS and KS n=284	IVDU n=51	Haemophilia n=76
	13-90%	65-96%	2-23%	0%

Non-HIV						
	KS n=50	High risk* n=166	IVDU n=38	Central African origin n=256	Blood donors n=557	Children <15 years n=265
	94-100%	5-12%	0%	35-100%	0-20%	2-8%

*: >1 sexually transmitted disease or unprotected sex. Refs.: [60;63;85-87].

The association between an active sexual life, the correlation with other sexually transmitted diseases, and demonstration of HHV-8 DNA in semen suggest that HHV-8 is transmitted through sexual contact and in particular through anal intercourse. Howeverk, the prevalence of HHV-8 DNA in tissues and bodily fluids is debated. In HIV-1-negative patients the detection rates of HHV-8 DNA in semen and the urogenital tract has varied in different investigations, perhaps reflecting different seroprevalences in the populations tested. A recent study searching for HHV-8 was able to detect HHV-8 DNA in skin lesions, normal skin, peripheral blood mononuclear cells, plasma, saliva and semen but not stool samples from patients with KS. Detection rates and HHV-8 copy number were highest in peripheral blood mononuclear cells and saliva samples, suggesting that oral transmission might also be of importance. In addition, while there is no evidence for association between plasma transfusion and KS, HHV 8 may possibly be transmitted through kidney transplants [64].

10. Toxoplasmosis

Cerebral toxoplasmosis is the most frequent infection of the central nervous system in AIDS patients in Europe. *Toxoplasma gondii* is an ubiquitous, obligate, intracellular protozoan widely distributed in human populations. Three life forms occur: The oocyst (in the intestine of cats), the tachyzoite (asexual invasive form) and the bradyzoit (persists in tissues during the latent stage of infection).

10.1 NATURAL HISTORY

Cats are the definitive hosts of *T. gondii* and the only animal that pass oocyst in their faeces. Major routes of transmission are oral or congenital. Oral infection is the most common with ingestion of tissue cysts from undercooked meat or oocysts from cat faeces or contaminated foods. World-wide seroprevalence for *T. gondii* varies between 8 and 90% with the highest prevalence in Central Europe and Sub-Saharan Africa.

In immunocompetent persons the absence of symptoms of infection is the rule. After ingestion, organisms spread from the gut by lymphatics and bloodstream throughout the body, where they multiply intracellularly. Termination of this stage depends upon both humoral and cell-mediated immunity. The predeliction of toxoplasma for brain tissue has been ascribed to low local immunity. Following primary infection, a latent phase with asymptomatic persistence of viable *T.gondii* in cyst form ensues. *T.gondii* apparently generates its own vesicle without fusing to any cellular vesicle, in this way isolating itself from the rest of the cell. However, in immunocompromised patients re-activation of chronic infection may occur with release of bradyzoites from cysts with development of toxoplasmosis. Alternatively, re-activation outside the brain with haematogenous spread and seeding in the brain has been suggested, supported by the observation of parasitaemia in 14-38% of toxoplasma encephalitis patients [66].

In AIDS patients most cases of toxoplasmosis is seen as a focal encephalitis. Toxoplasmic encephalitis (TE) almost invariably results from re-activation of latent infection. TE occurs at an advanced stage of HIV infection with CD4 counts <100 cells/µl. Before HAART, between 20 to 47% of AIDS patients seropositive for *T. gondii* would ultimately develop TE. Toxoplasma-seropositive patients with CD4 counts of <100/µl should be given prophylaxis against TE. Trimethoprimsulfamethoxazole, the first choice agent for PCP prophylaxis, also has some protection against toxoplasmosis [67].

Information on the immune response to *T.gondii* is mainly derived from animal experiments. Humoral immunity results in limited protection against less virulent strains of *T.gondii* but not against virulent strains. In murine models, interferon-γ and tumour necrosis factor-α are important in the development and/or progression of toxoplasmic encephalitis. In addition, interleukin-4 is protective against development of TE by preventing formation of *T.gondii* cyst and proliferation of tachyzoites in the brain. Interferon-γ enhances survival of mice and limits severity of encephalitis. Lymphocytes from AIDS patients have impaired production of interferon-γ and interleukin-2 in response to stimulation with toxoplasma antigens [68].

11. Concluding Remarks

The occurrence of the HIV epidemic has focused research on infections associated with immunodeficiency. Results of the research can be used not only for the benefit of HIV-infected patients, but also a variety of other types of immunodeficiencies, including those iatrogenically induced. Effective, cost-effective prophylactic regimens are now available for many of the most prominent pathogens, excluding tuberculosis and *Cytomegalovirus*.

The introduction of a long-lasting anti-HIV response by combining three or more anti-retroviral drugs is, however, the most consistent and logical way to prevent the occurrence of opportunistic infections. HAART can increase the number of circulating CD4 T cells even in patients with advanced AIDS. The increased number of CD4 T lymphocytes is both a result of regeneration of T cells and of redistribution of T cells sequestered during periods of active HIV-1 replication. Importantly, after effective therapy, the repertoire of T-cell receptors for both CD4 and CD8 lymphocytes remains skewed despite a greater number of circulating T cells. The increase in T cells appears to be composed of an initial redistribution of memory cells followed by a slow increase in naive T cells [70]. Hence, the ability of a patient's immune system to recognise certain antigens or pathogens may be altered or absent despite an overall increase in the number of circulating T cells. This could suggest that a CD4 count of *i.e.* 200 in a patient with a previously low CD4 cell count on antiviral therapy may be less protective than a similar CD4 count in a untreated patient. However, experience from the clinics suggest that the risk of contracting opportunistic infections diminish considerably in patients responding to triple-combination anti-retroviral therapy, even in those patients with a severely impaired immune system. To what extent the altered T-cell repertoire may lead to new patterns of opportunistic infections and AIDS-related malignancies remains to be determined.

12. References

1. Kaslow, R.A., Phair, J.P., Friedman, H.B., Lyter, D., Solomon, R.E., Dudley, J., Polk, B.F. and Blackwelder, W.: Infection with the human immunodeficiency virus: clinical manifestations and their relationship to immune deficiency. A report from the Multicenter AIDS Cohort Study, *Ann. Intern. Med.* **107** (1987), 474.
2. Morgan, D., Maude, G.H., Malamba, S.S., Okongo, M.J., Wagner, H.U., Mulder, D.W. and Whitworth, J.A.: HIV-1 disease progression and AIDS-defining disorders in rural Uganda, *Lancet* **350** (1997), 245.
3. Russian, D.A. and Kovacs, J.A.: Pneumocystis carinii in Africa: an emerging pathogen?, *Lancet* **346** (1995), 1242.
4. Lundgren, J.D., Barton, S.E., Lazzarin, A., Danner, S., Goebel, F.D., Pehrson, P., Mulcahy, F., Kosmidis, J., Pedersen, C., Philips, A.N. and AIDS in Europe Study Group: Factors associated with the development of Pneumocystis carinii pneumonia in 5,025 European patients with AIDS, *Clin. Infect. Dis.* **21** (1995), 106.
5. Moore, R.D. and Chaisson, R.E.: Natural history of opportunistic disease in an HIV-infected urban clinical cohort, *Ann. Intern. Med.* **124** (1996), 633.
6. Palella, F.J.J., Delaney, K.M., Moorman, A.C., Loveless, M.O., Führer, J., Satten, G.A., Aschman, D.J. and Holmberg, S.D.: Declining morbidity and mortality among patients with advanced human immunodeficiency virus infection. HIV Outpatient Study Investigators, *New Engl. J. Med.* **338** (1998), 853.
7. Mocroft, A., Vella, S., Benfield, T.L., Chiesi, A., Miller, G., Gargalianos, P., Monforte, A., Yust, I., Bruun, J.N., Philips, A.N. and Lundgren, J.D.: Changing pattern of mortality across Europe in patients infected with HIV-1, *Lancet* **352** (1998), 1725.

8. Kirk, O., Mocroft, A., Katzenstein, T.L., Lazzarin, A., Antunes, F., Francioli, P., Brettle, R.P., Parkin, J.M., Gonzales-Lahoz, J. and Lundgren, J.D.: Changes in use of antiretroviral therapy in regions of Europe over time. EuroSIDA Study Group, *AIDS* **12** (1998), 2031.
9. Sepkowitz, K.A.: Effect of HAART on natural history of AIDS-related opportunistic disorders, *Lancet* **351** (1998), 228.
10. Aberg, J.A., Yajko, D.M. and Jacobson, M.A.: Eradication of AIDS-related disseminated *mycobacterium avium* complex infection after 12 months of antimycobacterial therapy combined with highly active antiretroviral therapy, *J. Infect. Dis.* **178** (1998), 1446.
11. Jacobson, M.A., Zegans, M., Pavan, P.R., O'Donnell, J.J., Sattler, F., Rao, N., Owens, S. and Pollard, R.: Cytomegalovirus retinitis after initiation of highly active antiretroviral therapy, *Lancet* **349** (1997), 1443.
12. Crump, J.A., Tyrer, M.J., Lloyd-Owen, S.J., Han, L.Y., Lipman, M.C. and Johnson, M.A.: Military tuberculosis with paradoxical expansion of intracranial tuberculomas complicating human immunodeficiency virus infection in a patient receiving highly active antiretroviral therapy, *Clin. Infect. Dis.* **26** (1998), 1008.
13. Denis, M. and Ghadirian, E.: *Mycobacterium avium* infection in HIV-1-infected subjects increases monokine secretion and is associated with enhanced viral load and diminished immune response to viral antigens, *Clin. Exp. Immunol.* **97** (1994), 76.
14. Chaisson, R.E., Gallant, J.E., Keruly, J.C. and Moore, R.D.: Impact of opportunistic disease on survival in patients with HIV infection, *AIDS* **12** (1998), 29.
15. Bentwich, Z., Kalinkovich, A. and Weisman, Z.: Immune activation is a dominant factor in the pathogenesis of African AIDS, *Immunol. Today* **16** (1995), 187.
16. Perneger, T.V., Sudre, P., Lundgren, J.D. and Hirschel, B.: Does the onset of tuberculosis in AIDS predict shorter survival? Results of a cohort study in 17 European countries over 13 years. AIDS in Europe Study Group, *Brit. Med. J.* **311** (1995), 1468.
17. Goletti, D., Weissman, D., Jackson, R.W., Graham, N.M., Vlahov, D., Klein, R.S., Munsiff, S.S., Ortona, L., Cauda, R. and Fauci, A.S.: Effect of *Mycobacterium tuberculosis* on HIV replication. Role of immune activation, *J. Immunol.* **157** (1996), 1271.
18. Orenstein, J.M., Fox, C. and Wahl, S.M.: Macrophages as a source of HIV during opportunistic infections, *Science* **276** (1997), 1857.
19. Nabel, G. and Baltimore, D.: An inducible transcription factor activates expression of human immunodeficiency virus in T cells, *Nature* **326** (1987), 711.
20. Osborn, L., Kunkel, S. and Nabel, G.J.: Tumor necrosis factor alpha and interleukin 1 stimulate the human immunodeficiency virus enhancer by activation of the nuclear factor kappa B, *Proc. Natl. Acad. Sci. USA* **86** (1989), 2336.
21. Staprans, S.I., Hamilton, B.L., Follansbee, S.E., Elbeik, T., Barbosa, P., Grant, R.M. and Feinberg, M.B.: Activation of virus replication after vaccination of HIV-1-infected individuals, *J. Exp. Med.* **182** (1995), 1727.
22. Edman, J.C., Kovacs, J.A., Masur, H., Santi, D.V., Elwood, H.J. and Sogin, M.L.: Ribosomal RNA sequence shows *Pneumocystis carinii* to be a member of the fungi, *Nature* **334** (1988), 519.
23. Hughes, W.T., Bartley, D.L. and Smith, B.M.: A natural source of infection due to *pneumocystis carinii*, *J. Infect. Dis.* **147** (1983), 595.
24. Peters, S.E., Wakefield, A.E., Sinclair, K., Millard, P.R. and Hopkin, J.M.: A search for *Pneumocystis carinii* in post-mortem lungs by DNA amplification, *J. Pathol.* **166** (1992), 195.
25. Chen, W., Gigliotti, F. and Harmsen, A.G.: Latency is not an inevitable outcome of infection with *Pneumocystis carinii*, *Infect. Immun.* **61** (1993), 5406.
26. Ribes, J.A., Limper, A.H., Espy, M.J. and Smith, T.F.: PCR detection of *Pneumocystis carinii* in bronchoalveolar lavage specimens: analysis of sensitivity and specificity, *J. Clin. Microbiol.* **35** (1997), 830.
27. Sinclair, K., Wakefield, A.E., Banerji, S. and Hopkin, J.M.: *Pneumocystis carinii* organisms derived from rat and human hosts are genetically distinct, *Mol. Biochem. Parasitol.*, 1991, p. 183.
28. Lee, C.H., Lu, J.J., Bartlett, M.S., Durkin, M.M., Liu, T.H., Wang, J., Jiang, B. and Smith, J.W.: Nucleotide sequence variation in *Pneumocystis carinii* strains that infect humans, *J. Clin. Microbiol.* **31** (1993), 754.

29. Lu, J.J., Bartlett, M.S., Shaw, M.M., Queener, S.F., Smith, J.W., Ortiz Rivera, M., Leibowitz, M.J. and Lee, C.H.: Typing of *Pneumocystis carinii* strains that infect humans based on nucleotide sequence variations of internal transcribed spacers of rRNA genes, *J. Clin. Microbiol.* **32** (1994), 2904.

30. Lee, C.H., Helweg-Larsen, J., Tang, X., Jin, S., Li, B., Bartlett, M.S., Lundgren, B., Olsson, M., Vermund, S.H., Lucas, S.B., Roux, P., Atzori, C., Matos, O. and Smith, J.W.: Update on *Pneumocystis carinii* f.sp. hominis typing based on nucleotide sequence variations in the internal transcribed spacer regions of rRNA genes, *J. Clin. Microbiol.* **36** (1998), 734.

31. Mei, Q., Turner, R.E., Sorial, V., Klivington, D., Angus, C.W. and Kovacs, J.A.: Characterization of major surface glycoprotein genes of human *Pneumocystis carinii* and high-level expression of a conserved region, *Infect. Immun.* **66** (1998), 4268.

32. Beck, J.M., Newbury, R.L. and Palmer, B.E.: *Pneumocystis carinii* pneumonia in SCID mice induced by viable organisms propagated *in vitro*, *Infect. Immun.* **64** (1996), 4643.

33. Harmsen, A.G., Chen, W. and Gigliotti, F.: Active immunity to *Pneumocystis carinii* reinfection in T-cell-depleted mice, *Infect. Immun.* **63** (1995), 2391.

34. Koziel, H., Eichbaum, Q., Kruskal, B.A., Pinkston, P., Rogers, R.A., Armstrong, M.Y.K., Richards, F.F., Rose, R.M. and Ezekowitz, R.A.B.: Reduced binding and phagocytosis of *Pneumocystis carinii* by alveolar macrophages from persons infected with HIV-1 correlates with mannose receptor downregulation, *J. Clin. Invest.* **102** (1998), 1332.

35. Benfield, T.L., Vestbo, J., Junge, J., Nielsen, T.L., Jensen, A.M.B. and Lundgren, J.D.: Prognostic value of interleukin-8 in AIDS-associated *Pneumocystis carinii* pneumonia, *Am. J. Respir. Crit. Care Med.* **151** (1995), 1058.

36. Pedersen, C., Benfield, T. and Lundgren, J.D.: HIV infection in Europe: an overview of epidemiology and pulmonary complications. In: *Human Immunodeficiency Virus and the Lung* (Rosen, M.J. and Beck, J.M., eds.), Marcel Dekker, New York, 1997, pp. 19-46.

37. Sepkowitz, K.A., Raffalli, J., Riley, L., Kiehn, T.E. and Armstrong, D.: Tuberculosis in the AIDS era, *Clin. Microbiol. Rev.* **8** (1995), 180.

38. Small, P.M., Hopewell, P.C., Singh, S.P., Paz, A., Parsonnet, J., Ruston, D.C., Schecter, G.F., Daley, C.L. and Schoolnik, G.K.: The epidemiology of tuberculosis in San Francisco. A population-based study using conventional and molecular methods [see comments], *New Engl. J. Med.* **330** (1994), 1703.

39. Britton, W.J. and Garsia, R.J.: Mycobacterial infections. In: *Clinical Immunology* (Bradley, J. and McCluskey, J., eds.), Oxford University Press, New York, 1997, pp. 483-498.

40. Orme, I.M., Andersen, P. and Boom, W.H.: T-cell response to *Mycobacterium tuberculosis*, *J. Infect. Dis.* **167** (1993), 1481.

41. Wilkinson, D., Squire, S.B. and Garner, P.: Effect of preventive treatment for tuberculosis in adults infected with HIV: systematic review of randomised placebo-controlled trials, *Brit. Med. J.* **317** (1998), 625.

42. Pierce, M., Crampton, S., Henry, D., Heifets, L., LaMarca, A., Montecalvo, M., Wormser, G.P., Jablonowski, H., Jemsek, J., Cynamon, M., Yangco, B.G., Notario, G. and Craft, J.C.: A randomized trial of clarithromycin as prophylaxis against disseminated *Mycobacterium avium* complex infection in patients with advanced acquired immunodeficiency syndrome, *New Engl. J. Med.* **335** (1996), 384.

43. Havlir, D.V., Dube, M.P., Sattler, F.R., Forthal, D.N., Kemper, C.A., Dunne, M.W., Parenti, D.M., Lavelle, J.P., White, A.J., Witt, M.D., Bozzette, S.A. and McCutchan, J.A.: Prophylaxis against disseminated *Mycobacterium avium* complex with weekly azithromycin, daily rifabutin, or both. California Collaborative Treatment Group [see comments], *New Engl. J. Med.* **335** (1996), 392.

44. Von, R.C., Waddell, R.D., Eaton, T., Arbeit, R.D., Maslow, J.N., Barber, T.W., Brindle, R.J., Gilks, C.F., Lumio, J. and Lahdevirta, J.: Isolation of *Mycobacterium avium* complex from water in the United States, Finland, Zaire and Kenya, *J. Clin. Microbiol.* **31** (1993), 3227.

45. Chin, D.P., Hopewell, P.C., Yajko, D.M., Vittinghoff, E., Horsburgh, C.R., Hadley, W.K., Stone, E.N., Nassos, P.S., Ostroff, S.M., Jacobson, M.A. *et al.*: *Mycobacterium avium* complex in the respiratory or gastrointestinal tract and the risk of *M.avium* complex bacteremia in patients with human immunodeficiency virus infection, *J. Infect. Dis.* **169** (1994), 289.

46. Pepose, J.S., Holland, G.N., Nestor, M.S., Cochran, A.J. and Foos, R.Y.: Acquired immune deficiency syndrome. Pathogenic mechanisms of ocular disease, *Ophthalmology* **92** (1985), 472.
47. Spector, S.A., Wong, R., Hsia, K., Pilcher, M. and Stempien, M.J.: Plasma cytomegalovirus (CMV) DNA load predicts CMV disease and survival in AIDS patients, *J. Clin. Invest.* **101** (1998), 497.
48. Shepp, D.H., Match, M.E., Ashraf, A.B., Lipson, S.M., Millan, C. and Pergolizzi, R.: Cytomegalovirus glycoprotein B groups associated with retinitis in AIDS, *J. Infect. Dis.* **174** (1996), 184.
49. Peterson, P.K., Gekker, G., Chao, C.C., Hu, S.X., Edelman, C., Balfour, H.H. and Verhoef, J.: Human cytomegalovirus-stimulated peripheral blood mononuclear cells induce HIV-1 replication via a tumor necrosis factor-alpha-mediated mechanism, *J. Clin. Invest.* **89** (1992), 574.
50. Pleskoff, O., Treboute, C., Brelot, A., Heveker, N., Seman, M. and Alizon, M.: Identification of a chemokine receptor encoded by human cytomegalovirus as a cofactor for HIV-1 entry, *Science* **276** (1997), 1874.
51. Webster, A., Lee, C.A., Cook, D.G., Grundy, J.E., Emery, V.C., Kernoff, P.B. and Griffiths, P.D.: Cytomegalovirus infection and progression towards AIDS in haemophiliacs with human immunodeficiency virus infection [see comments], *Lancet* **2** (1989), 63.
52. Lundgren, J.D., Pedersen, C., Clumeck, N., Gatell, J.M., Johnson, A.M., Ledergerber, B., Vella, S., Phillips, A. and Nielsen, J.O.: Survival differences in European patients with AIDS, 1979-89. The AIDS in Europe Study Group, *Brit. Med. J.* **308** (1994), 1068.
53. Hermans, P., Lundgren, J., Sommereijns, B., Pedersen, C., Vella, S., Katlama, C., Luthy, R., Pinching, A.J., Gerstoft, J., Pehrson, P. and Clumeck, N.: Epidemiology of AIDS-related Kaposi's sarcoma in Europe over 10 years. AIDS in Europe Study Group, *AIDS* **10** (1996), 911.
54. Beral, V., Peterman, T.A., Berkelman, R.L. and Jaffe, H.W.: Kaposi's sarcoma among persons with AIDS: a sexually transmitted disease?, *Lancet* **335** (1990), 123.
55. Foreman, K.E., Friborg Jr., J., Kong, W.P., Woffendin, C., Polverini, P.J., Nickoloff, B.J. and Nabel, G.J.: Propagation of a human herpesvirus from AIDS-associated Kaposi's sarcoma [see comments], *New Engl. J. Med.* **336** (1997), 163.
56. Boshoff, C., Schulz, T.F., Kennedy, M.M., Graham, A.K., Fisher, C., Thomas, A., McGee, J.O., Weiss, R.A. and O'Leary, J.J.: Kaposi's sarcoma-associated herpesvirus infects endothelial and spindle cells, *Nature Med.* **1** (1995), 1274.
57. Ensoli, B., Gendelman, R., Markham, P., Fiorelli, V., Colombini, S., Raffeld, M., Cafaro, A., Chang, H.K., Brady, J.N. and Gallo, R.C.: Synergy between basic fibroblast growth factor and HIV-1 Tat protein in induction of Kaposi's sarcoma, *Nature* **371** (1994), 674.
58. Russo, J.J., Bohenzky, R.A., Chien, M.C., Chen, J., Yan, M., Maddalena, D., Parry, J.P., Peruzzi, D., Edelman, I.S., Chang, Y. and Moore, P.S.: Nucleotide sequence of the Kaposi sarcoma-associated herpesvirus (HHV8), *Proc. Natl. Acad. Sci. USA* **93** (1996), 14862.
59. Gill, P.S., Lunardi Ishkandar, Y., Louie, S., Tulpule, A., Zheng, T., Espina, B.M., Besnier, J.M., Hermans, P., Levine, A.M., Bryant, J.L. and Gallo, R.C.: The effects of preparations of human chorionic gonadotropin on AIDS-related Kaposi's sarcoma, *New Engl. J. Med.* **335** (1996), 1261.
60. Miller, G., Rigsby, M.O., Heston, L., Grogan, E., Sun, R., Metroka, C., Levy, J.A., Gao, S.J., Chang, Y. and Moore, P.: Antibodies to butyrate-inducible antigens of Kaposi's sarcoma-associated herpesvirus in patients with HIV-1 infection, *New Engl. J. Med.* **334** (1996), 1292.
61. Lennette, E.T., Blackbourn, D.J. and Levy, J.A.: Antibodies to human herpesvirus type 8 in the general population and in Kaposi's sarcoma patients, *Lancet* **348** (1996), 858.
62. Simpson, G.R., Schulz, T.F., Whitby, D., Cook, P.M., Boshoff, C., Rainbow, L., Howard, M.R., Gao, S.J., Bohenzky, R.A., Simmonds, P., Lee, C., De Ruiter, A., Hatzakis, A., Tedder, R.S., Weller, I.V., Weiss, R.A. and Moore, P.S.: Prevalence of Kaposi's sarcoma-associated herpesvirus infection measured by antibodies to recombinant capsid protein and latent immunofluorescence antigen, *Lancet* **348** (1996), 1133.

63. Gao, S.J., Kingsley, L., Hoover, D.R., Spira, T.J., Rinaldo, C.R., Saah, A., Phair, J., Detels, R., Parry, P., Chang, Y. and Moore, P.S.: Seroconversion to antibodies against Kaposi's sarcoma-associated herpesvirus-related latent nuclear antigens before the development of Kaposi's sarcoma, *New Engl. J. Med.* **335** (1996), 233.
64. LaDuca, J.R., Love, J.L., Abbott, L.Z., Dube, S., Freidman-Kien, A.E. and Poiesz, B.J.: Detection of human herpesvirus 8 DNA sequences in tissues and bodily fluids, *J. Infect. Dis.* **178** (1998), 1610.
65. Regamey, N., Tamm, M., Wernli, M., Witschi, A., Thiel, G., Cathomas, G. and Erb, P.: Transmission of human herpesvirus 8 infection from renal-transplant donors to recipients, *New Engl. J. Med.* **339** (1998), 1358.
66. Araujo, F., Slifer, T. and Kim, S.: Chronic infection with *Toxoplasma gondii* does not prevent acute disease or colonization of the brain with tissue cysts following reinfection with different strains of the parasite, *J. Parasitol.* **83** (1997), 521.
67. Subauste, C.S., Wong, S. and Remington, J.S.: AIDS-associated toxoplasmosis. In: *The Medical Management of AIDS* (Sande, M.A. and Volberding, P.A., eds.), W.B. Saunders, Philadelphia, 1997, pp. 343-362.
68. Pelloux, H., Ricard, J., Bracchi, V., Markowicz, Y., Verna, J.M. and Ambroise-Thomas, P.: Tumor necrosis factor alpha, interleukin-1 alpha, and interleukin-6 mRNA expressed by human astrocytoma cells after infection by three different strains of *Toxoplasma gondii*, *Parasitol. Res.* **80** (1994), 271.
69. Suzuki, Y., Yang, Q., Yang, S., Nguyen, N., Lim, S., Liesenfeld, O., Kojima, T. and Remington, J.S.: IL-4 is protective against development of toxoplasmic encephalitis, *J. Immunol.* **157** (1996), 2564.
70. Pakker, N.G., Notermans, D.W., De Boer, R.J., Roos, M.Th.L., De Wolf, F., Hill, A., Leonard, J.M., Danner, S.A., Miedema, F. and Schellekens, P.Th.A.: Biphasic kinetics of peripheral blood T cells after triple combination therapy in HIV-1 infection: a composite of redistribution and proliferation, *Nature Med.* **4** (1998), 208.
71. Griffiths, P.D.: Herpesviruses and AIDS, *Scand. J. Infect. Dis. Suppl.* **100** (1996), 3.
72. Kolls, J.K., Beck, J.M., Nelson, S., Summer, W.R. and Shellito, J.: Alveolar macrophage release of tumor necrosis factor during murine *Pneumocystis carinii* pneumonia, *Am. J. Respir. Cell Mol. Biol.* **8** (1993), 370.
73. Benfield, T.L., Van Steenwijk, R., Nielsen, T.L., Dichter, J.R., Lipschik, G.Y., Jensen, B.N., Junge, J., Shelhamer, J.H. and Lundgren, J.D.: Interleukin-8 and eicosanoid production in the lung during moderate to severe *Pneumocystis carinii* pneumonia in AIDS: a role of interleukin-8 in the pathogenesis of *P.carinii* pneumonia, *Respir. Med.* **89** (1995), 285.
74. Gazzinelli, R.T., Sher, A., Cheever, A., Gerstberger, S., Martin, M.A. and Dickie, P.: Infection of human immunodeficiency virus 1 transgenic mice with *Toxoplasma gondii* stimulates proviral transcription in macrophages *in vivo*, *J. Exp. Med.* **183** (1996), 1645.
75. Clouse, K.A., Robbins, P.B., Fernie, B., Ostrove, J.M. and Fauci, A.S.: Viral antigen stimulation of the production of human monokines capable of regulating HIV-1 expression, *J. Immunol.* **143** (1989), 470.
76. Zhang, Y., Nakata, K., Weiden, M. and Rom, W.N.: *Mycobacterium tuberculosis* enhances human immunodeficiency virus-1 replication by transcriptional activation at the long terminal repeat, *J. Clin. Invest.* **95** (1995), 2324.
77. Davis, M.G., Kenney, S.C., Kamine, J., Pagano, J.S. and Huang, E.S.: Immediate-early gene region of human cytomegalovirus transactivates the promoter of human immunodeficiency virus, *Proc. Natl. Acad. Sci. USA* **84** (1987), 8642.
78. Gendelman, H.E., Phelps, W., Feigenbaum, L., Ostrove, J.M., Adachi, A., Howley, P.M., Khoury, G., Ginsberg, H.S. and Martin, M.A.: Transactivation of the human immunodeficiency virus long terminal repeat sequence by DNA viruses, *Proc. Natl. Acad. Sci. USA* **83** (1986), 9759.
79. Biegalke, B.J. and Geballe, A.P.: Sequence requirements for activation of the HIV-1 LTR by human cytomegalovirus, *Virology* **183** (1991), 381.
80. Lusso, P., Ensoli, B., Markham, P.D., Ablashi, D.V., Salahuddin, S.Z., Tschachler, E., Wong-Staal, F. and Gallo, R.C.: Productive dual infection of human $CD4^+$ T lymphocytes by HIV-1 and HHV-6, *Nature* **337** (1989), 370.

81. Kenney, S., Kamine, J., Markovitz, D., Fenrick, R. and Pagano, J.: An Epstein-Barr virus immediate-early gene product transactivates gene expression from the human immunodeficiency virus long terminal repeat, *Proc. Natl. Acad. Sci. USA* **85** (1988), 1652.
82. Keely, S.P., Baughman, R.P., Smulian, A.G., Dohn, M.N. and Stringer, J.R.: Source of *Pneumocystis carinii* in recurrent episodes of pneumonia in AIDS patients, *AIDS* **10** (1996), 881.
83. Tsolaki, A.G., Miller, R.F., Underwood, A.P., Banerji, S. and Wakefield, A.E.: Genetic diversity at the internal transcribed spacer regions of the rRNA operon among isolates of *Pneumocystis carinii* from AIDS patients with recurrent pneumonia, *J. Infect. Dis.* **174** (1996), 141.
84. Latouche, S., Poirot, J.L., Bernard, C. and Roux, P.: Study of internal transcribed spacer and mitochondrial large-subunit genes of *Pneumocystis carinii hominis* isolated by repeated bronchoalveolar lavage from human immunodeficiency virus-infected patients during one or several episodes of pneumonia, *J. Clin. Microbiol.* **35** (1997), 1687.
85. Tural, C., Romeu, J., Sirera, G., Andreu, D., Conejero, M., Ruiz, S., Jou, A., Bonjoch, A., Ruiz, L., Arno, A. and Clotet, B.: Long-lasting remission of cytomegalovirus retinitis without maintenance therapy in human immunodeficiency virus-infected patients, *J. Infect. Dis.* **177** (1998), 1080.
86. Mocroft, A.J., Lundgren, J.D., d'Armino, M.A., Ledergerber, B., Barton, S.E., Vella, S., Katlama, C., Gerstoft, J., Pedersen, C. and Phillips, A.N.: Survival of AIDS patients according to type of AIDS-defining event. The AIDS in Europe Study Group, *Intern. J. Epidemiol.* **26** (1997), 400.
87. Ghassemi, M., Andersen, B.R., Reddy, V.M., Gangadharam, P.R., Spear, G.T. and Novak, R.M.: Human immunodeficiency virus and *Mycobacterium avium* complex coinfection of monocytoid cells results in reciprocal enhancement of multiplication, *J. Infect. Dis.* **171** (1995), 68.

CHAPTER 10

A - AIDS-RELATED NON-HODGKIN'S LYMPHOMAS

MARIE JOSÉ KERSTEN[1] and M.H.J. VAN OERS[2]

[1]*Department of Medical Oncology, Netherlands Cancer Institute/ Antoni van Leeuwenhoekhuis, Amsterdam, and* [2]*Department of Hematology, Academic Medical Center, University of Amsterdam, Amsterdam, The Netherlands*

1. Epidemiology

Non-Hodgkin's lymphomas (NHL) are monoclonal expansions of B- or T-lymphocytes. B-cell lymphomas have long been recognised to arise more frequently in patients with either congenital (*e.g.* X-linked lymphoproliferative disease [1]) or acquired cellular immunodeficiency (*e.g.* solid organ or bone marrow-transplant recipients) [2]. The first cases of NHL occurring in homosexual males were described in 1982 [3], only one year after the first reports on *Pneumocystis carinii* pneumonia and Kaposi's sarcoma in this patient group, and well before the human immunodeficiency virus (HIV) was identified as the cause of the immunodeficiency. Whereas primary central nervous system lymphoma (PCNSL) was considered one of the initial criteria for a diagnosis of AIDS, systemic NHL was included as an AIDS-indicator diagnosis in the revised case definition of AIDS proposed by the Centers for Disease Control in 1985.

The incidence of NHL is equally increased in all HIV-risk groups, such as homosexual or bisexual males, intravenous drug users [4], hemophiliacs [5] and transfusion recipients [6]. In contrast to Kaposi's sarcoma (KS), there appears to be no geographic variation in incidence. In the San Francisco City Clinic Cohort, the frequency of NHL as initial AIDS-defining diagnosis was 4.2%, whereas the lifetime incidence was 12.6% [7]. In another large prospective observational study the incidence rate was approximately 1.6% per year [8]. In HIV-1-infected individuals, the risk of NHL is approximately 100-fold increased when compared to the general population [9]. Since NHL is often a late complication of AIDS, the incidence was expected to increase because of increasing life expectancy due to anti-retroviral therapy and improving management of opportunistic infections [10]. Whereas the incidence of KS has dropped by approximately 60% since the advent of highly active anti-retroviral therapy (HAART), the incidence of AIDS-related lymphomas (ARL) is still increasing (A.M. Levine, personal communication).

H. Schuitemaker and F. Miedema (ed.), Aids Pathogenesis, 177–194.

Although the frequency of Hodgkin's disease, in particular the mixed cellularity type, is somewhat increased in HIV-infected persons [11], Hodgkin's disease is not included in the case definition of AIDS. Other infrequently reported lymphoproliferations in patients with AIDS include multiple myeloma [12], peripheral T-cell lymphoma, T-lymphoblastic lymphoma and multicentric Castleman's disease [13].

2. Clinical presentation

When compared to high-grade B-cell NHL in the general population, ARL have several distinctive features [14-17]. The median age at presentation is 35-42 years. 40-60% of the patients have a prior diagnosis of AIDS. The majority of patients has widespread (Ann Arbor stage III or IV) disease already at presentation, and most patients have B symptoms (fever, weight loss). About 70-97% of patients have extranodal disease, often at unusual localisations, such as the heart, muscle, soft tissues, anus and rectum. Especially striking is the high incidence of central nervous system (CNS) involvement: whereas CNS localisations occur in only 2% of NHL in the general population, CNS involvement (either primary CNS lymphoma or leptomeningeal localisation) is seen in 17-43% of ARL. This is also true for other immunodeficiency-related lymphomas, such as post-transplant lymphoproliferative disease. Presenting symptoms are headache, seizures, confusion, lethargy or focal neurologic deficits [18], but the diagnosis is often made at autopsy. On a CT scan one or two slightly hypodense mass lesions are seen, whereas in toxoplasmosis, the main differential diagnosis, usually more lesions are seen. Although the demonstration of EBV-DNA (Epstein-Barr virus) by PCR in cerebrospinal fluid has been shown to be a highly sensitive and specific marker for AIDS-related PCNSL [19], and also PET-scanning (positron emission tomography) may be a useful new diagnostic tool, a definite diagnosis can be made only by brain biopsy [20]. The incidence of bone marrow, gastrointestinal and liver involvement in ARL is also high.

The prognosis of ARL is extremely poor: the median survival reported in various studies is only 1.5-8 months. In a recent epidemiologic survey of survival in AIDS patients according to type of AIDS-defining event, median survival after NHL was only 5 months [21]. The most important adverse prognostic factors are a CD4-cell count less than $0.2x10^9/l$[22], a prior diagnosis of AIDS and poor performance status. Other factors influencing survival are the presence of extranodal disease, bone-marrow and/or CNS involvement, and constitutional symptoms [17,22]. AIDS-related Hodgkin's disease tends to display more aggressive clinical and biologic features than Hodgkin's disease in the immunocompetent host, and response to treatment is often poor[23].

3. Histopathology

In contrast to lymphomas in the general population, the lymphomas found in patients with AIDS are almost invariably of intermediate or high-grade malignancy according to the Working Formulation. Morphologically, the majority (60-70%) is classified as diffuse large cell lymphomas which are either polymorphic centroblastic or immunoblastic according to the Kiel classification. They often show pronounced tumour cell polymorphism and multiple bizarre nuclei (Fig. 1).

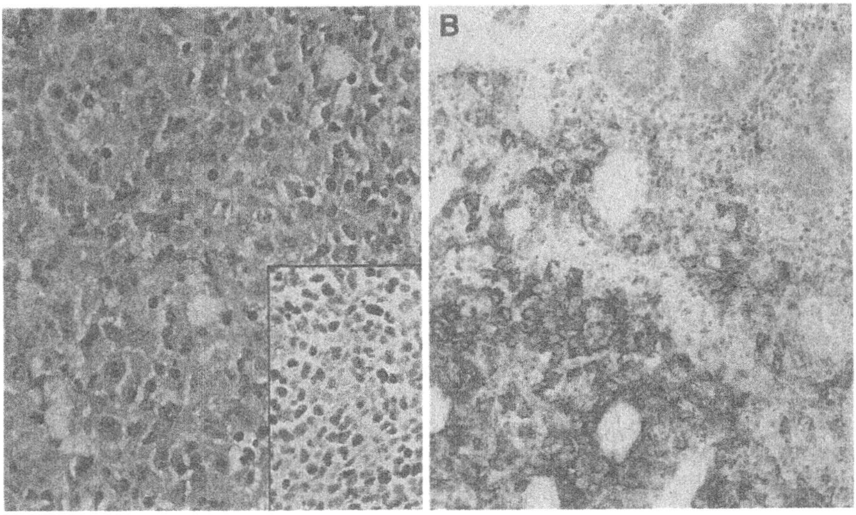

Fig. 1a - This tissue section shows diffuse infiltration of the large bowel with large atypical lymphoid cells with pleiomorphic nuclei with multiple nucleoli (haematoxylin-eosin, 400x magnification). The insert shows expression of EBER (Epstein-Barr virus-related small RNA's) in the nuclei (RNA *in-situ* hybridisation technique).
Fig. 1b - The tumour cells all stain positive with a monoclonal antibody against CD20 (a B-cell antigen).

In the recently developed REAL-classification (revised European/American classification of lymphoid neoplasms) they are classified as diffuse large cell lymphomas (DLCL). 30-40% of the tumours, however, is of the high-grade small non-cleaved cell variety (Burkitt- or non-Burkitt-type). Interestingly, this extremely high proportion of Burkitt-type tumours is not seen in immunosuppressed transplant patients.

Based on the presence of either surface immunoglobulin, B-cell-specific membrane markers or immunoglobulin gene rearrangement, virtually all ARL are found to be of B-cell origin [15,16]. Although also oligo- and polyclonal tumours have been described, most lymhomas are monoclonal [15,24]. The expression of EBV-related antigens is discussed under Pathogenesis.

4. Pathogenesis

The pathogenesis of lymphomas in the face of HIV-1 infection is still incompletely understood, but a multi-step model is presumed. The first phase is characterised by polyclonal B-cell expansion (clinically corresponding to persistent generalised lymphadenopathy or PGL). Factors important in this phase are thought to be latent EBV infection in combination with disruption of immunosurveillance, chronic antigen stimulation and cytokine deregulation, all resulting from HIV-1 infection. In the second phase, multiple genetic lesions accumulate, leading to clonal evolution. This model is supported by a study by Pelicci, demonstrating the existence of multiple clonal B-cell expansions in biopsies from patients with PGL, possibly representing precursor lesions of lymphoma [24]. Moreover, the presence of EBV in PGL biopsies has been associated with an increased risk of concurrent or subsequent development of NHL [25]. Although the pathogenesis for the two major histologic subgroups (Burkitt lymphomas and DLCL) in ARL may differ, several of these factors are shared. They will first be discussed separately, after which a model of lymphomagenesis in AIDS is presented.

4.1. HIV-1

There is no evidence that HIV-1 itself is directly involved in lymphomagenesis. Although HIV-1 has been shown to be able to infect peripheral B cells from EBV-seropositive, HIV-seronegative individuals, HIV-1 sequences are generally not detectable in lymphoma tissue by Southern blot analysis. PCR analysis has shown levels of HIV-1 that can be explained by the presence of infiltrating T cells. HIV-1 does, however, contribute to lymphomagenesis indirectly by inducing polyclonal B-cell expansion, reactivation of EBV infection, chronic antigen stimulation, and cytokine deregulation.

4.2. DISRUPTED IMMUNOSURVEILLANCE

The risk of NHL (especially DLCL) increases with the duration of HIV infection and the decline in CD4-cell count [26]. The immune deficiency presumably contributes to B-cell expansion via reactivation of EBV. This would be analagous to other states of cellular immune deficiency associated with increased risk of NHL. *In vitro*, defective regulation of EBV infection in patients with AIDS has been shown, causing an increase in the number of EBV-infected B cells in the peripheral blood [27]. One cross-sectional study concerning EBV-specific T-cell responses in HIV-1 infection showed severely decreased EBV-specific cytotoxic T-cell (CTL) activity in patients with AIDS and AIDS-related complex [28], but two other studies comparing EBV- with HIV-1-specific T-cell responses suggested preservation of EBV-specific CTL in advanced HIV-1 infection [29,30]. We have studied EBV-load and EBV-CTL precursor frequencies longitudinally in long-term asymptomatic HIV-1-infected individuals, patients with opportunistic infections and patients with NHL, and found in the patients with NHL, prior to the onset of lymphoma, a decrease in EBV-CTLp and an increase in EBV-load [31].

4.3. CHRONIC ANTIGENIC STIMULATION

B-cell hyperplasia and hypergamma-globulinaemia (sometimes even an oligo- or monoclonal paraprotein can be demonstrated) are often found already early in the course of HIV-1 infection and can be caused by HIV-1 itself [32]. Two AIDS-related Burkitt lymphoma cell lines have been shown to produce high-affinity IgM antibodies which are self-reactive with i/I determinants on red blood cells [33]. Other ARL cell lines have been demonstrated to be reactive to the HIV-1 gp160 determinant [34]. Thus, a process of B-cell clonal selection driven by a self antigen or by a viral antigen could contribute to the development of NHL.

4.4. CYTOKINE DYSREGULATION

Cytokine dysregulation is one of the key features of HIV-1-infection [35]. Cytokines including IL-1 (interleukin-1), IL-2, IL-4, IL-6, IL-7, IL-10, IL-12, IFNγ (interferon-γ), and TNF (tumor necrosis factor) are important in B-cell proliferation and differentiation, and some of these cytokines are also considered to be involved in lymphomagenesis. HIV-1 infection has been shown to induce monocyte production of IL6 [36], a paracrine growth factor in multiple myeloma, non-HIV-associated NHL and also AIDS-related DLCL [37]. Using a bio-assay, elevated levels of IL-6 have been shown to be associated with an increased risk of NHL in HIV-infected patients [26]. B-cell lines from patients with AIDS and BL constitutively secrete large quantities of IL-10, a potent B-cell growth factor [38]. IL-10 (both viral and human) can also contribute to lymphomagenesis by interfering with the generation of CTL and down-regulating HLA class I, thereby enabling tumour cells to escape (EBV-specific) immunosurveillance [39].

4.5. EBV INFECTION

The Epstein-Barr virus is a ubiquitous human gamma herpesvirus affecting more than 90% of the population. Essentially, all HIV-1-infected persons are EBV-seropositive. Apart from being the causative agent of infectious mononucleosis, EBV has been implicated in the pathogenesis of nasopharyngeal carcinoma, endemic Burkitt's lymphoma, lymphomas in patients with congenital or acquired cellular immune deficiency, and Hodgkin's disease. In HIV-1-infected individuals it also causes oral hairy leukoplakia. EBV shows tropism for epithelial cells in the oropharynx, and B cells. The 172-kb double-stranded DNA genome encodes approximately 100 proteins [40], which are only expressed when the virus is in lytic cycle, *i.e.* during primary infection in the oropharyngeal epithelium.

After primary infection, an NK-cell (natural killer cell) reaction and antibody response are elicited, but more important for control of the virus is the EBV-specific cytotoxic T-cell response [41]. However, the virus is never completely eradicated from the body, and will remain latent presumably in B cells. During latency, the EBV DNA circularises to form an episomal nuclear structure, which encodes only a limited number of proteins [42]. They include 6 EBV nuclear antigens (EBNA 1, 2, 3A, 3B, 3C and LP (leader protein), and 3 latent membrane proteins (LMP1 and LMP2A/2B). In addition, in all latently infected cells two small Epstein-Barr RNA's (EBER 1/2) and BARFO are transcribed that do not translate into a protein, and whose function is therefore unknown. Of these latent antigens EBNA1 is essential for maintenance of the episome [43]; this is the only antigen expressed in all latently infected cells. EBNA2 (through transactivation of LMP1 expression and up-regulation of CD21) and LMP1 (through up-regulation of bcl2, which protects cells from apoptosis [44]) are considered to be important for transformation. Other gene products more abundantly expressed in the lytic cycle but possibly playing a role in B-cell transformation are BHRF1 (a functional bcl2 homologue) and BCRF1 (an IL-10 homologue, which can induce B-cell expansion). Using a sensitive RNA *in situ* hybridisation technique, EBER (EBV- small RNA's) can be detected in up to 80% of systemic DLCL, in 90-100% of primary central nervous system lymphomas [45] and AIDS-associated Hodgkin's disease, but in only 30-40% of Burkitt-type NHL [46]. The fact that several studies have found evidence for clonal EBV infection in ARL, indicates that EBV integration occurred before clonal B-cell expansion and thus supports a causative role for EBV in lymphomagenesis [47].

In vitro, three distinct forms of viral latency have been described, each characterised by expression of a different array of latent proteins which is based on the use of different promoters. *Latency type I* (expression of only EBNA1) is found in resting EBV-infected B cells, believed to be the pool of latent infection, and in Burkitt-lymphoma; *latency type II* (expression of EBNA1, LMP1,2A 2B) is found in nasopharyngeal carcinoma and Hodgkin's disease; and *latency type III* (full expression of latent antigens) is found during primary EBV infection, in lymphoblastoid cell lines, and in post-transplant lymphoproliferative disease and some ARL. In ARL the expression of latent antigens differs between BL and DLCL: whereas BL usually exhibit a type I latency phenotype, in DLCL type II and type III latency patterns are found [48]. The pattern of expression of latent antigens has important implications for immune recognition: whereas EBNA1 thusfar has never been recognised as a target for cytotoxic T cells, EBNA3A/B/C and LMP2 are highly immunogenic [49]. This explains why tumours with a type III latency phenotype can only occur in immunosuppressed individuals (*e.g.* PTLD), whereas Burkitt-lymphomas can evade immunosurveillance.

Based on allelic polymorphism of the nuclear antigens EBNA 2, 3A, 3B and

3C, two subtypes of EBV are recognised: type 1 and 2 or A and B. Whereas type 1 EBV is more prevalent in Western countries, type 2 is mainly found in Africa and New Guinea [50]. As in endemic Burkitt lymphoma, both type 1 and type 2 EBV strains are found in ARL, whereas in EBV-positive NHL in immunocompetent individuals in Western countries only type 1 is detected [51]. However, type 2 EBV does not appear to play a causative role in ARL pathogenesis [52].

4.6. GENETIC LESIONS

In ARL, a number of genetic lesions have been demonstrated involving dominantly acting oncogenes as well as tumour suppressor genes.

Oncogenes: - c-myc. Deregulation of c-myc is seen in 100% of AIDS-BL but only in 20% of DLCL [53]. C-myc can cause tumourigenic conversion in lymphoblasts, and is responsible for down-regulation of the adhesion molecules LFA-1 (leukocye function antigen) and ICAM-1 (intercellular adhesion molecule) which are necessary for immune recognition requiring intercellular contact. In endemic and sporadic BL, different molecular mechanisms can lead to chromosomal translocations involving the c-myc oncogene on chromosome 8, and one of the immunoglobulin genes on chromosome 14 (Ig_H), 2 ($Ig_{L\kappa}$) or 22 ($Ig_L\lambda$). ARL resemble sporadic BL in this respect, in addition to sharing a more immature phenotype with these lymphomas.

- bcl1 and bcl2 are infrequently associated with ARL.
- bcl6 is a newly identified oncogene which codes for a member of the zinc-finger family of transcription factors. Bcl6 rearrangements occur in 20% of DLCL, but are consistently absent in BL. c-Myc and bcl-6 rearrangements in DLCL are mutually exclusive [54].
- ras-mutations, although never present in NHL in immunocompetent hosts, are found in 15% of both AIDS-related BL and DLCL [53].

Tumor suppressor genes: - p53: inactivation of p53 through point mutation of one allele and deletion of the other is detected in 60% of AIDS-BL but never in DLCL. They occur in both EBV-positive and EBV-negative tumuor clones [53,55]. The frequent association of c-myc and p53 mutations in BL suggests a synergistic role for these genetic lesions.

- 6q-deletions clustering to 6q27 and 6q21-23 occur in 20-25% of ARL and represent the putative site of two distinct tumour suppressor genes [56].

4.7. OTHER VIRUSES

Recently, DNA sequences have been identified in Kaposi's sarcoma lesions which are derived from a novel virus, named Kaposi's sarcoma herpes virus (KSHV) or human herpesvirus 8 (HHV-8) [57]. This virus bears great sequence homology to other herpesviruses (especially EBV). It has been found not only in AIDS-associated KS but also in classic KS. Although extensively searched for, HHV-8 sequences have not been found in ARL, except for a rare variant termed body-cavity-based-lymphoma (BCBL) or primary effusion lymphoma. These BCBL present as lymphomatous effusions growing in the pleural, pericardial and peritoneal cavities [58]. The HHV-8 therefore, although able to infect B cells in addition to endothelial cells and spindle cells, is associated with specific types of malignancies and immunosuppression rather than being a general feature of HIV-1-related proliferative disorders. The neoplastic cells in these BCBL often also harbour EBV, suggesting a synergistic role for both viruses in this disorder.

Other viruses, such as CMV, HHV-6, and HTLV-1, have never been implicated in the pathogenesis of ARL.

4.8. ZIDOVUDINE TREATMENT

Several studies have suggested that use of zidovudine or other anti-retroviral drugs may be etiologically linked to development of NHL [59]. However, in a recent population-based case control study of patients with ARL in Los Angeles, no increased risk of lymphoma was found in those patients treated with zidovudine [60]. Zidovudine probably merely functions to prolong survival, providing additional time for NHL to develop.

In conclusion, several factors are thought to contribute to lymphomagenesis in HIV-1-infected individuals, and in the two major histologic subgroups different pathways appear to exist. BL tend to occur in younger, less immunocompromised individuals. They are monoclonal and are nearly always associated with c-myc rearrangements and often with loss of the p53 tumour suppressor gene. Only 30-40% are EBV-positive, and when EBV-positive, they only express EBNA1 and thus do not express target antigens for EBV-specific CTL. Chronic antigenic stimulation and cytokine deregulation caused by HIV-1 itself and by other infections leads to increased B-cell proliferation, rendering these cells more susceptible to genetic hits. This clinical setting is comparable to the relative immunodeficiency and antigenic stimulation seen in chronic malarial infestation in those regions of Africa where BL is endemic [61].

DLCL, on the other hand, are more similar to NHL seen in other immunodeficiency states, in that they are often EBV-positive and can express transforming latent antigens, such as LMP1 and EBNA2. They tend to occur later in HIV-1 infection at lower CD4 cell counts, and are thought to arise because of uncontrolled EBV-driven proliferation of B lymphocytes. They are less often associated with specific genetic lesions. This model is illustrated in Fig. 2 (adapted from Gaidano *et al* [62]).

5. Treatment

5.1. TREATMENT OF SYSTEMIC NHL

Since patients with ARL often present with disseminated disease, combination chemotherapy is in general the treatment of choice. In view of the aggressive nature of the tumours in this patient group, intensive treatment appears warranted. However, HIV-1-infected patients often have limited bone marrow reserve and are at high risk of (oppor-

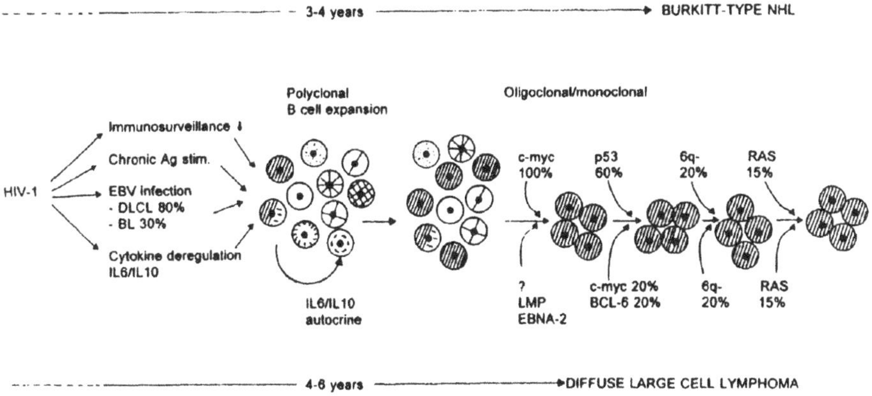

In this figure, a multi-step model of lymphomagenesis in HIV-1 infection is shown, in which several factors are thought to be important. In the upper part, factors considered to contribute to the development of Burkitt-type NHL are shown, and in the lower part those important for DLCL. The first phase of lymphomagenesis is characterised by polyclonal B-cell expansion (clinically corresponding to persistent general lymphadenopathy or PGL, previously known as AIDS-related complex or ARC). Factors important in this phase are thought to be disruption of immunosurveillance, chronic antigen stimulation and cytokine deregulation, all resulting from HIV-1 infection and EBV infection. In the second phase, multiple genetic lesions accumulate, leading to clonal evolution. In the figure, emergence of a single clone (identified by oblique stripes) is depicted (adapted from Gaidano *et al.* [62]).

tunistic) infections. In fact, early retrospective analyses have shown that more intensive treatment was associated with shorter survival [16], and one trial employing high dose methotrexate, cyclophosphamide and cytarabine was stopped early because of the extremely high incidence of opportunistic infection [63]. These findings have led to the development of other or modified treatment regimens, such as low dose M-BACOD.

In Table 1 [16,63,64-72,74], the majority of prospective therapy trials is reviewed. To make the studies more comparable, the percentages of patients with known adverse prognostic factors, such as low CD4 cell count, poor performance status and a prior diagnosis of AIDS, are listed. No results are as yet available of trials employing different treatment regimens in patients that were stratified according to risk factors (risk-adapted treatment). In most trials, patients with relatively good prognosis were enrolled (less than 25% of patients had a prior diagnosis of AIDS and/or a Karnofsky score of <70%). As can be seen in table 1, both dose-intensive regimens such as MACOP-B and LNH84 and low-dose regimens, such as modified M-BACOD, resulted in complete remission rates of 43-67%. Thus, contrary to what would be expected in these highly malignant tumours, lower dose regimens may be as effective as high-dose regimens. However, in a prospective study by the French-Italian Coope-

Table 1. Results of prospective chemotherapy trials in patients with AIDS-related NHL

Author	Regimen[#]	n[@]	CD4[^]	Stage III/IV[&]	Prior AIDS[&]	PS <70[*]	CR[†]	Opp. inf.[~]	Adverse progn.[!]	Survival months[¶]	NHL cause of death[§]
Gill 1987	M-BACOD AraC/MTX-	13 9	0.36 0.17	100 88	n.a.	(70) (90)	54 33	8 78	PS	11 6	n.a.
Kaplan 1989	COMET-A-	38 (33)	0.16	87	21	(75)	58	34	CD4,PS AIDS	5.2	36 (71)[!]
Levine 1991	M-BACOD Low +AZT-	42 (35)	57 0.15	68	23	18	43	21	CD4,stIVA IDS,BM	5.6 (1-43)	50
Kaplan 1991	CHOP+GMCSF -GM-CSF-	16 10	0.22 0.29	48 70	20 10	 (83)	67 67	n.a.	n.a.	10 9	35 50
Sawka 1992	MACOP-B-	30	76 0.06	77	47	(80)	33	24	CD4	8.1	59
Tirelli 1992	CHVmP-VB- + AZT	37 (29)	0.03	69	32	100	14	43	CD4,CNS	3.5	n.a.
Remick 1993	CCPE oral regimen	18	0.07	72	28	28	39	n.a.	CD4,PS	7 (1-36)	40
Gisselbrecht 1993	LNH84-	141	27	63	15	25	63	n.a.	CD4,PS, AIDS	9.3 (1-45+)	50
Walsh 1993	M-BACOD- + GM-CSF	17 (16)	n.a.	62	19	0 (83)	50	19	n.a.	14	n.a.
Sparano 1994	CDE-	21	0.08	76	14	(70)	62	10	n.a.	18.0	33
Schurmann 1995	MACOP-B-	8	62 0.08	87	75	75	50	n.a.	n.a.	4 (1-86)	43

Levine 1996	M-BACOD- low +ddC+G-CSF	28 (25)	54 0.16	76	20	n.a.	56	12	n.a.	8.1 (3-61+)	n.a.
Kaplan	M-BACOD low	94	0.1	71	53	18	41	22	CD4	7.8	70
1997	M-BACOD high	81	0.1	64	48	17	52	23		6.9	57

[#]Abbreviations used: n.a., not available. Chemotherapy regimens: M-BACOD: methotrexate, bleomycin, doxorubicin, cyclophosphamide, vincristine sulfate, dexamethasone; AraC/MTX: high dose cytarabine and methotrexate; COMET-A, cyclophosphamide, vincristine sulphate, methotrexate, etoposide, cytarabine; AZT, zidovudine; it araC, intrathecal prophylaxis with cytarabine; CHOP: cyclophosphamide, doxorubicin, vincristine, prednisone; MACOP-B: methotrexate, doxorubicin, cyclophosphamide, vincristine, bleomycin, prednisone; GM-CSF: granulocyte-monocyte colony stimulating factor; CHVmP: cyclophosphamide, doxorubicin, teniposide, prednisone, vincristine, bleomycin; CCPE oral regimen: CCNU, cyclophosphamide, etoposide, procarbazine; LNH84: doxorubicin, cyclophosphamide, vindesine, bleomycin, prednisolon, high dose methotrexate, ifosfamide, etoposide, L-asparaginase, cytarabine; CDE: cyclophosphamide, doxorubicin, etoposide; ddC, zalcitabine.

-Treatment regimens that included prophylactic intrathecal administration of either methotrexate or cytarabine

[@]nr. of patients enrolled in the trial (nr. of evaluable patients)

[^]% of patients with CD4 cell counts less than $0.2x10^9/l$ or median CD4 cell count ($x10^9/l$)

[&]% of patients presenting with stage III or IV NHL, or % of patients with a prior diagnosis of AIDS

[*]% of patients with a Karnofsky performance score <70%, or median performance score (%)

[†]% of patients attaining a complete remission

[~]% of patients developing an opportunistic infection during chemotherapy

[']Independent prognostic factors predictive of survival

[¶]Survival in months (median and range)

[§]% of patients in whom NHL was the major cause of death; [1]Although NHL was considered the major cause of death in 36% of this patient group, active NHL was present at the time of death in 71% of patients.

rative Study Group, a 50% dose-reduced regimen CHVmP-Vincristine-bleo resulted in only 4 complete remissions in 29 evaluable patients [67]. The median survival in complete responders in all studies is approximately 15 months but overall survival does not exceed 12 months in any but one study. In this study by Sparano, an infusional CDE regimen containing cyclophosphamide, doxorubicin and etoposide was used [71]. This resulted in an estimated survival of 18 months. An update on this study has not yet been published, but a similar study in which didanosine was added to chemotherapy resulted in comparable survival [73]. Durable disease-free remissions of 3-5 years in selected patients have been reported in all studies, indicating that treatment of ARL is not hopeless per se. However, most complete responders do relapse, and in more than half of the patients NHL is the major cause of death. Opportunistic infections and other HIV-related syndromes (wasting) are responsible for the other deaths.

Thus, both complete remission rates and relapse rates in ARL compare unfavourably to those found in non-HIV-related NHL. Since the advent of *haematopoietic growth factors*, two prospective studies were published using rhu-GM-CSF (recombinant human granulocyte-macrophage colony-stimulating factor). In a study by Kaplan *et al.* patients receiving CHOP were randomised to GM-CSF or no GM-CSF. Although there was a significant reduction in neutropenia and febrile episodes, there was no difference in complete remission rates or survival [65]. A randomised trial comparing full-dose M-BACOD with GM-CSF with low-dose M-BACOD without GM-CSF also shows comparable results for both treatment arms with respect to complete remission rates (52 versus 41%) and median survival (7.0 versus 7.8 months) [74] and thus no clear advantage of the administration of GM-CSF is demonstrated. On the other hand, treating HIV-infected patients with GM-CSF is not without risk, since GM-CSF (but not G-CSF) has been shown to increase HIV-1 replication [75]. In fact, in the study by Kaplan HIV-1 p24 antigen levels increased in the GM-CSF arm, whereas they decreased in the control group receiving only CHOP [65]. G-CSF can be used safely to sustain haematopoiesis during chemotherapy in ARL [76].

One of the newer chemotherapeutic agents used with some success in relapsed or refractory ARL is MGBG (mitoguazone dihydrochloride). MGBG single agent therapy has resulted in a 23% response rate, with some durable remissions [77]. Treatment of ARL with *biologic response modifiers,* such as IL-2 or anti-IL-6 monoclonal antibody, has resulted in clinically significant tumour responses in anecdotal cases only. The use of IL-2 has also been proposed in patients with complete or partial remission to augment host response.

Although the use of *zidovudine* has been epidemiologically linked to the emergence of NHL, *in-vitro* zidovudine has not only anti-viral but also anti-neoplastic activity, and its use has been associated with regression of lymphoma in a few case reports. Several studies have included anti-retroviral therapy in their regimens in an attempt also to slow down progression of the underlying HIV infection [64,67]. In a study recently published by Levine *et al.*, zalcitabine (ddC) was used in combination with low-dose M-BACOD, which did not result in increased (neuro)toxicity [72].

Although the effectiveness of *CNS prophylaxis* with intrathecal administration of either methotrexate or cytarabine has never been studied in a randomised trial, it is incorporated in most treatment regimens (Table 1), and appears to reduce the incidence of relapse in the CNS.

Since opportunistic infections are a frequent cause of morbidity and mortality in NHL patients, the use of PCP (*Pneumocystis carinii* pneumonia) prophylaxis (either low-dose trimethoprim-sulfamethoxazole or aerosolised pentamidine) is recommended in all patients receiving chemotherapy, regardless of their CD4 cell count.

5.2. TREATMENT OF CNS-LYMPHOMA

Patients with AIDS who develop PCNSL are usually severely immunocompromised with CD4 cell counts below 0.05×10^9/l and they often have a prior diagnosis of AIDS. Their median survival is only 2-6 months. In view of difficult accessibility for tissue diagnosis, the standard practice for HIV-1-infected patients with CNS masses has been to treat them empirically with anti-toxoplasmosis treatment for approximately 2 weeks, followed by a diagnostic biopsy in those patients who do not respond. Treatment of CNS lymphomas with radiotherapy and/or dexamethasone can result in clinical and radiologic improvement, but survival is not prolonged [78]. One study reported remissions of 11-16 months in four patients treated with whole brain irradiation and chemotherapy, which shows that selected patients with AIDS and PCNSL may benefit from more intensive treatment [79].

Lymphomatous meningitis, occurring in 20-25% of patients both with systemic and primary CNS lymphoma, is usually treated palliatively with intrathecal administration of cytarabine or methotrexate, often via an Ommaya reservoir [80].

In conclusion, the optimal treatment of ARL remains to be defined. The introduction of highly active anti-retroviral treatment (resulting in improved immune function, decreased antigenic stimulation and cytokine dysregulation, and possibly decreased EBV viral load) might be expected to result in a lower incidence of ARL. However, preliminary observations suggest that this is not the case (A.M. Levine, personal communication). Because immune deficiency is not a prerequisite for the occurrence of Burkitt-type lymphoma in patients with AIDS, HAART may also result in a shift from diffuse large cell lymphomas to the Burkitt-type lymphomas. Because of the disappointing results of combination chemotherapy, advances may depend on the introduction of new treatment modalities, such as monoclonal antibodies targeting the B-cell antigen CD20, either unconjugated or radiolabelled with I^{131} (Iodine) or Y^{90} (Yttrium) for targeted delivery of radiation. In EBV-positive lymphomas, in analogy to the situation in patients with post-transplant lymphoproliferative disease, the feasibility of treatment with either autologous or allogeneic *in-vitro* expanded EBV-specific cytotoxic T cells is currently being explored.

6. References

1. Gatti, R.A. and Good, R.A.: Occurrence of malignancy in immunodeficiency diseases: a literature review, *Cancer* **28** (1971), 89-98.
2. Penn, I., Hammond, A., Brett Scheider, L. and Starzl, T.E.: Malignant lymphomas in transplantation patients, *Transplant. Proc.* **1** (1969), 106-111.

3. Centers for Disease Control. Diffuse, undifferentiated non-Hodgkin's lymphoma among homosexual males. United States, *MMWR* **31** (1982), 277-285.
4. Monfardini, S., Vaccher, E., Foa, R., and Tirelli, U.: AIDS associated non-Hodgkin's lymphoma in Italy: intravenous drug users versus homosexual men. *Ann. Oncol.* **1** (1990), 203-211.
5. Ragni, M., Belle, S.H., Jaffe, R.A., Duerstein, S.L., Bass, D.C., McMillan, C.W., Lovrien, E.W., Aledort, L.M., Kisker, C.T. and Kingsley, L.: Acquired immunodeficiency syndrome-associated non-Hodgkin's lymphomas and other malignancies in patients with hemophilia, *Blood* **81** (1993), 1889-1897.
6. Beral, V., Peterman, T., Berkelman, R. and Jaffe, H.: AIDS-associated non-Hodgkin's lymphoma, *Lancet* **337** (1991), 805-809.
7. Katz, M.H., Hessol, N.A., Buchbinder, S.P., Hirozawa, A., O'Malley, R. and Holmberg, S.D.: Temporal trends of opportunistic infections and malignancies in homosexual men with AIDS, *J. Infect. Dis.* **170** (1994), 198-202.
8. Moore, R.D., Kessler, H., Richman, D.D., Flexner, C., and Chaission, R.E.: Non-Hodgkin's lymphoma in patients with advanced HIV infection treated with zidovudine, *JAMA* **265** (1991), 2208-2211.
9. Reynolds, P., Saunders, L.D., Layefsky, M.E., and Lemp, G.F.: The spectrum of acquired immunodeficiency syndrome-associated malignancies in San Francisco, 1980-1987, *Am. J. Epidemiol.* **137** (1993), 19-30.
10. Pluda, L.M., Yarchoan, R. and Jaffe, E.S.: Development of non-Hodgkin's lymphoma in a cohort of patients with severe HIV infection on longterm antiretroviral therapy, *Ann. Intern. Med.* **113** (1990), 276-282.
11. Rabkin, C.S. and Yellin, F.: Cancer incidence in a population with a high prevalence of infection with human immunodeficiency virus type 1, *J. Natl.Cancer Inst.* **86** (1994), 1711-1716.
12. Carbone, A., Tirelli, U., Vaccher, E., Volpe, R., Gloghini, A., Bertda, G., DeRe, V., Rossi, C., Boiocchi, M. and Monfardini, S.: A clinicopathologic study of lymphoid neoplasms associated with human immunodeficiency virus infection in Italy, *Cancer* **68** (1991), 842-852.
13. Oksenhendler, E., Duarte, M., Soulier, J., Cacoub, P., Welker, Y., Cadranel, J., Cazals-Hatem, D., Autran, B., Clauvel, J. and Raphael, M.: Multicentric Castleman's disease in HIV-infection: a clinical and pathological study of 20 patients, *AIDS* **10** (1996), 61-67.
14. Ziegler, J.L., Beckstead, J.A., Volberding, P.A., Abrams, D.I. and Levine, A.M.: Non-Hodgkin's lymphoma in 90 homosexual men. Relation to generalized lymphadenopathy and the acquired immunodeficiency syndrome, *New Engl. J. Med.* **311** (1984), 565-570.
15. Knowles, M.D., Chamulak, G.A. and Subar, M.: Lymphoid neoplasia associated with the acquired immunodeficiency syndrome (AIDS): the New York University Medical Center experience with 105 patients (1981-1986), *Ann. Intern. Med.* **108** (1988), 744-753.
16. Kaplan, L.D., Abrams, D.I., Feigal, E., McGrath, M., Kahn, J., Neville, P., Ziegler, J. and Volberding, P.A.: AIDS-associated non-Hodgkin's lymphoma in San Francisco, *JAMA* **261** (1989), 719-724.
17. Pedersen, C., Gerstoft, J., Lundgren, J.D., Skinhoj, P., Bottzauw, J., Geisler, C., Hamilton-Dutoit, S.J., Thorsen, S., Lisse, I., Ralfkiaer, E. and Pallesen, G.: HIV-associated lymphoma: histopathology and association with Epstein-Barr virus genome related to clinical, immunological and prognostic features, *Eur. J. Cancer* **27** (1991), 1416-1423.
18. Rosenblum, M.L., Levy, R.M., Bredesen, D.A., So, Y.T., Wara, W. and Ziegler, J.L.: Primary central nervous system lymphoma in patients with AIDS, *Ann. Neurol.* **23 (suppl)** (1988), S13-S16.
19. De Luca, A., Antinori, A., Cingolani, A., Larocca, L.M., Linzalone, A., Ammassari, A., Scerrati, M., Roselli, R., Tamburrini, E. and Ortona L.: Evaluation of cerebrospinal fluid EBV-DNA and IL-10 as markers for *in-vivo* diagnosis of AIDS-related primary central nervous system lymphoma, *Brit. J. Haematol.* **90** (1995), 844-849.
20. Bishburg, E., Eng, R.H.K., Slim, J., Perez, G. and Johnson, E.: Brain lesions in patients with acquired immunodeficiency syndrome, *Arch. Intern. Med.* **149** (1989), 941-943.

21. Mocroft, A.J., Lundgren, J.D., d'Armino Monforte, A., Ledergerber, B., Barton, S.E., Vella, S., Katlama, C., Gerstoft, J., Pedersen, C. and Phillips, A.N.: Survival of AIDS patients according to type of AIDS-defining event. The AIDS in Europe Study Group, *Intern. J. Epidemiol.* **26** (1997), 400-407.
22. Levine, A.M., Sullivan-Halley, J., Pike, M.C., Rarick, M.U., Loureiro, C., Bernstein-Singer, M., Willson, E., Brynes, R., Parker, J., Rasheed, S. and Gill, P.S.: Human Immunodeficiency virus-related lymphoma. Prognostic factors predictive of survival, *Cancer* **68** (1991), 2466-2472.
23. Rubio, R.: Hodgkin's disease associated with human immunodeficiency virus infection. A clinical study of 46 cases. Cooperative Study Group of Malignancies associated with HIV-infection of Madrid, *Cancer* **73** (1994), 2400-2407.
24. Pelicci, P.G., Knowles, D.M., Jarlin ,Z.A., Wieczorek, R., Luciw, P., Dina, D., Basilico, C. and Dalla-Favera, R.: Multiple monoclonal B cell expansions and c-myc rearrangements in acquired immune deficiency syndrome-related lymphoproliferative disorders: implications for lymphoma-genesis, *J. Exp. Med.* **164** (1986), 2049-2076.
25. Shibata, D., Weiss, L.M., Nathwani, B.N., Brynes, R.K. and Levine, A.M.: Epstein-Barr virus in benign lymph node biopsies from individuals infected with the human immunodeficiency virus is associated with concurrent or subsequent development of non-Hodgkin's lymphoma, *Blood* **77** (1991), 527-1533.
26. Pluda, J.M., Venzon, D.J., Tosato, G., Lietzau, J., Wyvill, K., Nelson, D.L., Jaffe, E.S., Karp, J.E., Broder, S. and Yarchoan, R.: Parameters affecting the development of non-Hodgkin's lymphoma in patients with severe human immunodeficiency virus infection receiving antiretroviral therapy, *J. Clin. Oncol.* **11** (1993), 1099-1107.
27. Birx, D.L., Redfield, R.R. and Tosato, G.: Defective regulation of Epstein-Barr virus infection in patients with acquired immunodeficiency syndrome (AIDS) or AIDS-related disorders, *New Engl. J. Med.* **314** (1986), 874-879.
28. Blumberg, R.S., Paradis, T., Byington, R., Henle, W., Hirsch, M.S. and Schooley, R.T.: Effects of Human Immunodeficiency Virus on the cellular immune response to Epstein-Barr Virus in homosexual men: characterization of the cytotoxic response and lymphokine production, *J. Infect. Dis.* **155** (1987), 877-890.
29. Carmichael, A., Jin, X., Sissons, P. and Borysiewicz, L.: Quantitative analysis of the human immunodeficiency virus type 1 (HIV-1)-specific Cytotoxic T Lymphocyte (CTL) response at different stages of HIV-1 infection: differential CTL response to HIV-1 and Epstein-Barr virus in late disease, *J. Exp. Med.* **177** (1993), 249-256.
30. Geretti, A.M., Dings, M.E.M., van Els, C.A.C.M., van Baalen, C.A., Wijnholds, F.J., Borleffs, J.C.C. and Osterhaus, A.D.M.E..: Human immunodeficiency virus type 1 (HIV-1)- and Epstein-Barr Virus-specific T lymphocyte precursors exhibit different kinetics in HIV-1-infected persons, *J. Infect. Dis.* **174** (1996), 34-45.
31. Kersten, M.J., Klein, M.R., Holwerda, A., Miedema, F. and van Oers, M.H.J.: EBV-specific T cell responses in HIV-1 infection: different kinetics in patients progressing to opportunistic infection or non-Hodgkin's lymphoma, *J. Clin. Invest.* **99** (1997), 1525-1533.
32. Schnittman, S.M., Lane, H.C. and Higgins, S.E.: Direct polyclonal activation of human B lymphocytes by the acquired immunodeficiency virus, *Science* **233** (1986), 1084-1086.
33. Riboldi , P., Gaidano, G., Schettino, E.W., Steger, T.G., Knowles, D.M., Dalla-Favera, R. and Casali, P.: Two AIDS-associated Burkitt's lymphomas produce specific anti-i IgM cold agglutinins utilizing somatically mutated V_H4-21 segments, *Blood* **83** (1994), 2952-2961.
34. Ng, V.L., Hurt, M.H., Fein, C.L., Khayam-Bashi, F., Marsh, J., Nunes, W.M., McPhaul, L.W., Feigal, E., Nelson, P., Herndier, B.G., Shiramizu, B., Reyes, G.R., Fry, K.E. and McGrath, M.S.: IgMs produced by two acquired immune deficiency syndrome lymphoma cell lines: Ig-binding specificiy and V_H gene putative somatic mutation analysis, *Blood* **83** (1994), 1067-1078.
35. Fauci, A.S., Pantaleo, G., Stanley, S. and Weissman, D.: Immunopathogenic mechanisms of HIV infection, *Ann. Intern. Med.* **124** (1996), 654-663.
36. Birx, D.L., Redfield, R.R., Tencer, K., Fowler, A., Burke, D.S. and Tosato, G.: Induction of Interleukin 6 during human immunodeficiency virus infection, *Blood* **76** (1990), 2303-2310.

37. Emilie, D., Coumbaras, J., Raphael, M., Devergne, O., Delecluse, H.J., Gisselbrechts, C., Michelis, J.F., van Damme, J., Taga, T., Kishimoto, T., Crevon, M.C. and Galanaud, P.: Interleukin-6 production in high-grade B lymphomas: correlation with the presence of malignant immunoblasts in acquired immunodeficiency syndrome and in human immunodeficiency virus-seronegative patients, *Blood* **80** (1992), 498-504.
38. Benjamin, D., Knobloch, T.J. and Dayton, M.A.: Human B-cell interleukin-10: B cell lines derived from patients with acquired immunodeficiency syndrome and Burkitt's lymphoma constitutively secrete large quantities of interleukin-10, *Blood* **80** (1992), 1289-1298.
39. Matsuda, M., Salazar, F., Petersson, M., Masucci, G., Hansson, J., Pisa, P., Zhang, Q.J., Masucci, M.G. and Kiessling, R.: Interleukin-10 pretreatment protects target cells from tumor- and allospecific cytotoxic T cells and downregulates class I expression, *J. Exp. Med.* **180** (1994), 2371-2376.
40. Baer, R., Bankier, A.T., Biggin, M.D., Deininger, P.L., Farrell, P.J., Gibson, T.J., Hatfull, G., Hudson, G.S., Satchwell, S.C., Sequin, C., Tuffnell, P.S. and Barrell, B.G.: DNA sequence and expression of the B95.8 Epstein-Barr virus genome, *Nature* **310** (1984), 207-211.
41. Rickinson, A.B.: Cellular immunological responses to the virus infection. Epstein M.A., Achong B.G., eds. In: *The Epstein-Barr Virus: Recent Advances*. New York, Wiley Medical Publications (1990), pp. 75-115.
42. Henderson, S.A., Huen, D. and Rowe, M. Epstein-Barr virus transforming proteins, *Semin. Virol.* **5** (1994), 391-399.
43. Reisman, D. and Sudgen ,B.: Transactivation of an Epstein-Barr viral transcriptional enhancer by the Epstein-Barr virus nuclear antigen 1, *Mol. Cell Biol.* **6** (1986), 3838-3846.
44. Henderson, S., Rowe, M., Gregory, C., Broom-Carter, D., Wang, F., Longnecker, R., Kieff, E. and Rickinson, A.: Induction of bcl-2 expression by Epstein-Barr virus latent membrane protein 1 protects infected B cells form programmed cell death, *Cell* **65** (1991), 1107-1115.
45. MacMahon, E.M.E., Glass, J.D., Hayward, S.D., Mann, R.B., Becker, P.X., Charache, P., McArthur, J.C. and Ambinder, R.F.: Epstein-Barr virus in AIDS-related primary central nervous system lymphoma, *Lancet* **338** (1991), 969-973.
46. Hamilton-Dutoit, S.J., Raphael, M., Audoin, J., Diebold, J., Lisse, I., Pedersen, C., Oksenhendler, E., Marelle, L. and Pallesen, G.: In situ demonstration of Epstein-Barr virus small RNA's (EBER 1) in Acquired Immunodeficiency Syndrome-related lymphomas: correlation with tumor morphology and primary site, *Blood* **82** (1993), 619-624.
47. Neri, A., Barriga, F., Inghirami, G., Knowles, D.M., Neequaye, J., Magrath, I.T. and Dalla-Favera, R.: Epstein-Barr virus infection precedes clonal expansion in Burkitt's and acquired immunodeficiency syndrome-associated lymphoma, *Blood* 77 (1991), 1092-1095.
48. Carbone, A., Tirelli, U., Gloghini, A., Volpe, R. and Boiocchi, M.: Human Immunodeficiency virus-associated systemic lymphomas may be subdivided into two main groups according to Epstein-Barr viral latent gene expression, *J. Clin. Oncol.* **11** (1993), 1674-1681.
49. Khanna, R., Burrows, S.R., Kurilla, M.G., Jacob, C.A., Misko, I.S., Sculley, T.B., Kieff, E. and Moss, D.J.: Localization of Epstein-barr virus cytotoxic T cell epitopes using recombinant vaccinia: implications for vaccine development, *J. Exp. Med.* **176** (1992), 169-176.
50. Zimber, U., Adlinger, H.K. and Lenoir, G.M.: Geographic prevalence of two types of Epstein-Barr virus, *Virology* **154** (1986), 56-66.
51. Goldschmidt, W.L., Bhatia, K., Johnson, F., Akar, N., Gitierrez, M.I., Shibata, D., Carolan, M., Levine, A. and Magrath, I.T.: Epstein-Barr virus genotypes in AIDS-associated lymphomas are similar to those in endemic Burkitt lymphoma, *Leukemia* **6** (1992), 875-878.
52. Van Baarle, D., Hovenkamp, E., Kersten, M.J., Klein, M.R., Miedema, F. and van Oers, M.H.J.: Direct EBV typing on PBMC: no association between Epstein-Barr virus type 2 infection or superinfection and the development of AIDS-related non-Hodgkin's lymphoma, *Blood*, in press 1999.
53. Ballerini, P., Gaidano, G., Gong, J.Z., Tassi, V., Saglio, G., Knowles, D.M. and Dalla-Favera, R: Multiple genetic lesions in acquired immunodeficiency-syndrome-related non-Hodgkin's lymphoma, *Blood* **81** (1993), 166-176.

54. Carbone, A., Gaidano, G., Gloghini, A., Larocca, L.M., Capello, D., Canzonieri, V., Antinor, A., Tirelli, U., Falini, B. and Dalla Favera, R.: Differential expression of bcl-6, CD138 and EBV-encoded latent membrane protein 1 identifies distinct histogenetic subsets of acquired immunodeficiency syndrome-related non-Hodgkin's lymphomas, *Blood* **91** (1998), 747-755

55. Gaidano, G., Parsa, N.Z., Tassi, V., Della-Latta, P., Chaganti, R.S.K., Knowles, D.M. and Dalla-Favera, R.: *In-vitro* establishment of AIDS-related lymphoma cell lines: phenotypic characterization, oncogene and tumour suppressor gene lesions, and heterogeneity in Epstein-Barr virus infection, *Leukemia* **7** (1993), 1621-1629.

56. Gaidano, G., Hauptschein, R.S., Parsa, N.Z., Offit, K., Rao, P.H., Lenoir, G., Knowles, D.M., Chaganti, R.S.K. and Dalla-Favera, R.: Deletions involving two distinct regions of 6q in B cell non-Hodgkin's lymphoma, *Blood* **80** (1992), 1781-1787.

57. Chang, Y., Cesarman, E. and Pessin, M.S.: Identification of herpesvirus-like DNA sequences in AIDS-associated Kaposi's sarcoma, *Science* **266** (1994), 1865-1869.

58. Cesarman, E., Chang, Y., Moore, P.S., Said, J.W. and Knowles, D.M.: Kaposi's sarcoma-associated herpesvirus-like DNA sequences in AIDS-related body-cavity-based lymphomas, *New Engl. J. Med.* **332** (1995), 1186-1191.

59. Pluda, J.M., Venzon, D.J., Tosato, G., Lietzau, J., Wyvill, K., Nelson ,D.L., Jaffe, E.S., Karp, J.E., Broder, S. and Yarchoan, R: Parameters affecting the development of non-Hodgkin's lymphoma in patients with severe human immunodeficiency virus infection receiving antiretroviral therapy, *J. Clin. Oncol.* **11** (1993), 1099-1107.

60. Levine, A.M., Bernstein, L., Sullivan-Halley, J., Shibata, D., Bauch Mahterian, S. and Nathwani, B.N.: Role of zidovudine antiretroviral therapy in the pathogenesis of acquired immunodeficiency syndrome-related lymphoma, *Blood* **86** (1995), 4612-4616.

61. Whittle, H.C., Brown ,J., Marsh, K, Greenwood, B.M., Seidelin, P., Tighe, H. and Wedderburn, L.: T-cell control of Epstein-Barr virus-infected B-cells is lost during *P. falciparum* malaria, *Nature* **312** (1984), 449-450.

62. Gaidano, G., Pastore, C., Lanza, C., Mazza, U. and Saglio, G.: Molecular pathology of AIDS-related lymphomas. Biologic aspects and clinicopathologic heterogeneity, *Ann. Hematol.* **69** (1994), 281-290.

63. Gill, P.S., Levine, A.M., Krailo, M., Rarick, M.U., Laureiro, C., Deyton, L., Meyer, P. and Rasheed, S.: AIDS-related malignant lymphoma: results of prospective treatment trials, *J. Clin. Oncol.* **5** (1987), 1322-1328.

64. Levine, A.M., Wernz, J.C., Kaplan, L.D., Rodman, R., Cohen, P., Metroka, C., Bennett, J.M., Rarick, M.U., Walsh, C., Kahn, J., Miles, S., Jehamann, W.C., Feinberg, J., Nathwani, B., Gill, P. and Misuyasu, R.: Low-dose chemotherapy with central nervous system prophylaxis and zidovudine maintenance in AIDS-related lymphoma. A prospective multi-institutional trial, *JAMA* **266** (1991), 84-88.

65. Kaplan, L.D., Kahn, J.O., Crowe, S., Northfelt, D., Neville, P., Grossberg, H., Abrams, D.I., Tracey, J., Mills, J. and Volberding, P.A.: Clinical and virologic effects of recombinant human granulocyte-macrophage colony- stimulating factor in patients receiving chemotherapy for human immunodeficiency virus-associated non-Hodgkin's lymphoma: results of a randomized trial, *J. Clin. Oncol.* **9** (1991), 929-940.

66. Sawka, C.A., Shepherd, F.A., Brandwein, J., Burkes, R.L., Sutton, D.M. and Warner, E.: Treatment of AIDS-related non-Hodgkin's lymphoma with a twelve week chemotherapy program, *Leukemia and Lymphoma* **8** (1992), 213-220.

67. Tirelli, U., Errante, D., Oksenhendler, E., Spina, M., Vaccher, E., Serraino, D., Gastaldi, R., Repetto, L, Rizzardini, G. and Carbone, A.: Prospective study with combined low-dose chemotherapy and zidovudine in 37 patients with poor-prognosis AIDS-related non-Hodgkin's lymphoma. French-Italian Cooperative Study Group, *Ann. Oncol.* **3** (1992), 843-847.

68. Remick, S.C., McSharry, J.J, Wolf, B.C., Blanchard, C.G., Eastman, A.Y., Wagner, H., Portuese, E., Wighton, T., Powell, D., Pearce, T., Horton, J. and Ruckdeschel, J.C.: Novel oral combination chemotherapy in the treatment of intermediate-grade and high-grade AIDS-related non-Hodgkin's lymphoma, *J. Clin. Oncol.* **11** (1993), 1691-1702.

69. Gisselbrecht, C., Oksenhendler, E., Tirelli, U., Lepage, E., Gabarre, J., Farcet, J.P., Gastaldi, R., Coiffier, B., Thyss, A., Raphael, M. and Monfardini, S.: Human immunodeficiency virus-related lymphoma treatment with intensive combination chemotherapy, *Am. J. Med.* **95** (1993), 189-196.

70. Walsh, C., Wernz, J.C., Levine, A., Rarick, M., Wilson, E., Melendez, D., Bonnem, E., Thompson, J. and Shelton, B.: Phase I trial of m-BACOD and Granulocyte-Macrophage Colony-stimulating factor in HIV-associated non-Hodgkin's lymphoma, *J. Acq. Immune Def. Syndr. Human Retrovir.* **6** (1993), 265-271.

71. Sparano, J.A., Wiernik, P.H., Strack, M., Leaf, A., Becker, N. and Valentine, E.S.: Infusional cyclophosphamide, doxorubicin and etoposide in human immunodeficiency virus- and human T cell leukemia virus type I-related non-Hodgkin's lymphoma: a highly active regimen, *Blood* **81** (1993), 2810-2815.

72. Levine, A.M., Tulpule, A., Espina, B., Boswell, W., Buckley, J., Rasheed, S., Stain ,S., Parker, J., Nathwani, B. and GIll, P.S.: Low dose methotrexate, bleomycin, doxorubicin, cyclophosphamide, vincristine, and dexamethasone with zalcitabine in patients with acquired immunodeficiency syndrome-related lympoma. Effect on human immunodeficiency virus and serum Interleukin 6 levels over time, *Cancer* **78** (1996), :517-526.

73. Sparano, J.A., Wiernik, P.H., Hu, X., Sarta, C., Schwartz, E.L., Soeiro, R., Henry, D.H., Mason, B., Ratech, H. and Dutcher, J.P.: Pilot trial of infusional cyclophosphamide, doxorubicin, and etoposide plus didanosine and filgrastim in patients with HIV-associated non-Hodgkin's lymphoma, *J. Clin. Oncol.* **14** (1996), 3026-3035.

74. Kaplan, L., Straus, D., Testa, M., von Roenn, J., Dezube, B.J., Cooley, T.P., Herndier, B., Northfelt, D.W., Huang, J., Tulpule, A. and Levine, A.M.: Low dose compared with standard dose mBACOD chemotherapy for non-Hodgkin's lymphoma associated with human immunodeficiency virus infection, *New Engl. J. Med.* **336** (1997), 1641-1648.

75. Koyanagi, Y., O'Brien, W.A., Zhao, J.Q., Golde, D.W., Gasson, J.C. and Chen, I.S.Y.: Cytokines alter production of HIV-1 from primary mononuclear phagocytes, *Science* **241** (1988), 1673-1675.

76. Kersten, M.J., Verduyn, T.J., Reiss, P., Evers, L.M., de Wolf, F. and van Oers, M.H.J.: Treatment of patients with AIDS-related non-Hodgkin's lymphoma with chemotherapy (CNOP) and r-hu-G-CSF: clinical outcome and effects on HIV-1 viral load, *Ann. Oncol.* **9** (1998), 1135-1138.

77. Levine, A.M., Tulpule, A., Tessman, D., Kaplan, L., Giles, F., Luskey, B.D., Scadden, D.T. and von Hoff, D.: Mitoguazone therapy in patients with refractory or relapsed AIDS-related lymphoma: results from a multicenter phase II trial, *J. Clin. Oncol.* **15** (1997), 1094-1103.

78. Donahue, B.R., Sullivan, J.W. and Cooper, J.S.: Additional experience with empiric radiotherapy for presumed human immunodeficiency virus-associated primary central nervous system lymphoma, *Cancer* **76** (1995), 328-332.

79. Chamberlain, M.C.: Long survival in patients with acquired immune deficiency syndrome-related primary central nervous system lymphoma, *Cancer* **73** (1994), 1728-1730.

80. Chamberlain, M.C. and Dirr, L.: Involved-field radiotherapy and intra-Ommaya methotrexate/cytarabine in patients with AIDS-related lymphomatous meningitis, *J. Clin. Oncol.* **11** (1993), 1978-1984.

CHAPTER 10

B - PATHOGENESIS OF KAPOSI'S SARCOMA

THOMAS F. SCHULZ

Molecular Virology Group, Department of Medical Microbiology and Genitourinary Medicine, The University of Liverpool, Liverpool

1. Historical Background

First described by the Austro-Hungarian dermatologist Moritz Kaposi in the last century [1], Kaposi's sarcoma (KS) was for many years considered a rare, slowly progressive tumour of elderly males of some Mediterranean and Eastern European countries. In this "classical" form of KS, lesions are mainly confined to the skin and localised preferentially on the extremities. In the early 1950s, it was found that KS comprised up to 8% of malignancies in some sub-Saharan regions, in particular in parts of Central Africa, such as Zaire (for a compilation of epidemiological studies, see [2]). This "endemic" form of KS occasionally affects children [3]. KS is also more common among immunosuppressed organ transplant recipients, in particular if they come from countries where other forms of HIV-negative KS are more frequent [4,5]. However, in developed countries KS is now most commonly associated with HIV infection, and this variant, termed "epidemic KS" is characterised by widely distributed lesions, visceral as well as involving lymph nodes, and a rapidly progressive course.

Epidemiological observations relating to all four clinical forms of KS had long pointed to the involvement of a transmissible agent in the pathogenesis of KS (reviewed in [6]). In the case of the HIV-negative forms of KS, the main arguments were the characteristic geographic distribution, the fact that immigrants to the US or UK from Mediterranean countries had a higher risk of 'classic' and post-transplant KS than the population of the host country, and the increase in 'classic' KS in several Scandinavian countries which preceded the AIDS epidemic (reviewed in [2,6]). For HIV-associated KS, the fact that it occurred much more frequently in homosexual or bisexual men than in other HIV transmission groups, that the risk of AIDS KS was related to the number of sexual partners and sexual lifestyle (receptive anal intercourse and in some studies rimming), and the observation that it occurred more frequently among homosexual men with links to countries or regions where AIDS KS was common, suggested that a sexually transmitted agent other than HIV was involved [7,8,9] (further references in [2]). The search for such an agent finally led to the discovery of

H. Schuitemaker and F. Miedema (ed.), Aids Pathogenesis, 195–208.

'Kaposi's sarcoma-associated herpesvirus' (KSHV) or human herpesvirus 8 (HHV8) [10], and the last few years have shown that this virus fits the pattern predicted for the 'Kaposi agent' very well [11].

2. The KS Lesion: Histological Features and Origin of the Tumour Cell

KS skin lesions develop from the early "patch" stage to the "plaque" and "nodular" stage (reviewed in [12]). The early patch stage is characterised by pink macular lesions. Histologically, these involve slit-like vascular spaces in the dermis, which are lined by flattened endothelial cells and lack any of the other histological features of small vessels. A cellular infiltrate of lymphocytes and plasma cells is often present and can create the impression of an inflammatory process.

Once the proliferation of these atypical vessels becomes more extensive and spreads to the whole of the dermis as well as the underlying layers of subcutaneous fat, the clinical appearance of the lesion is that of a plaque or papule. The accompanying cellular infiltrate is much more pronounced and there are many extravasated erythrocytes and their remnants, hyaline deposits. With the progression of KS lesions to nodules and tumours, a spindle-like morphology of the abnormal endothelial cells predominates the histological appearance and numerous interweaving spindle cells, surrounding extravasated erythrocytes and hyaline deposits make up the characteristic appearance of these advanced lesions. These "KS spindle cells" represent the neoplastic component and the hallmark of advanced KS lesions. Most of the available evidence suggests an endothelial origin of these spindle cells (either vascular or lymphatic endothelium), but cells from venous lymphatic junctions, fibroblasts, smooth muscle cells and dermal dendrocytes have all been proposed as progenitors of KS spindle cells (reviewed [13]).

Thus endothelial cell markers, such as the EN-4 antigen, Ulex europaeus agglutinin (UEA-1), the BMA 120 antigen, factor VIII-related antigen, and OKM 5 antigen have been detected on KS spindle cells [13,14,15,16]. The absence on KS spindle cells of the Pal-E antigen, expressed on blood vessels, but not lymphatic endothelial cells, may indicate a relationship with lymphatic endothelium [14]. KS spindle and endothelial cells express LAM-1 (leukocyte adhesion molecule 1) and thrombomodulin which are markers of lymphokine-activated endothelial cells [18], in keeping with the notion that KS endothelial spindle cells are activated by growth factors (see below). Short- term cultures established from KS lesions also yielded cells with an endothelial surface marker profile in many cases (reviewed in [13]) and spindle-shaped cells showing a moderate expression of endothelial antigens were also cultured from peripheral blood of KS patients [19]. However, other cell types, such as monocytes/macrophages and dermal dendrocytes have also been suggested as precursors for KS spindle cells [13,14,20].

3. The Presence and Expression of KSHV in KS Lesions

Following the discovery of KSHV in KS lesions [10], it soon became apparent that KSHV could be detected in virtually all cases of AIDS KS [10,21,22, 23], classical Mediterranean KS [21, 24, 25], post-transplant KS [21, 25] and African endemic KS

[25,26]. KS lesions contain mainly circular viral DNA, characteristic of a latent viral persistence, but variable amounts of linear viral genomes have also been described [27,28]. PCR *in situ*, conventional *in-situ* hybridisation and immunohistological studies also indicate that the majority of KS spindle cells harbour latently persisting virus (Fig. 1), but that some infected spindle cells, as well as infiltrating mononuclear cells, are undergoing lytic viral replication [29,30,31,32,33]. Herpesvirus particles have also been seen by electron microscopy [34]. It is thus possible that KSHV persists in KS spindle cells in a latent state for long time periods with only a few cells ever switching into lytic replication. An alternative interpretation of these results would be that lytically infected cells disappear so quickly from KS tissue by lysis or apoptosis that they are only rarely documented. Which of these scenarios applies is currently the focus of debate.

The majority of spindle cells only appear to express a limited set of viral genes, as expected during viral latency. These include a set of three genes encoded on alternatively spliced multi-cistronic mRNAs transcribed from the same latent promoter [35]. Of these, open reading frame (orf) 73 encodes the latent nuclear antigen, LNA [32] (Fig. 1), orf72 a homologue of a D-type cyclin, v-cyc [36], and orf71/K13 a homologue of apoptosis inhibitors of the FLICE inhibitory proteins, v-FLIP [37] (further references in [11].

V-cyc associates with cyclin-dependent kinase (cdk) 6 to phosphorylate the retinoblastoma protein RB, the physiological target of D-type cyclins [38,39]. However, it has a wider spectrum of activity than cellular D-type cyclins and can also mediate phosphorylation, by cdk 6, of histone H1, the dual specificity phosphatase cdc25a, the transcription factor Id2, and the cdk inhibitor $p27^{Kip}$ [38,40,41]. Phosphorylated $p27^{Kip}$ is targeted for degradation. As a result, v-cyc can mediate not only the entry of resting fibroblasts from the G1 into the S phase of the cell cycle (through phosphorylation of pRB), but also, by removing $p27^{Kip}$–mediated inhibition of cyclin E/cdk2, G2/M progression [40-42]. However, on its own v-cyc does not appear to have transforming properties, at least as assessed by tumorigenicity studies in nude mice inoculated with v-cyc–transfected fibroblasts.

The role of v-FLIP may be to inhibit apoptosis of latently infected cells [37]. The function of LNA is not yet known in any detail, although features suggestive of an activator of gene expression are beginning to emerge.

The other viral mRNA expressed in KS spindle cells is T0.7, transcribed from orf K12 [43], which is thought to encode a short hydrophobic membrane protein, 'kaposin', reported to have transforming properties in rodent fibroblasts [44]. However, it is likely that other transcripts also arise from this gene locus and their functional importance and abundance, relative to that of T0.7, is still under investigation.

Of the many other KSHV genes with suggestive homologies to cellular proteins involved in growth control and differentiation (see [11] for review), the chemokine homologue, v-MIP I, and the bcl-2 homologue, v-bcl2, have been shown to be expressed in a few KS spindle cells and infiltrating inflammatory cells. The majority of these viral "cell homologous" genes are only expressed in KSHV-infected lymphoma cells in which lytic viral replication is thought to take place (reviewed in [45]). By analogy, and on the basis of RT-PCR studies, it is thought that v-MIP I, v-bcl 2, as well as v-MIP II, v-MIP III, v-GCR, K1, could also be expressed in a few virus-producing KS cells. V-MIPI and II have angiogenic proper-

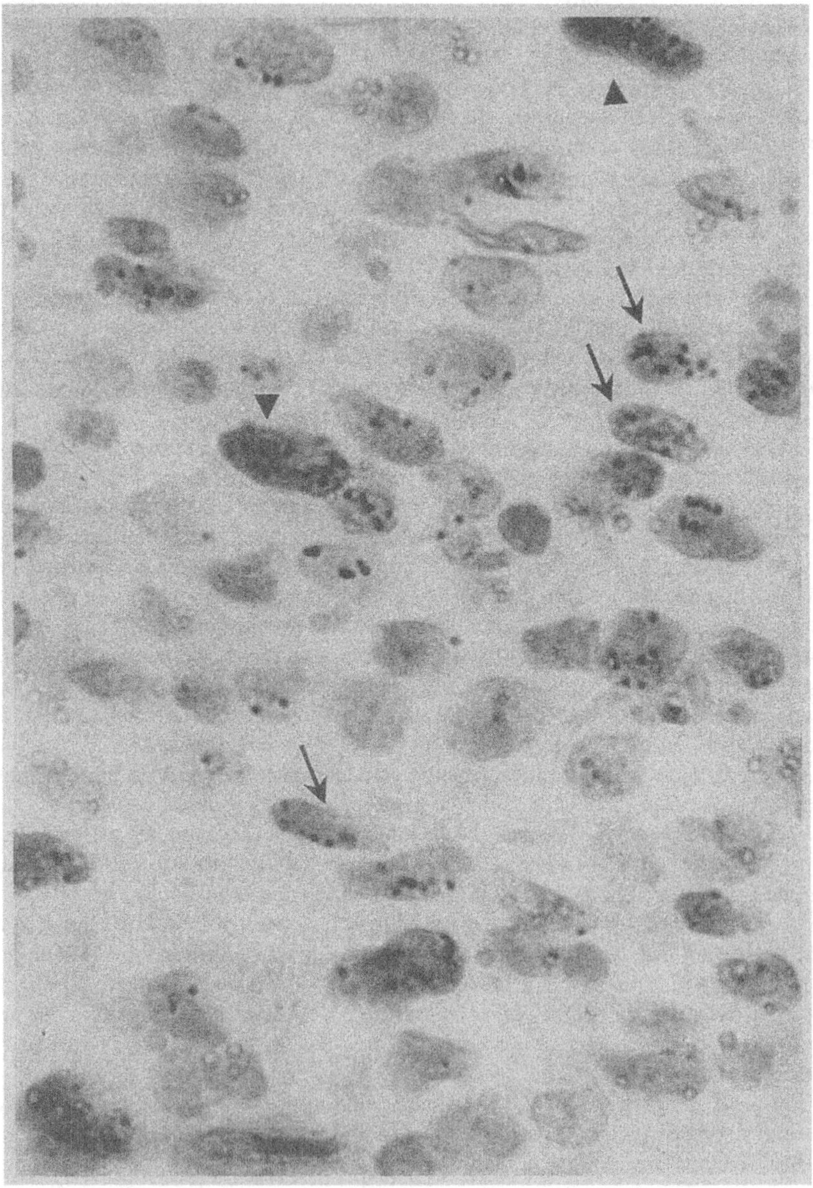

Fig. 1 - **Expression of KSHV LNA in KS spindle cells. KS tissue sections were stained with an affinity-**purified human antibody to LNA [32]. Nuclei of KS spindle cells express this antigen, either in a speckled or diffuse pattern (arrows), demonstrating the presence and latent persistence of KSHV in KS tumour cells.

ties on the chorioallantoic membrane and it appears possible that their release from a few virus-producing cells could be involved in the pathogenesis of this vascular tumour [51].

In the course of these studies it has also become apparent that the viral gene expression pattern in KS tissue is subtly different from that in the two lymphoid neoplasias associated with KSHV infection, *i.e.* primary effusion lymhoma (PEL) and some cases of the plasma cell variant of Multicentric Castleman's Disease (MCD). In these two conditions, tumour cells also express the viral IL-6 homologue [33] which has been shown to bind to the β chain of the IL-6 receptor and to promote growth of plasma cell lines *in vitro* (reviewed in [11]).

In addition to the orf K12/T0.7 encoded 'kaposin' [43,44], three other viral genes have been reported to have transforming properties in rodent fibroblasts or primary T cells. These include the type I membrane glycoprotein encoded by orf K1, the viral homologue of a chemokine receptor (v-GCR; orf74) and the viral homologue of an interferon regulatory factor (v-IRF 1; orf K9) [47-49]. The cytoplasmic domain of the K1 glycoprotein engages cytoplasmic signalling pathways through an ITAM domain, involving Syk, PI 3kinase, and c-vav, and intracellular Ca^{++} mobilisation [50]. V-GCR induces the secretion of VEGF in transfected cells [48], and v-IRF1 inhibits interferon-γ-mediated signalling [49,51]. However, there is currently no evidence that any of these are expressed *in vivo* in latently infected KS spindle cells, PEL lymphoma cells, or lymphoid cells in MCD, and their role in the pathogenesis of KS therefore remains to be elucidated. *In-vitro* studies in PEL cell lines suggest that v-GCR and K1 are only expressed after induction of lytic viral replication, whereas v-IRF 1 shows basal expression during the latent state [52]. In view of the tissue-specific expression of some viral genes (*e.g.* v-IL-6; see above), care should be taken in extrapolating from the expression pattern in cultured lymphoma cells to KS lesions *in vivo*. However, it is conceivable that the role of K1 and vGCR may be confined to the lytic state.

Recent *in-vitro* studies also suggest that infection of cultured primary endothelial cells with KSHV leads to a latent stage infection in most, and lytic infection in a minority of these cells, *i.e.* a pattern similar to that observed in KS tumours [53]. Latently infected spindle cells have an increased life span *in vitro* and grow in soft agar [53]. KSHV-infected endothelial cell cultures release VEGF which promotes the growth of uninfected cells, highlighting the possibility that paracrine mechanisms could play a role in the pathogenesis of KS. KSHV-infected cell cultures also have increased telomerase activity, considered to be a key step in cellular transformation [53].

4. HIV-1 Infection as a Co-factor in the Pathogenesis of KS

While epidemiological studies (see below) show that KSHV is a prerequisite for the development of KS, there is clearly a need for other co-factors. HIV-1 infection appears to be a particularly strong co-factor, illustrated by the high rate of AIDS KS in homosexual men who are dually infected with KSHV and HIV-1 (see below), and by the dramatic effect of highly active retroviral therapy (HAART) on AIDS KS [54-56]. The rapid disappearance of AIDS KS lesions after HAART could suggest that the role of HIV-1 in promoting KS is not confined to immune suppression, as is the case in transplant recipients.

4.1 INFLAMMATORY CYTOKINES

Initiated before the discovery of KSHV, a number of *in-vitro* studies have suggested that inflammatory cytokines and the Tat protein of HIV-1 may play a role in the generation of cells with features of KS spindle cells. These studies therefore point towards possible interactions of inflammatory cytokines and perhaps HIV-1 Tat with KSHV.

Treatment of endothelial cells with the inflammatory cytokines TNF- α, IL-1β and γ-interferon induces basic fibroblast growth factor (bFGF), a spindle-shaped morphology and a spindle cell surface marker profile [57]. Endothelial cells treated in this way induce bFGF-mediated angiogenic lesions in nude mice with some similarities to KS [58,59]. These inflammatory cytokines, in particular γ-interferon, are expressed in CD8 T cells and macrophages-infiltrating KS lesions, as well as peripheral blood mononuclear cells of these patients [60,61]. Inflammatory cytokines and platelet-derived growth factor have also been reported to induce cultured KS cells to produce VEGF, another angiogenic factor which is also induced by transfection with vGCR (see above), and which may synergise with bFGF to induce angiogenesis and oedema [48,62].

4.2 HIV-1 TAT

Early experiments had suggested that the HIV-1 Tat protein could induce KS-like lesions in nude mice when administered together with bFGF [58]. Although many epidemiological studies clearly indicate that HIV-1 infection alone is neither sufficient nor necessary for the development of KS, HIV-1 infection is clearly an important co-factor (see below), and HIV-1 Tat may therefore play a role in the pathogenesis of AIDS KS. HIV-1 Tat binds to the heparin-sulphate proteoglycans of extracellular matrix. This could displace bFGF which would then be able to act on its target cells [63]. HIV-1 Tat has also been reported to potentiate angiogenesis by molecular mimicry of extracellular matrix proteins containing an RGD integrin-binding domain and by a heparin-binding effect [64]. Furthermore, HIV-1 Tat has been shown to bind to the VEGF (Flk/KDR) receptor and to activate its signalling pathway [65]. Other studies have reported that HIV-1 Tat enhances the adhesiveness and chemotaxis of monocytes [66], and may reactivate KSHV infection [67].

4.3 CLONALITY OF KS TUMOURS

Whether KS nodules are in fact of monoclonal origin, or whether they represent an oligoclonal or polyclonal proliferation of KSHV-infected endothelial cells akin to EBV-driven oligoclonal B-cell proliferations in immunosuppressed individuals is still unclear. While one study found evidence, based on X-chromosome inactivation in individual KS lesions, for monoclonality [68], others disagree [69,70].

5. Epidemiological Clues to the Role of KSHV in the Pathogenesis of KS

Thus cell and molecular biology studies have highlighted a number of ways in which KSHV infection of endothelial cells could lead to their proliferation and atypical differentiation, and also induce angiogenesis. However, the strongest support for KSHV being the infectious cause of KS comes from epidemiological studies. In addition, epidemiological observations also indicate clearly that KSHV rarely causes disease in the immunocompetent host, but more frequently in the context of immune suppression. HIV-1 infection is an extremely strong co-factor, and the mechanisms involved may go beyond immune suppression.

Many PCR-based and sero-epidemiological studies have confirmed and extended the strong association between all forms and KS and infection with KSHV. These studies have recently been reviewed by several authors [11,71-73].

Unlike many other herpesviruses KSHV is not ubiquitous in most countries. Owing to remaining uncertainties about serological assays [74] (discussed in [71]) the exact prevalence rates of KSHV infection in many countries are still controversial. KSHV infection appears to be infrequent in the general population of Northern Europe, the USA, and parts of Asia, but more frequent in some Mediterranean countries (Italy, Greece) and widespread in most parts of sub-Saharan Africa [75-79] (reviewed in [11,71-73]). This distribution of KSHV reflects the geographic variations in the incidence of HIV-negative KS, with one important exception (see below). In some countries, seroprevalence rates, as determined by assays using defined antigens, show marked regional variation, and in Italy this corresponds to the regional differences in classic KS incidence [78,79]. In spite of KSHV seroprevalence rates of 25–35%, as determined by assays using defined KSHV antigens [78,79], population-based incidence rates of 'classic' KS are only in the order of 1 – 3/100,000 in parts of Southern Italy, Sicily and Sardinia [80]. Similar KSHV strains appear to circulate in Northern and Southern Europe. As defined on the basis of variability in the K1 gene of KSHV (see above), two major groups of KSHV, termed A and C, are found in different parts of Europe and in some cases isolates in Southern Europe are closely to those encountered in Northern Europe [81]. Although the available data are still limited, this distribution pattern probably indicates that no particular KSHV strains predominate in the endemic regions of Europe and that KSHV strains have been transmitted across Europe for some time [81]. These comparisons of KSHV prevalence rates and KSHV variability with incidence rates for 'classic' KS indicate that only a small proportion, perhaps fewer than 0.1%, of KSHV-infected, but immuncompetent, individuals in Southern Europe ever develop 'classic' KS.

In contrast to the good correlation between 'classic' KS rates and KSHV seroprevalence in Europe, there is no such agreement between KSHV seroprevalence in Africa, conservatively estimated to be in the order of 40–70 % in most parts of sub-Saharan Africa [75,77,82 –84], and the increased incidence of 'endemic' KS in East and Central Africa. Initial studies have also failed to find evidence for the existence of different KSHV variants, as assessed by variability in the K1 gene, in East and Central Africa, compared to West Africa [81,85,86]. KSHV strain variation may therefore not be the explanation for the higher rates of endemic KS in East and Central Africa. However, further studies are probably required, in view of the evidence for recombination among KSHV genomes and the fact that variability in the K1 locus does not always reflect variability in other regions of the genome [85-87].

Among those at risk for HIV infection in Western countries, KSHV is much more common (20–35% in most studies) among homosexual men, than in patients with haemophilia, intravenous drug users, or heterosexually infected women [75-77,88-90]. This distribution mirrors exactly that of the postulated 'KS agent', predicted to be more common among homosexual men [7]. Also as predicted [6-9], KSHV prevalence among homosexual men increases with the number of sexual partners [88,90], other STDs [90], and contact (in the early 1980s) with men from countries where AIDS KS is common [88]. Risk factors for KSHV transmission, as deduced from seroepidemiological studies in several prospective cohorts, also include HIV-1 infection, receptive anal intercourse [88], and oro-genital contact [91]. KSHV DNA has been detected infrequently in semen and saliva from KSHV-infected individuals (see [71] for review), providing a possible explanation for the route of transmission deduced from seroepidemiological studies.

While transmission associated with sexual contact, is thus likely to represent the main route of transmission among homosexual men, the evidence for sexual transmission among female sex workers is still limited, and is unlikely to explain the high KSHV seroprevalence rates in parts of Southern Europe and, in particular, Africa. First indications are that KSHV transmission in Africa occurs in young children, mostly after the age of 2 but before puberty, perhaps following routes similar to hepatitis-B virus [82] or from mother to child [83]. Transmission through sexual contact may also occur [84], but, unlike among homosexual men of Western countries, may not represent the main route of transmission in Africa.

HIV-1 infection clearly has a dramatic effect on the development of KS in a KSHV-infected individual with at least 50% of KSHV and HIV-infected homosexual men developing AIDS over a period of 10 years [89,90]. A recent study measured progression rates to AIDS KS from the point of infection with KSHV and HIV-1, and found that those who contracted KSHV infection after a pre-existing HIV-1 infection progressed faster (median 5 years) than those who were infected with KSHV before HIV-1 (median 10 years) [89]. This result could indicate that HIV-1 infection promotes KSHV replication and/or its seeding into endothelial cells and the establishment of a latent infection, either by interfering with a primary T-cell response to KSHV, or as a result of inflammatory cytokines promoting KSHV replication (see above).

A recent case-control study from West Africa points to the possibility that HIV-2 infection may be a much less efficient co-factor than HIV-1 infection, since AIDS KS in The Gambia, West Africa, appears to be much more common among HIV-1-infected, than among HIV-2-infected individuals [92]. A direct comparison with the results obtained among homosexual men is however not possible, since the patients investigated in the West African study are likely to have been infected with KSHV in their childhood.

6. Summary

The epidemiological case for KSHV being the cause of Kaposi's sarcoma is very strong: (*i*) all patients with KS are infected with KSHV; (*ii*) cross-sectional case-control studies have shown a strong link association between KS and KSHV infection; (*iii*) prospective studies among HIV-infected individuals have estimated the median interval between KSHV infection and the appearance of tumours to be around 5–10 years. In contrast, KS occurs only in a very small proportion of immunocompetent, HIV-negative individuals in Southern Europe. HIV-1 infection is, therefore, a potent co-factor in KS pathogenesis, partly through immunosuppression, but other features (inflammatory cytokines, Tat) may also contribute. KSHV established a latent infection in endothelial spindle cells, the neoplastic component of KS, and several viral genes have the potential to dysregulate the cell cycle or affect intracellular signalling. Other viral genes, expressed only during lytic viral replication which occurs in a few cells in KS lesions, may also play a role in the pathogenesis of KS.

7. References

1. Kaposi, M.: Idiopathisches multiples Pigmentsarkon der Haut, *Arch. Dermatol. Syphil.* **4** (1872), 742-749.
2. IARC monographs on the evaluation of carcinogenic risks to humans: *Human Immunodeficiency Viruses and Human T-cell Lymphotropic Viruses* **67** (1996), 76.
3. Matondo, P. and Zumla, A.: The spectrum of African Kaposi's sarcoma: is it consequential upon diverse immunological responses?, *Scand. J. Infect. Dis.* **28** (1996), 225-230.
4. Farge, D.: Kaposi's sarcoma in organ transplant recipients. The Collaborative Transplantation Research Group of Ile de France, *Eur. J. Med.* **2** (1993), 339-343.
5. Penn, I.: Kaposi's sarcoma in transplant recipients, *Transplantation* **64** (1997), 669-673
6. Beral, V.: Epidemiology of Kaposi's sarcoma. In: *Cancer, HIV and AIDS, Cancer Surveys* (V. Beral, H.W. Jaffe, and R.A. Weiss, eds.), Cold Spring Harbor Laboratory, Cold Spring Harbor, New York, **10** (1991), pp. 5–22.
7. Beral, V., Peterman, T. A., Berkelman, R.L. and Jaffe, H.W.: Kaposi's sarcoma among persons with AIDS: a sexually transmitted infection?, *Lancet* **335** (1990), 123–128.
8. Beral, V., Bull, D., Jaffe, H., Evans, B., Gill, N., Tillett, H. and Swerdlow, A.J.: Risk of Kaposi's sarcoma in AIDS patients in the British Isles; is it increased if sexual partners come from the USA or Africa?, *Brit. Med. J.* **324** (1991), 624-625.
9. Archibald, C.P., Schechter, M.T., Craib, K.J.P. *et al.*: Risk factors for Kaposi's sarcoma in the Vancouver lymphadenopathy-AIDS study, *J. Acq. Immune Def. Syndr. Human Retrovir.* **3 (suppl. 1)** (1990), S18-S23.
10. Chang, Y., Cesarman, E., Pessin, M.S., Lee, F., Culpepper, J., Knowles, D.M. and Moore, P.S.: Identification of herpesvirus-like DNA sequences in AIDS-associated Kaposi's sarcoma, *Science* **266** (1994), 1865–1869.
11. Schulz, T.F.: Epidemiology of Kaposi's sarcoma-associated herpesvirus (KSHV/HHV8), *Adv. Cancer Res.*, in press.
12. Cockerell, C.J.: Histopathological features of Kaposi's sarcoma in HIV- infected individuals. In: *Cancer, HIV and AIDS* Beral,V., Jaffe, H.W. and Weiss, R.A., eds.), *Cancer Surveys* **10** (1991), 39-52.
13. Roth,W.K., Brandstetter, H. and Stürzl, M.: Cellular and molecular features of HIV-associated Kaposi's sarcoma, *AIDS* **6** (1992), 895-913.
14. Rappersberger, K., Tschachler, E., Zonzit, E. *et al.*: Endemic Kaposi's sarcoma in human immunodeficicency virus type I-seronegative persons: demonstration of retrovirus-like particles in cutaneous lesions, *J. Invest. Dermatol.* **95** (1991), 371-381.

15. Nadji, M., Morales, A.R., Ziegler-Weissman, J., Penneys, N.S.: Kaposi's sarcoma: immunohistological evidence for an endothelial origin, *Arch. Pathol.* Lab. Med. **105** (1981), 274-275.
16. Modlin, R.L., Hofman, F.M., Kempf, R.A., Taylor, C.R., Conant, M.A. and Rea, T.H.: Kaposi's sarcoma in homosexual men: an immunohistochemical study, *J. Am. Acad. Dermatol.* **8** (1983), 620-627.
17. Roth, W.K., Werner, S., Risau, W., Remberger, K. and Hofschneider, P.H.: Cultured, AIDS-related Kaposi's sarcoma cells express endothelial cell markers and are weakly malignant *in vitro*, *Int. J. Cancer* **42** (1988), 767-773.
18. Zhang, Y.-M., Bachmann, S., Hemmer, C., van Lunzen, J., von Stemm, A., Kem, P., Dietrich, M., Ziegler, R., Waldherr, R. and Nawroth, P.P.: Vascular origin of Kaposi's sarcoma: expression of leukocyte adhesion molecule-1, thrombomodulin, and tissue factor, *Am. J. Pathol.* **144** (1994), 51–59.
19. Browning, P.J., Sechler, J.M.G., Kaplan, M., Washington, R.H., Gendelman, R., Yarchoan, R., Ensoli, B. and Gallo, R.C.: Identification and culture of Kaposi's sarcoma-like spindle cells from the peripheral blood of human immunodeficiency virus-1-infected individuals and normal controls, *Blood* **84** (1994), 2711–2720.
20. Nickoloff, B.J. and Griffiths, C.E.M.: Factor XIIIa-expressing dermal dendrocytes in AIDS-associated cutaneous Kaposi's sarcoma, *Science* **243** (1989), 1736-1737.
21. Boshoff, C., Whitby, D., Hatzioannou, T., Fisher, C., van der Walt, J., Hatzakis, A., Weiss, R.A. and Schulz, T.F.: Kaposi's sarcoma-associated herpesvirus in HIV-negative Kaposi sarcoma, *Lancet* **345** (1995), 1043–1044.
22. Huang, Y.Q., Li, J.J., Kaplan, M.H., Poiesz, B.J., Katabira, E., Zhang, W.C., Feiner, D. and Friedman-Kien, A.: Human herpesvirus-like nucleic acid in various forms of Kaposi's sarcoma, *Lancet* **345** (1995), 759-761.
23. Lebbé, C., de Cremoux, P., Rybojad, M., Costa da Cunha, C., Movel, P. and Calvo, F.: Kaposi's sarcoma and new herpes virus, *Lancet* **345** (1995), 1180.
24. Moore, P.S. and Chang, Y.: Detection of herpesvirus-like DNA sequences in Kaposi's sarcoma in patients with and without HIV infection *New Engl. J. Med.* **332** (1995), 1181–1185.
25. Buonaguoro, F.M., Tomesello, M.L., Beth-Giraldo, E., Hatzakis, A., Mueller, N., Downing, R., Biryamvaho, B., Sempala, D.K. and Giraldo, G.: Herpes virus-like DNA sequences detected in endemic, classic, iatrogenic and epidemic Kaposi's sarcoma (KS) biopsies, *Int. J. Cancer* **65** (1996), 25–28.
26. Schalling, M., Eukman, M., Kaaya, E.E., Linde, A. and Biberfeld, P.: A role for a new Herpesvirus (KSHV) in different forms of Kaposi's sarcoma, *Nature Med.* **1** (1995), 707-708.
27. Decker, L.L., Shankar, P., Khan, G.: The Kaposi sarcoma-associated herpesvirus (KSHV) is present as an intact latent genome in KS tissue but replicates in the peripheral blood mononuclear cells of KS patients, *J. Exp. Med.* **184** (1996), 283–288.
28. Russo, J.J., Bohenzky, R.A., Chien, M.-C., Chen, J., Yan, M., Maddalena, D., Parry, J.P., Peruzzi, D., Edelman, I.S., Chang, Y. and Moore, P.S.: Nucleotide sequence of Kaposi's sarcoma-associated herpesvirus (HHV 8) *Proc. Natl. Acad. Sci. USA* **93** (1996), 14862–14868.
29. Boshoff, C., Schulz, T.F., Kennedy, M.M., Graham, A.K., Fisher, C., Thomas, A., McGee, J.O., Weiss, R.A., and O'Leary, J.J.: Kaposi's sarcoma- associated herpesvirus infects endothelial and spindle cells, *Nature Med.* **1** (1995), 1274-1278.
30. Staskus, K.A., Zhong, W., Gebhard, K., Herndier, B., Wang, H., Renne, R., Beneke, J., Pudney, J., Anderson, D.J., Ganem, D. and Haase, A.T.: Kaposi's sarcoma-associated herpesvirus gene expression in endothelial (spindle) tumor cells, *J. Virol.* **71** (1997), 715–719.
31. Davis, M.A., Stürzl, M., Blasig, C., Schreier, A., Guo, H.G., Reitz, M., Opalenik, S.R. and Browning, P.J.: Expression of human herpesvirus 8-encoded cyclin D in Kaposi's sarcoma spindle cells, *J. Natl. Cancer Inst.* **89** (1997), 1868–1874.
32. Rainbow, L., Platt, G.M., Simpson, G.R., Sarid, R., Gao, S.-J., Stoiber, H., Herrington, C.S., Moore, P.S. and Schulz, T.F.: The 222-234 kd nuclear protein (LNA) of Kaposi's sarcoma-associated herpesvirus (Human herpes virus 8) is encoded by orf73 and a component of the latency-associated nuclear antigen, *J. Virol.* **71** (1997), 5915-5921.
33. Moore, P.S., Boshoff, C., Weiss, R.A. and Chang, Y.: Molecular mimicry of human cytokine and cytokine response pathway genes by KSHV, *Science* **274** (1996), 1739-1744.
34. Orenstein, J.M., Alkan, S., Blauvelt, A., Jeang, K.-T., Weinstein, M.D., Ganem, D. and Herndier, B.: Visualization of human herpesvirus type 8 in Kaposi's sarcoma by light and transmission electron microscopy, *AIDS* **11** (1997), F35-745.

35. Dittmer, D., Lagunoff, M., Renne, R., *et al.*: A cluster of latently expressed genes in Kaposi's sarcoma-associated herpesvirus, *J. Virol.* **72** (1998), 8309-8315.
36. Chang,Y., Moore, P.S., Talbot, S.J., Boshoff, Zarkowska, C.T., Godden-Kent, D., Paterson, H., Weiss, R.A. and Mittnacht, S.: Cyclin encoded by KS herpesvirus, *Nature* **382** (1996), 410.
37. Thome, M., Schneider, P., Hofman, K., Fickenscher, H., Meinl, E., Neipel, F., Mattmann, C., Burns, K., Bodmer, J.-L., Schröter, M., Scaffidl, C., Krammer, P.H., Peter, M.E. and Tschopp, J.: Viral FLICE-inhibitory proteins (FLIPs) prevent apoptosis induced by death receptors, *Nature* **386** (1997), 517-521.
38. Godden-Kent, D., Talbot, S.J., Boshoff, C., Chang, Y., Moore, P.S., Weiss, R.A., and Mittnacht, S.: The cyclin encoded by Kaposi's sarcoma-associated herpesvirus (KSHV) stimulates cdk6 to phoshorylate the retinoblastoma protein and Histone H1, *J. Virol.* **71** (1997), 4193-4198.
39. Li, M.T., Lee, H.R., Yoon, D.W., Albrecht, J.C., Fleckenstein, B., Neipel, F. and Jung, J.U.: Kaposi's sarcoma-associated herpesvirus encodes a functional cyclin, *J. Virol.* **71** (1997), 1984-1991.
40. Ellis, M., Chew, Y.P., Fallis, L., Freddersdorf, S., Boshoff, C., Weiss, R.A., Lu, X. and Mittnacht, S.: Degradation of $p27^{Kip}$ cdk inhibitor triggered by Kaposi's sarcoma virus cyclin-cdk6 complex, *EMBO J.* **18** (1999), 644–653
41. Mann, D.J., Child, E.S., Swanton, C., Laman, H. and Jones, N.: Modulation of $p27^{Kip1}$ levels by the cyclin encoded by Kaposi's sarcoma-associated herpesvirus, *EMBO J.* **18** (1999), 654-663.
42. Swanton, C., Mann, D.J., Fleckenstein, B., Neipel, F., Peters, G. and Jones, N.: Herpes viral cyclin/cdk6 complexes evade inhibition by CDK inhibitor proteins, *Nature* **390** (1997), 184-187.
43. Zhong, W., Wang, H., Herndier, B. and Ganem, D.: Restricted expression of Kaposi's sarcoma-associated herpesvirus (human herpesvirus 8) genes in Kaposi's sarcoma, *Proc. Natl. Acad. Sci. USA* **93** (1996), 6641–6646.
44. Muralidhar, S., Pumfery, A.M., Hassani, M., *et al.*: Identification of Kaposin (open reading frame K12) as a human herpesvirus 8 (Kaposi's sarcoma-associated herpesvirus) transforming gene, *J. Virol.* **72** (1998), 4980-4988.
45. Biberfeld, P., Ensoli, B., Sturzl, M., and Schulz, T.F.: Kaposi sarcoma-associated herpes-virus/human herpesvirus 8, cytokines, growth factors and HIV in the pathogenesis of Kaposi's sarcoma, *Curr. Opinion Inf. Dis.* **11** (1998), 97–105.
46. Boshoff, C., Endo, Y., Collins, P.D., Takeuchi, Y., Reeves, J.D., Schweickart, V.L., Siani, M.A., Sasaki, T., Williams, T.J., Gray, P.W., Moore, P.S., Chang, Y. and Weiss, R.A.: Angiogenic and HIV-inhibitory functions of KSHV-encoded chemokines, *Science* **278** (1997), 290-294.
47. Lee, H., Veazy, R., Williams, K., Li, M., Guo, J., Neipel, F., Fleckenstein, B., Lackner, A., Desrosiers, R.C. and Jung, J.U.: Deregulation of cell growth by the K1 gene of Kaposiis sarcoma-associated herpesvirus, *Nature Med.* **4** (1998), 435–440.
48. Bais, C., Santomasso, B., Coso, O., Arvanitakis, L., Reaka, E.G., Gutkind, J.S., Asch, A.S., Cesarman, E., Gerhengorn, M.C. and Mesri, E.A.: G-protein-coupled receptor of Kaposi's sarcoma-associated herpesvirus is a viral oncogene and angiogenesis activator, *Nature* **341** (1998), 86-89.
49. Gao, S.J., Boshoff, C., Jayachandra, S., Weiss, R.A., Chang, Y. and Moore, P.S.: KSHV orf K9 (vIRF) is an oncogene which inhibits the interferon signaling pathway, *Oncogene* **15** (1997), 1979–1985.
50. Lee, H., Guo, J., Li, M., Choi, J.-K., DeMaria, M., Rosenzweig, M. and Jung, J.U.: Identification of an immunoreceptor tyrosine-based activation motif of K1-transforming protein of Kaposi's sarcoma-associated herpesvirus, *Mol. Cell. Biol.* **18** (1998), 5219–5228.
51. Zimring, J.C., Goodbourn, S. and Offermann, M.K.: Human herpesvirus 8 encodes an interferon regulatory factor (IRF) homolog that represses IRF-1- mediated transcription, *J. Virol.* **72** (1998), 701–707.
52. Sarid, R., Flore, O., Bohenzky, R.A., Chang, Y. and Moore, P.S.: Transcriptional mapping of the Kaposi's sarcoma-associated herpesvirus (KSHV) genome in the body cavity-based lymphoma cell line BC-1, *J. Virol.* **72** (1998), 1005-1012.
53. Flore, O., Rafii, S., Ely, S., O'Leary, J.J., Hyjek, E.M. and Cesarman, E.: Transformation of primary endothelial cells by Kaposi's sarcoma-associated herpesvirus, *Nature* **394** (1998), 588–592.
54. Conant, M.A., Opp, K.M., Poretz, D. and Mills, R.G.: Reduction of Kaposi's sarcoma lesions following treatment of AIDS with ritonavir, *AIDS* **11** (1997), 1300–1301.

55. Blum, L., Pellet, C., Agbalika, F., Blanchard, G., Morel, P., Calvo, F. and Lebbé, C.: Complete remission of AIDS-related Kaposi's sarcoma associated with undetectable human herpesvirus-8 sequences during anti-HIV protease therapy, *AIDS* **11** (1997), 1653–1655.
56. Murphy, M., Armstrong, D., Sepkowitz, K.A., Ahkami, R.N. and Myskowski, P.L.: Regression of AIDS-related Kaposi's sarcoma following treatment with HIV-protease inhibitor, *AIDS* **11** (1997), 261–262.
57. Fiorelli, V., Gendelman, R., Samaniego, F., Markham, P.D. and Ensoli, B.: Cytokines from activated T-cells induce normal endothelial cells to acquire the phenotypic and functional features of AIDS-Kaposi's sarcoma spindle cells, *J. Clin. Invest.* **95** (1995), 1723–1734.
58. Ensoli, B., Gendelman, R., Markham, P., Fiorelli, V., Colombini, S., Raffeld, M., Cafaro, A., Chang, H.K., Brady, J.N. and Gallo, R.C.: Synergy between basic fibroblast growth factor and HIV-1 tat protein in induction of Kaposi's sarcoma, *Nature* **371** (1994a), 674-680.
59. Samaniego, F., Markham, P.D., Gendelman, R., Gallo, R.C. and Ensoli, B.: Inflammatory cytokines induce endothelial cells to produce and release fibroblast growth factor and to promote Kaposi's sarcoma-like lesions in nude mice, *J. Immunol.* **158** (1997), 1887–1894.
60. Sirianni, M.C., Vincenzi, L., Fiorelli, V., Topino, S., Scala, E., Uccini, S., *et al.*: γ-interferon production in peripheral blood mononuclear cells (PBMC) and tumour-infiltrating lymphocytes from Kaposi's sarcoma patients: correlation with the presence of human herpesvirus-8 in PBMC and lesional macrophages, *Blood* **91** (1998), 968-976.
61. Fiorelli, V., Gendelman, R., Sirianni, M.C., Chang, H.K., Colombini, S., Markham, P.D. *et al.*: γ-interferon produced by $CD8^+$ T cells infiltrating Kaposi's sarcoma induces spindle cells with angiogenic phenotype and synergy with HIV-1 Tat protein: an immune response to HHV-8 infection, *Blood* **91** (1998), 956-967.
62. Samaniego, F., Markham, P.D., Gendelman, R., Watanabe, Y., Kao, V., Kowalski, K., *et al.*: Vascular endothelial growth factor and basic fibroblast growth factor are expressed in Kaposi's sarcoma and synergise to induce angiogenesis, vascular permeability and KS lesion development: induction by inflammatory cytokines, *Am. J. Pathol.* **152** (1998), 1433-1443.
63. Chang, H.Q., Samaniego, F., Nair, B.C., Buonaguro, L. and Ensoli, B.: HIV-1 Tat protein exits from intact cells via a leaderless secretory pathway and binds extracellular matrix-associated heparan sulphate proteoglycans through its basic region, *AIDS* **11** (1997), 1421–1431.
64. Barillari, G., Fiorelli, V., Gendelman, R., Colombini, S., Samaniego, F., Morris, B.C., *et al.*: HIV-1 Tat protein enhances angiogenesis and Kaposi's sarcoma (KS) development triggered by inflammatory cytokines (IC) or bFGF by engaging the v3 integrin, *J. Acq. Immune Def. Syndr. Human Retrovir.* **114** (1997), A33.
65. Albini, A., Soldi, R., Giunciuglio, D., Giraudo, E., Benelli, R., Primo, L., *et al.*: The angiogenesis induced by HIV-1 Tat protein is mediated by the Flk-1/KDR receptor on vascular endothelial cells, *Nature Med.* **2** (1996), 1371– 1375.
66. Lafrenie, R.M., Wahl, L.M., Epstein, J. S., Hewlett, I.K., Yamada, K.M., and Dhawan, S.: HIV-1 Tat protein promotes chemotaxis and invasive behaviour by monocytes, *J. Immunol.* **157** (1996), 974–977.
67. Harrington, W., Sieczkowski, L., Sosa, C., Chan-a-Sue, S., Cai, J.P., Cabral, L. and Wood, C.: Activation of HHV-8 by HIV-1 Tat, *Lancet* **349** (1997), 774.
68. Rabkin, C.S., Janz, S., Lash, A., Coleman, A.E., Musaba, E., Liotta, L., *et al.*: Monoclonal origin of multicentric Kaposi's sarcoma lesions, *New Engl. J. Med.* **336** (1997), 988-993.
69. Gill, P., Tsai, Y., Rao, A.P. and Jones, A.P.: Clonality in Kaposi's sarcoma, *New Engl. J. Med.* **337** (1997), 570–571.
70. Delabesse, E., Oksenhendler, E., Lebbé, C., Verola, O., Varet, B. and Turhan, A.G.: Molecular analysis of clonality in Kaposi's sarcoma, *J. Clin. Pathol.* **50** (1997), 664–668.
71. Schulz, T.F. : Epidemiology of Kaposi's sarcoma-associated herpesvirus/ Human herpesvirus 8, *Adv. Cancer Res.*, in press.
72. Boshoff, C. and Weiss, R.A.: Kaposi's sarcoma–associated herpesvirus, *Adv. Cancer Res.* **75** (1998), 57–86.
73. Olsen, S.J. and Moore, P.S.: Kaposi's sarcoma-associated herpesvirus (KSHV/HHV8) and the etiology of KS. In: *Molecular Immunology of Herpesviruses*, (H. Freidman, P. Medvecky and M. Bendinelli, eds.), New York: Plenum Press, 1999, in press.
74. Rabkin, C.S., Schulz, T.F., Whitby, D., Lennette, E.T., Magpanty, L.I., Chatlynne, L. and Biggar, R.J.: Interassay correlation of human herpesvirus 8 (HHV-8) serologic tests, *J. Infect. Dis.* **178** (1998), 304-309.

75. Gao, S.J., Kingsley, L., Li, M., Zheng, W., Parravicini, C., Ziegler, J., Newton, R., Rinaldo, C.R., Saah, A., Phair, J., Detels, R., Chang, Y. and Moore, P.S.: Seroprevalence of KSHV antibodies among North Americans, Italians, and Ugandans with and without Kaposi's sarcoma, *Nature Med.* **2** (1996b), 925–928.

76. Kedes, D.H., Operskalski, E., Busch, M., Kohn, R., Flood, J. and Ganem, D.: The seroprevalence of human herpesvirus 8 (HHV 8): Distribution of infection in Kaposi's sarcoma risk groups and evidence for sexual transmission, *Nature Med.* **2** (1996), 918–924.

77. Simpson, G.R., Schulz, T.F., Whitby, D., Cook, P.M., Boshoff, C., Rainbow, L., Howard, M.R., Gao, S.-J., Bohenzky, R.A., Simmonds, P., Lee, C., de Ruiter, A., Hatzakis, A., Tedder, R.S., Weller, I.V.D., Weiss, R.A. and Moore, P.S.: Prevalence of Kaposi's sarcoma-associated herpes-virus infection measured by antibodies to recombinant capsid protein and latent immuno-fluorescence antigen, *Lancet* **348** (1996), 1133-1138.

78. Whitby, D., Luppi, M., Barozzi, P., Boshoff, C., Weiss, R. A. and Torelli, G.: Human herpesvirus 8 seroprevalence in blood donors and patients with lymphoma from different regions of Italy, *J. Natl. Cancer Inst.* **90** (1998), 395–397.

79. Calabrò, M.L., Sheldon, J., Favero, A., Simpson, G.R., Fiore, J.R., Gomes, E., Angarano, G., Chieco-Bianchi, L. and Schulz, T.F.: Seroprevalence of Kaposi's sarcoma-associated herpesvirus (KSHV/HHV8) in different regions of Italy, *J. Hum. Virol.* **1** (1998), 207–213.

80. Geddes, M., Franceschi, S., Barchielli, A., Falcini, F., Carli, S., Cocconi, G., Conti, E., Crosignani, P., Gafà, L., Giarelli, L., Vercelli, M. and Zanetti, R.: Kaposi's sarcoma in Italy before and after the AIDS epidemic, *Brit. J. Cancer* **69** (1994), 333–336.

81. Cook, P.M., Whitby, D., Calabro, M.L., Luppi, M., Kakoola, D.N., Hjalgrim, H., Ariyoshi, K., Ensoli, B., Davison, A.J. and Schulz, T.F.: Variability and evolution of Kaposi's sarcoma-associated herpesvirus in Europe and Africa, submitted for publication.

82. Mayama, S., Cuevas, L., Sheldon, J., Smith, D., Okong, P., Silvel, B. and Schulz, T.F.: Prevalence of Kaposi's sarcoma-associated herpesvirus (Human herpesvirus 8) in a young Ugandan population, *Int. J. Cancer* **77** (1998), 817-820.

83. Bourboulia, D., Whitby, D., Boshoff,C., *et al.*: Serologic evidence for mother-to-child transmission of Kaposi's sarcoma–associated herpesvirus infection, *JAMA* **280** (1998), 31-32.

84. Bestetti, G., Renon, G., Mauclère, P., Ruffié, A., Mbopi Kéou, F.X., Eme, D., Parravicini, C., Corbellino, M., de Thé, G. and Gessain, A.: High seroprevalence of human herpesvirus-8 in pregnant women and prostitutes from Cameroon, *AIDS* **12** (1998), 541-543.

85. Nicholas, J., Zong, J., Alcendor, D.J., Ciufo, D., Poole, L.J., Sarisky, R.T., Chiou, C.J., Zhang, X., Wan, X., Guo, H.G., Reitz, M.S. and Hayward, G.S.: Novel organizational features, captured cellular genes and strain variability within the genome of KSHV/HHV8, *Monogr. Natl. Cancer Inst.* **23** (1998), 79-88.

86. Zong, J.-C., Ciufo, D.M., Alcendor, D.J., Wan, X., Nicholas, J., Browning, P., Rady, P., Tyring, S.K., Orenstein, J., Rabkin, C., Su, I.-J., Powell, K.F., Croxson, M., Foreman, K.E., Nickoloff, B.J., Alkan, S. and Hayward, G.S.: High level variability in the ORF K1 membrane protein gene at the left end of the Kaposi's sarcoma-associated herpesvirus (HHV 8) genome defines four major virus subtypes and multiple clades in different human populations, J. Virol., in press

87. Glenn, M.A., Rainbow, L., Auradé, F., Davison, A. and Schulz, T.F.: Identification of a spliced KSHV/HHV8 gene that encodes a protein with similarities to latent membrane proteins 1 and 2A of Epstein-Barr virus, submitted for publication.

88. Melbye, M., Cook, P.M., Hjalgrim, H., Begtrup, K., Simpson, G.R., Biggar, R.J., Ebbesen, P. and Schulz, T.F.: Transmission of human herpesvirus 8 (HHV 8) among homosexual men follows the pattern of a 'Kaposi's sarcoma agent', *Int. J. Cancer* **77** (1998), 543-548.

89. Renwick, N., Halaby, T., Weverling, G.J., Simpson, G.R., Coutinho, R.A., Lange, J.M.A., Schulz, T.F. and Goudsmit, J.: Seroconversion for Kaposi's sarcoma-associated herpesvirus is highly predictive of KS development in HIV-1-infected individuals, *AIDS* **12** (1998), 2481–2488.

90. Martin, J.N., Ganem, D.E., Osmond, D.H., Page-Shafer, K.A., Macrae, D. and Kedes, D.H.: Sexual transmission and the natural history of human herpesvirus 8 infection, *New Engl. J. Med.* **338** (1998), 948–954.

91. Dukers, N.H.T.M., Renwick, N., Prins, M., Geskus, R.B., Schulz, T.F., Weverling, G.-J., Coutinho, R.A. and Goudsmit, J.: Risk factors for HHV8 seropositivity and HHV8 seroconversion in a cohort of homosexual men, Am. J. Epidemiol, in press.

92. Ariyoshi, K., Schim van der Loeff, M., Corrah, T., Cham, F., Cook, P.M., Whitby, D., Weiss, R.A., Schulz, T.F. and Whittle, H.: Kaposi's sarcoma and human herpesvirus 8 (HHV8) in HIV-1 and HIV-2 infection in The Gambia, *J. Hum. Virol.* **1** (1998), 193-199.

CHAPTER 11

AIDS DEMENTIA COMPLEX

PETER PORTEGIES AND ROELIEN H. ENTING

Department of Neurology, Academic Medical Centre, Amsterdam, The Netherlands.

1. Definition

HIV-associated dementia was originally described in 1983 and termed subacute encephalitis. The term AIDS dementia complex (ADC) has been introduced by Navia *et al.* in 1986 in two papers in the Annals of Neurology, in which they describe the clinical and neuropathological characteristics of 46 AIDS patients who suffered unexplained mental impairment [1,2]. This progressive dementia was often associated with motor and behavioural abnormalities. Based on the characteristic psychomotor slowing, the authors classified the disease as a subcortical dementia. Progressive encephalopathy is the term for this condition in children. In 1987, HIV-encephalopathy was added to the list of AIDS-defining illnesses. In 1991, the American Academy of Neurology AIDS Task Force developed definitional criteria for HIV-1-associated dementia and for the less severe forms of cognitive impairment (Table 1) [3]. In this report HIV-1-associated cognitive/motor complex was introduced, which can be separated in two categories: 1] a more severe form: HIV-1-associated dementia complex, and 2] a less severe form: HIV-1-associated minor cognitive/motor disorder. The severe form is sufficient for a diagnosis of AIDS, the less severe form is not. It is still unknown whether the severe and minor forms of this complex are the same entity, or whether patients with the minor form will invariably progress to a severe form. Although HIV-1-associated cognitive/motor complex is not identical to the AIDS dementia complex, it is consistent with it. The more official term HIV-1-associated cognitive/motor complex is less frequently used than AIDS dementia complex or ADC. Several authors simply use HIV-dementia.

For established cases of ADC a staging scheme, based on functional severity of cognitive and motor abnormalities, has been developed by Price and Brew (Table 2) [4]. This Memorial Sloan Kettering or MSK-rating scale is generally accepted and has proven to be very useful in interventional studies. For most ADC treatment studies patients with ADC MSK stage 1 are eligible to enrol into the study.

H. Schuitemaker and F. Miedema (ed.), Aids Pathogenesis, 209–220.

TABLE 1. Criteria for clinical diagnosis of HIV-1-associated cognitive/motor complex

I. Sufficient for diagnosis of AIDS

A. HIV-1-associated cognitive/motor complex

Probable (must have each of the following)

1. Acquired abnormality in at least two of the following cognitive abilities (present for at least one month): attention/concentration, speed of processing of information, abstraction/reasoning, visuospatial skills, memory/learning, and speech/language. The decline should be verified by reliable history and mental status examination. In all cases, when possible, history should be obtained from an informant, and examination should be supplemented by neuropsychological testing. Cognitive dysfunction causing impairment of work or activities of daily living (objectively verifiable or by report of a key informant). This impairment should not be attributable solely to severe systemic illness.

2. At least one of the following:

   a. Acquired abnormality in motor function or performance verified by clinical examination (*e.g.* slowed rapid movements, abnormal gait, limb incoordination, hyperreflexia, hypertonia, or weakness), neuropsychological tests (*e.g.* fine motor speed, manual dexterity, perceptual motor skills), or both.

   b. Decline in motivation or emotional control or change in social behaviour. This may be characterised by any of the following: change in personality with apathy, inertia, irritability, emotional lability, or new onset of impaired judgment characterised by socially inappropriate behaviour or disinhibition.

3. Absence of clouding of consciousness during a period long enough to establish the presence of 1#.

4. Evidence of another etiology, including active CNS opportunistic infection or malignancy, psychiatric disorder (*e.g.* depressive disorder), active alcohol or substance use, or acute or chronic substance withdrawal, must be sought from history, physical and psychiatric examination, and appropriate laboratory and radiologic investigation (*e.g.* lumbar puncture, neuro-imaging). If another potential etiology (*e.g.* major depression) is present, it is not the cause of the above cognitive, motor, or behavioural symptoms and signs.

Possible (must have one of the following):

1. Other potential etiology present (must have each of the following):

   a. As above (see *Probable*) #1, 2 and 3.

   b. Other potential etiology is present but the cause of #1 is uncertain.

2. Incomplete clinical evaluation (must have each of the following):

   a. As above (see *Probable*) #1, 2 and 3.

   b. Etiology cannot be determined (appropriate laboratory or radiologic investigation not performed).

TABLE 2. Memorial Sloan Kettering Rating Scale

Stage 0 - **Normal**

Normal mental and motor function.

Stage 0.5 **Equivocal/subclinical**

Either minimal or equivocal symptoms of cognitive or motor dysfunction characteristic of ADC, or signs (snout response, slowed extremity movements), but without impairment of work or capacity to perform activities of daily living (ADL). Gait and strength are normal.

Stage 1 - **Mild**

Unequivocal evidence (symptoms, signs, neuropsychological test performance) of functional intellectual or motor impairment characteristic of ADC, but able to perform all but the more demanding aspects of work or ADL. Can walk without assistance.

Stage 2 - **Moderate**

Cannot work or maintain the more demanding aspects of daily life, but able to perform basic activities of self care. Ambulatory, but may require a single prop.

Stage 3 - **Severe**

Major intellectual incapacity (cannot follow news or personal events, cannot sustain complex conversation, considerable slowing of all output), or motor disability (cannot walk unassisted, requiring walker or personal support, usually with slowing and clumsiness of arms as well.

Stage 4 - **End stage**

Nearly vegetative. Intellectual and social comprehension and responses are at a rudimentary level. Nearly or absolutely mute. Paraparetic or paraplegic with double incontinence.

2. Epidemiology

ADC develops in 15-20% of untreated patients in the advanced stages of HIV-infection [5]. A CDC-analysis over the period 1987-1991 revealed an incidence of 7.3% [6]. Most patients with ADC have already been diagnosed with AIDS, but in 3% ADC is the AIDS-defining event. The introduction of anti-retroviral drugs in 1986-1987 has had a substantial influence on the epidemiology of ADC. With zidovudine (ZDV) use, the incidence has declined to less than 5%, and ADC has become a rare complication in AIDS patients on continous ZDV-therapy [5,7]. Although conflicting data have been reported by several groups in the USA, who did not detect significant changes in ADC-incidence, most published data demonstrate that

ADC has become an infrequent clinical complication. Widespread ZDV-use is a major factor in this change, but other factors, such as a changing neuro-virulence, may have contributed as well. In 11 placebo-controlled ZDV-trials with HIV-infected individuals at various stages of their infection, ADC occurred in 12/4403 ZDV-users, compared to 20/3292 in the placebo-users (relative risk: 0.54 [0.29-1.04]). With the introduction of HAART (highly active anti-retroviral therapy) ADC has remained a rare complication.

The epidemiology of HIV-associated minor cognitive-motor disorder is uncertain. Reliable data on its incidence, prevalence and natural history are not available. Even its relationship to ADC is unclear.

3. Clinical features

Clinically, ADC is characterised by disturbances in cognition, motor performance and behaviour [1]. The earliest symptoms usually consist of difficulties with concentration and memory. Patients lose track of conversation while speaking to people. Difficulty in reading is common and is usually due to difficulty in concentrating (so that frequent re-reading is required), rather than difficulty in understanding the printed words. Patients have to keep lists to maintain their daily schedules. Many patients complain of slowness in thinking. The patient takes long pauses before answering questions and exhibits a general torpidity. Complex tasks become increasingly difficult to complete. Abulia, defined as "loss of will", is often striking. They become apathetic and lose interest in everything. As a consequence they may become socially withdrawn, which can be mistaken for depression. Motor symptoms include clumsiness, sloppy handwriting, tremor, poor balance, unsteadiness of gait, and slowness of rapidly alternating movements. Organic psychosis may develop in some patients. Cortical symptoms, such as aphasia, alexia and agraphia, are lacking. Saccadic and pursuit eye movements are often slowed and inaccurate. Fine finger movements are slowed, snout response is common, and deep tendon reflexes are brisk. With time increasing psychomotor slowing may progress to severe dementia with akinetic mutism, paraparesis, and incontinence. The clinical and neuropsychological abnormalities in ADC are compatible with what has been called subcortical dementia.

Without treatment, ADC is rapidly progressive, with a mean survival of about 6 months. In one study it has been suggested that patients with ADC can be grouped in rapid-progressors and slow-progressors [8].

4. Diagnosis

Essential for the diagnosis of ADC are the presence of well-documented cognitive decline and the exclusion of other neurological complications of HIV-infection. Therefore, CSF examination and imaging studies of the brain are mandatory [3].

4.1 IMAGING STUDIES

Cortical atrophy, enlarged ventricles and diffusely decreased attenuation of the deep white matter on CT-scan in patients with ADC have been described in several studies. Atrophy and enlarged ventricles may be present in patients who are not demented. Bilateral areas of increased signal intensity on T2-weighted MR images, usually in the periventricular white matter and centrum semiovale in patients with ADC have also been reported in a number of studies [5]. Three patterns of white matter abnormalities on MRI may be found: 1] patchy (localised involvement with ill-defined margins), 2] punctate lesions (small foci less than 1 cm in diameter) and 3] diffuse (widespread involvement of a large area). The diffuse pattern is the most frequent pattern in ADC [9]. Quantitative MRI studies have shown selective atrophy in basal ganglia structures and posterior cortex.

4.2 CEREBROSPINAL FLUID

CSF analysis in patients with suspected ADC is mandatory to rule out other conditions that could explain the clinical findings. Besides excluding other CNS complications, CSF analysis may reveal abnormalities that may support the clinical diagnosis of ADC.

4.2.1 *HIV-specific CSF markers*

HIV-isolation from the CSF has been reported in patients with and without ADC. The virus can be isolated as free virus from the CSF and from cells present in the CSF. By polymerase chain reaction (PCR), HIV-RNA load can be determined in the CSF. HIV-RNA levels in the CSF correlate with ADC and its severity [10]. CNS opportunistic infections may also increase HIV-RNA levels and these infections must be excluded first. HIV p24 core-protein is detectable in 40-50% of patients with ADC [11]. Intrathecal synthesis of HIV-1 antibodies has been demonstrated in all stages of HIV infection, in patients with and without neurological syndromes, and is a poor marker for ADC.

4.2.2 *Non-HIV-1-specific CSF parameters*

A mildly elevated protein in the CSF is present in 60% of ADC patients and a mild mononuclear pleocytosis in nearly 25%, and is not very helpful diagnostically [5,12]. Increased CSF levels of several aspecific markers of immune activation (beta-2-microglobulin, neopterin and quinolinic acid) and several cytokines (tumour necrosis factor alfa, interleukin-1-beta, interleukin-1-alpha, interleukin-6) have been reported in ADC. All these markers are increased in opportunistic infections of the CNS as well, so these have to be ruled out.

4.3 *Neuropsychological assessment*

Neuropsychological assessment in patients with ADC may be helpful in confirming the clinical diagnosis of ADC. It may be useful in quantifying disease severity, in following the course of ADC and in documenting its response to anti-retroviral treatment. Many different test-batteries have been used. A compact set on which a group of experts agreed is shown in Table 3. The characteristic abnormalities in patients with ADC include difficulty with complex sequencing, slowed psychomotor speed, impaired fine and rapid motor movement, and reduced verbal fluency [13]. Cortical dysfunction is usually less prominent. In short: "slowing and loss of precision in both mentation and motor control" [4].

TABLE 3. Neuropsychological testbattery for ADC

1.	Symbol Digit Modalities Test
2.	Rey Auditory Verbal Learning Test
3.	Grooved Pegboard
4.	Verbal Fluency
5.	Trail Making A and B
6.	Cal Cap Reaction Time

5. Neuropathology

The principal histopathological abnormalities are most prominent in the subcortical structures, notably in the central white matter, basal ganglia, thalamus, brain stem, and spinal cord [14]. The most common of these abnormalities is diffuse pallor of the white matter, which is usually accompanied by astrocytic reaction, perivascular lymphocytes, and brown-pigmented macrophages. Multinucleated cells are found in a subgroup of patients with more severe clinical disease. Substantional neuronal loss has been demonstrated as well, which may play a secondary role.

6. Pathogenesis

6.1 VIRAL ENTRANCE

Some studies have suggested that HIV may reach the brain relatively early in HIV infection, possibly at the time of or shortly after the primary infection. This is based on the following observations. Some patients have clinical neurological syndromes at the time of the seroconversion: meningitis, meningo-encephalitis, and myelopathy. These neurological syndromes tend to have a good prognosis with recovery within weeks, although persistent neurological deficit has been reported. In some patients it was possible to isolate HIV from the CSF or to demonstrate HIV p24 antigen in the CSF at the time of seroconversion. Finally, HIV-specific neuropathological abnormalities in the brain have been described in patients who died of other causes than HIV disease, relatively early in HIV-infection [15].

One theory for viral entrance is that infected monocytes bring the virus into the brain: the Trojan horse mechanism [16,17]. It is unclear what triggers these infected monocytes to leave the capillaries in the brain while trafficking through the brain. However the infected monocyte crosses the endothelium and settles as an infected perivascular macrophage. It has been suggested that from these macrophages the virus is spread through the brain by cell-to-cell contact between macrophages and microglia cells. Other possibilities for viral entry are: cell-free virus crossing the endothelium or penetration of the parenchyma from the CSF. This is supported by the observation that endothelial cells and cells in the choroid plexus exhibit high levels of productive infection.

Based on the limited evidence available, it is impossible to determine whether early entrance of HIV in the brain occurs in every infected individual. Recent CSF studies in asymptomatic HIV-infected individuals, however, demonstrated that the majority had detectable levels of HIV-RNA, suggesting that HIV replication takes place in the CNS (not necessarily in the parenchyma) in the majority of HIV-infected hosts at early stages [18]. All these data together suggest that the virus is present and replicating in the CNS in the majority of (if not all) infected individuals early after the primary infection.

6.2 INFECTED CELLS

By *in-situ* hybridisation and immunocytochemistry it has been demonstrated that monocytes/macrophages and microglia cells are the the principal, if not exclusive, cell types in the CNS showing productive infection *in vivo* [19]. These studies indicate that the major site of virus accumulation is within macrophages in the perivascular spaces and in the multinucleated giant cells (fused macrophages/microglia cells). Infection of cells of neuro-ectodermal origin (neurons, astrocytes and oligodendrocytes) is rare in CNS HIV-1 infection. Infection of astrocytes and oligodendrocytes has been reported. Only rarely has a cell resembling a neuron been positive. Infection of capillary endothelial cells has been suggested by a few studies. Nevertheless, most studies have confirmed that the macrophage/microglia cell is the major cell type in the brain infected by HIV-1.

6.3 MECHANISMS OF NEURONAL DYSFUNCTION

Viral latency is defined as presence of viral genome in infected cells without replication. Based on the recent CSF studies (mentioned above), showing detectable HIV-RNA levels in most infected individuals, it is likely that in the majority viral replication does occur and that true latency is rare. In most patients HIV replication is controlled and kept at low levels by the host immune defenses (systemically and locally) [17]. In advanced HIV infection a progressive systemic immunodeficiency leads in a minority of patients to enhanced replication in the brain. This enhanced replication in microglia cells is the first step in a complex cascade. Viral proteins (gp120, tat, nef) are released in the brain tissue. Microglia and astrocytes are stimulated by these proteins to produce cytokines and other inflammatory mediators: tumour necrosis factor-alfa, nitric oxide (NO), neopterin, beta-2-microglobulin, quinolinic acid and GM-CSF [20-22]. These cell-encoded signals further enhance HIV replication in microglia cells, and, together with the increasing amount of viral proteins, have a toxic effect on the final target cell: the neuron. So, initially there is neuronal dysfunction, later on there may be additional neuronal loss. This largely hypothetical model of pathogenesis, in which HIV replication in the CNS is the essential step, makes it understandable that with inhibition of viral replication by anti-retroviral drugs, the process is to a certain extent reversible. Other investigators have postulated that toxins produced in the systemic infection may play a role in CNS dysfunction.

7. Intervention

7.1 ZIDOVUDINE

Zidovudine is a nucleoside analogue that was introduced as an anti-retroviral drug in 1986. It penetrates well into the CSF with mean CSF concentrations between 0.15-0.43 μmol/l [23]. The neuroprotective efficacy of zidovudine has been demonstrated in several clinical studies.

7.2 THERAPEUTIC EFFICACY

A therapeutic efficacy of ZDV in patients with HIV dementia has been demonstrated in several studies. The first report was published in 1987: Yarchoan *et al.* describe neurological and neuropsychological improvement in 3 out of 4 patients with HIV dementia who were treated with ZDV [24]. The authors suggested that the disease was at least partially reversible. This was further supported by the study of Schmitt *et al.* They performed neuropsychological examination at baseline and after 8 and 16 weeks in the first placebo-controlled trial with ZDV. Of 134 ZDV users 60 patients were impaired at baseline, and showed improvement after 16 weeks. In the placebo-arm (n=128) no improvement could be demonstrated in those who were impaired at baseline (n=69) [25]. Although these patients were not classified as having HIV dementia, retrospectively the majority of these impaired patients would fulfill the CDC-criteria for HIV dementia. In another study of 30 patients with HIV dementia, 25 showed improvement after 6 months of ZDV treatment. In some of these patients the benefit was transient [26]. In the placebo-controlled trial by Sidtis *et al.*, 25 patients on ZDV (12 on 1000 mg daily; 13 on 2000 mg daily) showed an improvement (on a neuropsychological testbattery) after 16 weeks of treatment, compared to 15 patients in the placebo-arm. After 16 weeks the placebo-arm patients were rerandomised to one of the treatment arms and also showed significant improvement by week 32 [27]. Noteworthy in this trial were the high doses of ZDV compared to the doses in use today, and the fact that all included patients had $CD4^+$ cell counts above 450.10^6/l. In HIV-infected children with encephalopathy (the equivalent of HIV dementia in adults), ZDV given intravenously was beneficial in a group of 21 patients [28]. In contrast with these results, a study by Nozyce *et al.* did not find improvement in neurodevelopmental functioning in 54 vertically HIV-infected children treated with oral ZDV [29].

In an observational study in London, long-term treatment with ZDV had a significant beneficial effect on global neuropsychological performance in patients with early symptomatic HIV infection (n=51) and AIDS (n=32) [30]. Although different dosages of zidovudine have not been compared for efficacy in ADC in clinical trials, it is generally accepted that the current dosages of 500-600 mg per day are sufficient. Up to date, factors (*e.g.* viral load, previous drug history, CD4 counts) that predict the responsiveness in a patient with ADC who starts treatment, are not identified.

7.3 PROPHYLACTIC EFFICACY

In the late 1980s, and shortly after the introduction of ZDV, a major decline in the incidence of ADC was observed at many AIDS centres in Europe and in the USA. ADC has become a rare disease in patients receiving ZDV treatment. In our group of patients in Amsterdam, the incidence declined from more than 50% before the introduction to less than 5% of patients with AIDS after the introduction of ZDV [7]. The preventive efficacy of ZDV has been confirmed in several other cohort studies and in at least 4 placebo-controlled trials with ZDV.

7.4 OTHER NUCLEOSIDE ANALOGUES

Didanosine improved neurocognitive functioning in a small group of children with AIDS, but it did not prevent the development of ADC in adults with advanced HIV infection [31,32]. Clinical neurological data on patients receiving zalcitabine, lamivudine and stavudine are not available. During chronic oral dosing regimens all nucleoside analogues reach comparable absolute CSF concentrations. When compared to IC_{50}-values, lamivudine and ZDV appear to have the most favourable CSF concentration-IC_{50} ratios.

7.5 PROTEASE INHIBITORS

The protease inhibitors are to a large extent protein-bound and for that reason do not penetrate well into the CSF. Indinavir is protein-bound for only 60%, and CSF levels are higher compared to the other PI's [23]. Clinical efficacy data in patients with ADC are not available for the PI's.

7.6 NON-NUCLEOSIDE REVERSE TRANSCRIPTASE INHIBITORS

The non-nucleoside reverse transcriptase inhibitor nevirapine penetrates relatively well into the CSF [23]. Clinical data on nevirapine are not available.

7.7 ADDITIONAL NEUROPROTECTIVE TREATMENT

Additional neuroprotection might be achieved by drugs that do not inhibit viral replication but interfere with other steps in the pathogenesis of ADC [17]. Several examples are: pentoxifylline inhibits the production and effects of TNF-alpha, memantine blocks the action of toxins acting at the N-methyl-D-aspartate receptor, nimodipine is a calcium-channel blocker, and peptide-T is a neurotrophic peptide. A small study with nimodipine failed to show clinical efficacy in patients with ADC. Some of the other drugs are at the point of entering clinical trials. So far neuroprotection in HIV infection by drugs other than antiretrovirals has not been demonstrated.

8. Prognosis

Although the prognosis in patients with ADC is generally poor, the course of ADC is highly variable, and may be influenced by anti-retroviral treatment. In the original series of Navia *et al.*, over one half of the (untreated) patients developed severe dementia within two months of onset of symptoms, although 20% had a protracted course (4-6 months). The mean survival, from the onset of symptoms, was 4.2 months (range: 1-9 months) [1]. In McArthur's series, 45% showed progression within 4 months, although 35% did not deteriorate in the following 8 months [12]. ZDV treatment may prolong survival. In our own series of patients with ADC, 20 patients with ADC who never used ZDV had a mean survival of 4 months, compared to 14.8 months in patients who started on ZDV after they were classified as having ADC [5]. It is obvious that survival in patients with ADC is influenced by a number of factors, but it is clear that anti-retroviral treatment is the key factor.

9. Conclusion

ADC is the result of uncontrolled HIV replication in the brain. The production of viral proteins and inflammatory molecules leads primarily to neuronal dysfunction, and secondary to neuronal loss. The most rational therapeutic approach is anti-retroviral treatment. To prevent the development of ADC in HIV-infected individuals, combination regimens should include at least one, but preferably several neuroprotective anti-retroviral drugs [33]. Suboptimal treatment may lead to the development of resistant strains in the CNS. Therefore, to suppress viral replication in the CNS, a combination of drugs that penetrate well into the CSF should be used. Lamivudine-ZDV or lamivudine-d4T-containing regimens can be recommended [18]. When more data on other drugs are available, these recommendations may be broadened to include other therapies. Approaches other than anti-retroviral treatment have been unsuccesful so far.

10. References

1. Navia, B.A., Jordan, B.D., Price, R.W., *et al.*: The AIDS dementia complex: I. Clinical features, *Ann. Neurol.* **19** (1986), 517-524.
2. Navia, B.A., Cho, E.-S., Petito, C.K. and Price, R.W.: The AIDS dementia complex: II. Neuropathology, *Ann. Neurol.* **19** (1986), 525-535.
3. Janssen, R.S., Cornblath, D.R., Epstein, L.G., *et al.*: Nomenclature and research case definitions for neurological manifestations of human immunodeficiency virus type-1 (HIV-1) infection. Report of a Working Group of the American Academy of Neurology AIDS Task Force. *Neurology* **41** (1991), 778-785.
4. Price, R.W. and Brew, B.J.: The AIDS dementia complex, *J. Infect. Dis.* **158** (1988), 1079-1083.
5. Portegies, P, Enting, R.H., de Gans, J., *et al.*: Presentation and course of AIDS dementia complex: 10 years of follow-up in Amsterdam, The Netherlands, *AIDS* **7** (1993), 669-675.
6. Janssen, R.S., Nwanyanwu, O.C., Selik, R.M. and Stehr-Green, J.K.: Epidemiology of human immunodeficiency virus encephalopathy in the United States, *Neurology* **42** (1992), 1472-1476.
7. Portegies, P., de Gans, J., Lange, J.M.A., *et al.*: Declining incidence of AIDS dementia complex after introduction of zidovudine treatment, *Brit. Med. J.* **299** (1989), 819-821.
8. Bouwman, F.H., Skolasky, R., Hes, D., Selnes, O., Glass, J., Nance, T., Royal, W., Dal Pan, G. and McArthur, J.C.: Variation in clinical progression of HIV-associated dementia, *Neurology* **50** (1998), 1814-1820.
9. Olsen, W.L., Longo, F.M., Mills, C.M. and Norman, D.: White matter disease in AIDS: findings at MR imaging, *Radiology* **169** (1988), 445-448.
10. Brew, B.J., Pemberton, L., Cunningham, P. and Law, M.G.: Levels of human immunodeficiency virus type 1 RNA in cerebrospinal fluid correlate with AIDS dementia stage, *J. Infect. Dis.* **175** (1997), 963-966.
11. Portegies, P., Epstein, L.G., Tjong A Hung, S., De Gans, J. and Goudsmit, J.: Human immunodeficiency virus type-1 antigen in cerebrospinal fluid. Correlation with clinical neurologic status, *Arch. Neurol.* **46** (1989), 261-264.
12. McArthur, J.C.: Neurologic manifestations of AIDS, *Medicine* **66** (1987), 407-437.
13. Dunbar, N. and Brew, B.: Neuropsychological dysfunction in HIV infection: a review, *J. Neuro-AIDS* **1** (1996), 73-102.
14. Sharer, L.R.: Pathology of HIV-1 infection of the central nervous system. A review, *J. Neuropath. Exp. Neurol.* **51** (1992), 3-11.
15. Gray, F., Lescs, M.C., Keohane, C., *et al.*: Early brain changes in HIV infection: neuropathological study of 11 HIV seropositive, non-AIDS cases, *J. Neuropath. Exp. Neurol.* **51** (1992), 177-185.
16. Epstein, L.G. and Gendelman, H.E.: Human immunodeficiency virus type 1 infection of the nervous system: pathogenetic mechanisms, *Ann. Neurol.* **33** (1993), 429-436.
17. Price, R.W.: Management of AIDS dementia complex and HIV-1 infection of the nervous system, *AIDS* **9 (suppl. A)** (1995), S221-S236
18. Foudraine, N.A., Hoetelmans, R.M.W., Lange, J.M.A. *et al.*: Cerebrospinal fluid HIV-1 RNA and drug concentrations after treatment with lamivudine plus zidovudine or stavudine, *Lancet* **351** (1998), 1547-1551.
19. Wiley, C.A., Schrier, R.D., Nelson, J.A., Lampert, P.W. and Oldstone, M.B.A.: Cellular localization of human immunodeficiency virus infection within the brains of acquired immunodeficiency patients, *Proc. Natl. Acad. Sci. USA* **83** (1986), 7089-7093.
20. Achim, C.L. and Wiley, C.A.: Inflammation in AIDS and the role of the macrophage in brain pathology, *Curr. Opin. Neurol.* **9** (1996), 221-225
21. Adamson, D.C., Wildemann, B., Saski, M., *et al.*: Immunologic NO synthase: elevation in severe AIDS dementia and induction by HIV-1 gp41, *Science* **274** (1996), 1917-1921.
22. Nottet, H.S.L.M. and Gendelman, H.E.: Unraveling the neuroimmune mechanisms for the HIV-1-associated cognitive/motor complex, *Immunol. Today* **16** (1995), 441-448.

23. Enting,R.H., Hoetelmans, R.M.W., Lange, J.M.A., Burger, D.M., Beijnen, J.H. and Portegies, P.: Antiretroviral drugs and the central nervous system, *AIDS* **12** (1998), 1941-1955.
24. Yarchoan, R., Berg, G., Brouwers, P., *et al.*: Response of human immunodeficiency virus-associated neurological disease to 3'-azido-3'-deoxythymidine, *Lancet* **1** (1987), 132-135.
25. Schmitt, F.A., Bigley, J.W., McKinnis, R., *et al.* Neuropsychological outcome of zidovudine (AZT) treatment of patients with AIDS and AIDS-related complex, *New Engl. J. Med.* **319** (1988), 1573-1578.
26. Tozzi, V., Narciso, P., Galgani, S., *et al.*: Effects of zidovudine in 30 patients with mild to end-stage AIDS dementia complex, *AIDS* **7** (1993), 683-692.
27. Sidtis, J.J., Gatsonis, C., Price, R.W., *et al.*: Zidovudine treatment of the AIDS dementia complex: results of a placebo-controlled trial, *Ann. Neurol.* **33** (1993), 343-349.
28. Pizzo, P.A., Eddy, J., Falloon, J., *et al.*: Effect of continuous intravenous infusion of zidovudine (AZT) in children with symptomatic HIV infection, *New Engl. J. Med.* **319** (1988), 889-896.
29. Nozyce, M., Hoberman, M., Arpadi, S., *et al.*: A 12-month study of the effects of oral zidovudine on neurodevelopmental functioning in a cohort of vertically HIV-infected inner-city children, *AIDS* **8** (1994), 635-639.
30. Baldeweg ,T., Catalan, J., Lovett, E., Gruzelier, J., Riccio, M. and Hawkins, D.: Long-term zidovudine reduces neurocognitive deficits in HIV-1 infection, *AIDS* **9** (1995), 589-596.
31. Butler, K.M., Husson, R.N., Balis, F.M., *et al.*: Dideoxyinosine in children with symptomatic human immunodeficiency virus infection, *New Engl. J. Med.* **324** (1991), 137-144.
32. Portegies, P., Enting, R.H., de Jong, M.D., *et al.*: AIDS dementia complex and didanosine, *Lancet* **344** (1994), 759.
33. Portegies, P.: HIV-1, the brain and combination therapy, *Lancet* **346** (1995), 1244-1245.

CHAPTER 12

ANTI-RETROVIRAL THERAPY AND RESISTANCE TO ANTI-RETROVIRAL DRUGS

JOEP M.A. LANGE[1] AND JULIO S.G. MONTANER[2]

[1]*National AIDS Therapy Evaluation Centre (NATEC) Academic Medical Centre, University of Amsterdam, The Netherlands and* [2]*Canadian HIV Trials Network, Centre for Excellence in HIV/AIDS St Paul's Hospital, University of British Columbia, Vancouver, Canada*

1. Introduction

The era of anti-retroviral therapy starts in 1986. Only three years after the now so designated human immunodeficiency virus (HIV) was identified as the causative agent of the acquired immunodeficency syndrome (AIDS), treatment with zidovudine (3'-azido-3'-deoxythymidine, AZT, ZDV), a nucleoside analogue inhibitor of the viral reverse transcriptase (RT), was shown to significantly reduce mortality in patients with advanced HIV-1 infection, which was paralelled by a rise in $CD4^+$ lymphocyte counts [1]. The agent also appeared to confer benefits in patients with less advanced HIV-1 infection [2-4]. Initial high hopes for this drug, however, led to subsequent disappointment when, with extended follow-up, its effects on morbidity and mortality proved to be of limited duration [5,6].

Zidovudine was soon followed by two other nucleoside analogue RT-inhibitors (nRTIs): zalcitabine (2',3'-dideoxycytidine, ddC) [7], and didanosine (2',3'-dideoxyinosine, ddI) [8-10]. Like zidovudine, these were initially almost exclusively used as monotherapy. Moreover, they were mainly used as salvage therapy in patients whose HIV-related disease progressed during zidovudine therapy or who were intolerant to zidovudine. As was the case with zidovudine, most benefical results were rather short-lived [11,12].

In the early 1990s, a new class of highly specific HIV-1 inhibitors -non-nucleoside RT-inhibitors (NNRTIs)- went into clinical development [13,14]. Their full development was, however, severely retarded by the extremely rapid appearance of viral drug resistance to these agents [15-18].

In the meantime, various exploratory clinical trials showed that therapy with combinations of two nRTIs was superior to monotherapy in suppressing viral replication and in restoring $CD4^+$ cell numbers [19-21]. Due to a widespread lack of understanding of the biology of HIV infection, general acceptance of the superiority of combination- over monotherapy, however, only came after several large clinical

H. Schuitemaker and F. Miedema (ed.), Aids Pathogenesis, 221–242.

endpoint studies were confirmatory of this [22,23]. Meanwhile, two additional nRTIs - lamivudine (the negative enantiomer of 2'-deoxy-3'-thiacytidine, 3TC) [24-28] and stavudine (2',3'-didehydro-3'-deoxythymidine, d4T) [29,30] had become available, providing an important extension of the anti-retroviral armamentarium.

If the appearance of zidovudine can be characterized as the birth of anti-retroviral therapy, a number of concomitant developments that occurred in the mid-1990s led to a dramatic change in outlook on the principles of anti-retroviral therapy. Until assays to measure HIV-1 RNA became available, quantification of the antiviral response could only be done in a crude way and in a subset of patients, by measuring levels of the viral core protein p24. [2]. Now, HIV-1 RNA assays enabled the direct assessment of antiviral drugs on HIV-1 replication in virtually every infected subject [31]. Seminal studies using HIV-1 RNA and potent inhibitors of HIV-1 replication led to the realization that generally there is continuous high level production of HIV-1 throughout the course of infection, driving relentless immune destruction [32-35]. The high viral turnover rate and the error-prone nature of RNA virus replication [36], explain why only combinations of multiple anti-retroviral agents will lead to durable suppression of HIV replication. Allowing for significant residual HIV replication, as was the case with the monotherapies and dual therapies that were used during the first eight years of anti-retroviral therapy, will invariably lead to selection and outgrowth of drug-resistant viruses and virological therapy failure: a powerful illustration of the fact how important a thorough understanding of Darwinian principles is for the optimal treatment of HIV infection [37,38]. An important development was the advent of a new class of anti-retrovirals: HIV-1 protease inhibitors (PIs), which appeared to have hitherto unsurpassed antiviral potency [39,40]. The unfolding understanding of viral dynamics, availability of HIV-1 RNA assays, and of multiple powerful drugs, including HIV PIs, led to a quantum leap in the treatment of HIV-1 infection. Triple- drug therapy, with two NRTIs and an HIV PI, proved vastly superior to dual therapy [41,42] and quickly became the standard of care in the industrialised world, which has led to a dramatic reduction in HIV-related morbidity and mortality [43-45].

A better understanding of the dynamics of HIV replication and of the importance of Darwinian evolution for development of antiviral drug resistance, led to a re-assessment of the value of NNRTIs, now in triple-drug combinations. It was demonstrated that if NNRTIs were taken as part of a potent combination regimen, which led to suppression of viral replication to minimal levels, the development of viral drug resistance against this potentially vulnerable class of agents might be prevented [46]. After a rather troublesome development, NNRTIs have now also made it to the market beside nRTIs and PIs. Recent studies have shown that regimens of 2nRTIs and an NNRTI, or even combinations of 3nRTIs may provide a viable alternative to combinations of 2nRTIs and a PI [47-50].

New agents belonging to the three classes of licensed anti-retrovirals, as well as agents with novel mechanisms of action are in preclinical or clinical development [51].

2. Anti-retroviral Agents

2.1 NUCLEOSIDE AND NUCLEOTIDE ANALOGUE REVERSE TRANSCRIPTASE INHIBITORS (nRTIs)

The oldest class of anti-retrovirals still forms the cornerstone of most anti-HIV drug regimens. Combinations of 2nRTIs and either one or two PIs or a NNRTI are standard of care [52]. In addition, triple-drug combinations of nRTIs only are being evaluated. [47,49,50].

nRTIs interfere with the virus life cycle through inhibition of DNA chain elongation during the process of reverse transcription and by competitive inhibition of RT [53]. They are pro-drugs which require intracellular phosphorylation to their triphosphate moieties, a process that is not equally efficient for different nucleoside analogues in different infected cell types [54,55].

Either ZDV or d4T is part of most anti-retroviral regimens currently used. Both are thymidine analogues, that appear to be antagonistic when used together, at least in ZDV-pretreated individuals [56,57]. Comparison between ZDV and d4T-containing double- and triple-drug regimens suggest that they have an equipotent antiviral effect [58-60]. Their side effect profile differs, however: ZDV is mainly associated with haematological toxicity, d4T with peripheral neuropathy [30,61]. As is the case for most anti-retrovirals, side effects tend to be more frequent and severe in patients with advanced HIV infection.

The other licensed nRTIs -ddC, ddI and 3TC- are mainly used in combination with either ZDV- or d4T-containing regimens. 3TC-containing regimens have become the most popular, due to the potency and excellent tolerability of this drug [62]. Moreover, combined ZDV/3TC treatment postpones the occurrence of ZDV-resistance and the 3TC resistance mutation at RT codon 184 may restore susceptibility to ZDV in case of existing ZDV resistance [63]. The latter may account for the durable success of ZDV/3TC/indinavir combination therapy in ZDV-pretreated patients [64]. Other than ZDV/3TC, the most popular nRTI cornerstones of multi-drug combinations are d4T/3TC [59,60] and d4T/ddI [47,65]. FTC ([1-beta-L-FTC] 2', 3'-dideoxy-5-fluoro-3'-thiacytidine) is a 3TC-like nRTI in clinical development that shows considerable promise [66].

Abacavir is a new nRTI, with great potency [67,68], that has been shown to confer similar antiviral potency as a PI when used in first-line triple-drug therapy [49,50]. Like most available anti-retrovirals, abacavir has its greatest potential in first-line therapy, as it has little effect against viruses that carry multiple RT resistance mutations [69]. In approximately 3% of patients, use of abacavir may be associated with a hypersensitivity reaction, that precludes re-challenge with the drug [68].

A considerable proportion of nRTI-associated toxicity has a mitochondrial origin and is the result of inhibition of DNA polymerase gamma by these agents. Manifestations of mitochondrial toxicity are neuropathy, myopathy, cardiomyopathy, pancreatitis, hepatic steathosis and lactic acidosis, with sometimes lethal consequences [70]. An extreme example of nRTI-associated mitochondrial toxicity was seen with fialuridine (FIAU), a nRTI that was under investigation for the treatment of hepatitis-B infections [71].

Adefovir dipivoxil (bis-POM PMEA), an oral pro-drug of PMEA (9-[2-(phosphonomethoxy)ethyl]adenine), is a nucleotide analogue with broad-spectrum antiviral activity against HIV, several herpes viruses and hepatitis-B virus [72]. It is currently in phase III trials for the treatment of HIV infection, but is associated with the development of renal tubular dysfunction [73]. PMPA ([R]-9-[2-phosphonyl-methoxypropyl]adenine), a related nucleotide, has shown great promise in the prevention and treatment of simian immunodeficiency (SIV) infections in rhesus macaques [74]. The pro-drug, bis-POC PMPA, has entered human studies and has displayed *in-vivo* anti-HIV activity [75].

2.2 NON-NUCLEOSIDE REVERSE TRANSCRIPTASE INHIBITORS (NNRTIs)

As indicated in the Introduction, the NNRTIs, after a difficult start, have made a striking comeback in the therapy of HIV-1 infection. NNRTIs interfere with the process of reverse transcription through non-competitive binding to the RT enzyme-template-primer complex [76]. Unlike nRTIs and protease inhibitors, the available NNRTIs have highly specific activity against HIV-1 and no activity against HIV-2 [13,77]. There are three NNRTIs on the market: nevirapine (NVP) [17,18,46], delavirdine (DLV) [78,79] and efavirenz (EFZ) [48,80]. The NNRTI loviride was found not to contribute in a noteworthy way to a combination of ZDV and 3TC, and further development of this compound was halted [62]. Several new NNRTIs are in clinical development [81-83]. A major toxicity associated with use of NNRTIs is rash, which necessitates cessation of therapy in a small subset of patients [46,48,74,75,76]. The start of efavirenz therapy may also be associated with nervous system symptoms, such as headache, dizziness, insomnia and fatigue [48,76]. Direct comparative studies employing different NNRTIs are not available yet.

2.3 PROTEASE INHIBITORS

As described in the Introduction, HIV PIs revolutionised the treatment of HIV infection. Recently, however, it has become evident that the use of these agents may be associated with a syndrome of peripheral lipodystrophy, hyperlipidaemia and insulin resistance which appears to become manifest in a large majority of treated patients [84]. Although protease inhibitors may not be the only culprits causing this syndrome [85], their obvious association with it is a major reason for the enthusiasm that so- called PI-sparing regimens of 2nRTIs and a NNRTI [47,48] or of 3nRTIs [47,49,50] have recently generated. HIV PIs are based on amino-acid sequences recognized and cleaved in HIV proteins by the HIV protease. They prevent cleavage of gag and gag-pol protein precursors in acutely and chronically infected cells, arresting maturation of virions and thereby blocking their infectivity [86].

Thus far, four PIs have been licensed: saquinavir (SQV), ritonavir (RTV), indinavir (IDV) and nelfinavir (NFV). SQV, RTV and NFV should be taken with (high-fat) meals, IDV on an empty stomach or with a light, low-fat snack [86]. The efficacy of SQV, the first available PI [87,88] is hampered by relatively poor oral bio-availability [86,89]. A new enhanced oral formulation appears to have improved bio-availability [90]. RTV is a potent PI [39,40,91], whose use in full dose is limited by poor tolerability [92]. It has an increasing role, however, in lower dosages, in combinations with other PIs, as a pharmacological enhancer of these drugs [93,94].

IDV is another potent PI [41,42], with a propensity to cause kidney stones as a distinguishing toxicity [95]. When taken as a single PI, IDV needs to be taken three times a day; in combination with RTV, however, it may be taken twice daily and does not have to be taken on an empty stomach [96,97]. NFV is a potent PI with a favourable toxicity profile, its major side effect being diarrhea [98]. Amprenavir is a novel PI in a very advanced stage of clinical development [99].

Combinations of two PIs, usually in combination with one or two nRTIs, are becoming increasingly popular. Encouraging results have been obtained with combinations of RTV/SQV [100,101], RTV/IDV [96,97], and SQV/NFV [102]. Given its potency in inhibiting hepatic cytochrome P-450 enzymes (which metabolise PIs), a tiny dose of RTV, that in it self does not have substantial anti-HIV activity, may suffice to increase levels of the other PI. It is not yet clear whether it would be more advantageous to employ RTV doses that may be expected to have anti-HIV activity.

Several new PIs, which may have favorable pharmacological and/or resistance profiles, are in clinical development [103-106]. A major advance would be separation of anti-HIV activity and induction of lipodystrophy and/or hyperlipidaemia.

2.4 ADDITIONAL APPROACHES

2.4.1 *Hydroxyurea*:

There is a growing interest in hydroxyurea (HU) as a potentially effective adjuvant to nRTIs in the treatment of HIV infection. HU has been used for many years to treat human malignancies. It inhibits ribonucleotide reductase, a rate-limiting cellular enzyme in the synthesis of deoxyribonucleotide triphosphates [107]. Most favorable effects have been seen in combination with ddI, where plasma HIV-RNA responses significantly improved when HU was added [108-112]. HU, however, through induction of lymphopenia, does negatively affect the $CD4^+$ cell rise that is usually seen with anti-retroviral therapy [112]. An alternative or additional mechanism of action of HU is inhibition of target cell availability [113]. A major advantage of HU is its low cost, which allows for the composition of anti-retroviral drug regimens that may be applicable in developing country settings [114].

2.4.2 *Interleukin-2*:

Interleukin-2 (IL-2) is a cytokine that regulates the proliferation and differentiation of lymphocytes. Intermittent intravenous or subcutaneous therapy in HIV-1-infected patients has led to striking increases in $CD4^+$ cell counts [115-117]. Whether this will translate into clinical benefit is still subject of investigation. Subjective side effects may be quite severe [115-117].

2.4.3 *Therapeutic vaccines*:

The possibility of preventing or slowing progression to AIDS by therapeutic immunisation with an HIV vaccine continues to provoke interest [118]. Although use of these vaccines has been associated with generation of new humoral and T-cell responses to HIV, clinical proof of efficacy is still lacking [119-122]. The ultimate hope is that in the era of multidrug therapy, where the HIV load can be brought down to minimal levels, immune responses generated by therapeutic vaccines will eventually be able to control viral replication in the absence of drug pressure [123]. This hope has not yet been substantiated by any facts, however.

2.5 NEW THERAPEUTIC TARGETS

Until quite recently, viral reverse transcriptase and protease were the only successful targets for HIV-drug discovery. An additional enzymatic target is viral integrase [124].

The *in vivo* antiviral effect of T-20, a peptide inhibitor of gp41-mediated virus entry of target cells, is proof of principle that successful drug development aimed at non-enzymatic targets is possible [125]. Other non-enzymatic targets include protease dimerisation, matrix zinc fingers and tat-TAR complexes [51,126, 127].

2.6 DRUG INTERACTIONS

Clinically relevant pharmacokinetic drug-drug interactions involving nRTIs are limited in number, but when PIs were introduced, drug-drug interactions gained a prominent place in the treatment of HIV infection. PIs may be subject to extensive cytochrome P-450 metabolism. In many cases, inhibitors of P-450 (such as ketoconazole) increase plasma concentrations of PIs, whereas P-450-inducers (such as rifampin) accelerate the clearance of PIs, leading to subtherapeutic plasma concentrations. PIs themselves may alter the pharmacokinetics of other drugs by acting as p-450 inhibitors or hepatic enzyme inducers. As has been indicated above, the potency of RTV in inhibiting cytochrome P-450 enzymes may be exploited in a positive manner to optimise exposure to other PIs [89,128].

The appearance of NNRTIs has further complicated the picture. NVP and EFZ have been shown to induce cytochrome P450 enzymes [76,129], while DLV inhibits them [130].

With all the potential for drug-drug interactions with currently available anti-retroviral therapies, it may be clear that a thorough knowledge of the clinical pharmacology of the drugs involved is a prerequisite for the appropriate and safe treatment of HIV infection. Accordingly, the clinical pharmacologist has become an invaluable player in the management of HIV-infected patients.

3. Drug resistance

Development of viral resistance to anti-retroviral drugs is an important cause of treatment failure and limits options for subsequent treatment [38,131]. Increasingly genotypic and phenotypic resistance assays are being used to guide therapeutic decisions [132-134].

With the high production rate of HIV and the continuous generation of genetically distinct viral variants, it can be assumed that replication-competent virus with every possible single drug mutation is likely to be generated daily. Double mutants are less likely to arise and the probability of three or more drug resistance mutations in the same genome is very low [131]. Thus the need for combinations of multiple anti-retroviral drugs to achieve long-term suppression of viral replication [37,38]. For some anti-retroviral drugs, such as 3TC or currently available NNRTIs, a single mutation can confer high level resistance. When these drugs are given in combinations only partially suppressing viral replication, drug-resistant mutants predominate within weeks [17,18,131, 135,136]. For other drugs, such as ZDV and most PIs, high-level resistance requires accumulation of 3 or more mutations in a single genome [137-140]. These highly resistant variants emerge more slowly, requiring months to predominate during sub-optimal viral suppression [138-143].

Fig. 1 lists common mutations associated with resistance to currently available anti-retrovirals. In general, there is good concordance between mutations seen in laboratory selection experiments and those in clinical isolates from patients. Some mutations associated with *in vitro* resistance are, however, rarely found in clinical isolates: examples are the d4T-resistance-associated V75T mutation [144] and the DLV-resistance-associated P236L mutation [145]. In fact, the genetic correlate of phenotypic d4T resistance is still unclear [146,147].

Some anti-retroviral drug-resistance-associated mutations directly affect viral enzymes and cause resistance via decreased drug binding, whereas others have indirect effects. Primary resistance mutations are generally selected early in the process of resistance mutation accumulation, are relatively inhibitor-specific and may have a discernable effect on virus drug susceptibility. Secondary mutations accumulate in viral genomes already containing one or more primary mutations. Many secondary mutations alone have little or no discernable effect on the extent of resistance, but they may be selected because they improve viral fitness [131].

Generally there is no cross-resistance among drugs belonging to different classes. The degree of cross-resistance within a class of drugs differs between classes. From this perspective, nRTIs offer the most options for second-line therapy after failure of first-line therapy and development of resistance to the drugs employed. Some mutations that may be selected during nRTI combination therapy, however, confer resistance to the whole class. One such multidrug resistance mutation is Q151M, associated with three or four additional mutations [148]. Recently, an insertion between codons 68 and 69 of RT has also been associated with multi-nRTI resistance [146,147,149,150]. Cross-resistance among NNRTI is nearly complete [131], and there is a large degree of cross-resistance among PIs, although primary PI-resistance mutations may differ [131,139].

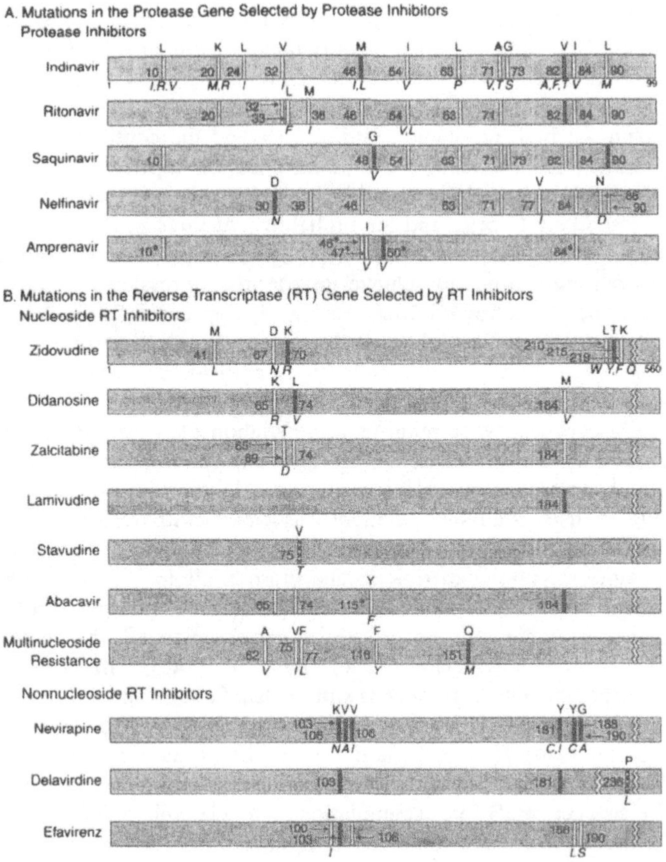

Fig. 1 - The most common human immunodeficiency virus 1 mutations selected by protease inhibitors (A), and nucleoside and non-nucleoside reverse transcriptase inhibitors (B) [31]. For each amino acid residue listed, the letter above the listing indicates the amino acid associated with the wild-type virus. The italicised letter below the residue indicates the substitution that confers drug resistance. The drug-selected mutations are categorised as "primary" (black bars) or "secondary" (white bars). (The black-and-white bar indicates a mutation selected *in vitro*, but rarely seen in specimens from patients in whom therapy fails). Primary mutations generally decrease inhibitor binding and are the first mutations selected. For indinavir, the mutations listed as primary may not be the first mutations selected, but they are selected in most patients' isolates in combination with other mutations. For zalcitabine, all mutations are listed as secondary because of inadequate clinical data to determine a common initial mutation. For nevirapine and delavirdine, each mutation can occur as either an initial or subsequent mutation and affect inhibitor binding. The asterisk indicates that the mutation has been reported *in vitro*, but relevance for clinical drug failure is uncertain. Amino acid abbreviations are as follows: A, alanine; C, cysteine; D, aspartate; E, glutamate; F, phenylalanine; G, glycine; H, histidine, I, isoleucine, K, lysine; L, leucine, M, methionine; N, asparagine; P, proline; Q, glutamine; R, arginine; S, serine; T, threonine; V, valine, W, tryptophan; Y, tyrosine. Multinucleoside resistance viruses have phenotypic resistance to most nucleoside reverse transcriptase inhibitors. Current listings are also available at http://hiv-web.lanl.gov/ or at http://www.viral-resistance.com.

This figure is from "Drug Resistance Testing in HIV-infected Adults by Hirsch *et al.* and published in JAMA 279, 1998, p. 1986. Permission for copyright was granted by American Medical Association.

Particular drug resistance mutations, selected by one drug, may suppress the phenotypic effects of mutations selected by other drugs. The effect of the 3TC resistance-associated M184V mutation on restoring phenotypic sensitivity to ZDV has already been mentioned [63]. The ddI resistance-associated L74V mutation has a similar effect [151]. M184V also increases susceptibility to the nucleotides adefovir dipivoxil and PMPA [152].

Transmission of HIV resistant to virtually all anti-retroviral agents available has already been observed [153].

4. Prevention of HIV Transmission

4.1 MOTHER-TO-CHILD TRANSMISSION

One of the greatest success stories of anti-retroviral therapy has been its impact on mother-to-child transmission of HIV-1. The pivotal ACTG 076 trial not only established that use of ZDV may reduce mother-to-child transmission of HIV to one-third of that in untreated mother-infant pairs [154], it also, at the time, was a much needed proof of principle of the potential of anti-retroviral therapy. No one could maintain any longer that anti-retroviral agents, such as ZDV, were only 'weak' drugs. It could be inferred from the ACTG 076 results that, if used properly, anti-retroviral agents could have an enormous impact on the burden of HIV disease.

The ZDV regimen employed in the ACTG 076 trial was, however, too complex and costly to be of use to most developing country settings, where the overwhelming burden of mother-to-child transmission of HIV occurs. Fortunately, simplified anti-retroviral regimens, investigated in clinical trials in developing countries, have now also been shown to significantly reduce mother-to-child transmission of HIV-1 [155-158]. It remains to be seen, however, to what extent such regimens will actually be implemented in the settings concerned, as this still requires a major effort. Furthermore, the absence of alternatives to breast feeding in particular areas, will reduce the effect of therapies targeted at the perinatal period [158].

Elective caesarian section appears to reduce the risk of mother-to-child transmission of HIV-1 independently of the effects of ZDV [159-161].

4.2 OCCUPATIONAL EXPOSURE

Prophylactic use of ZDV by health care workers following percutaneous exposure to HIV, has been shown to reduce the risk of infection [162]. Current guidelines recommend use of two or more anti-retroviral drugs for a period of four weeks following high-risk occupational exposure [52,163].

4.3 SEXUAL TRANSMISSION

The male condom is a very effective barrier against HIV transmission. Nevertheless, there is great need for female-controlled methods of prevention of sexual HIV transmission [164]. The female condom may be one such method [165], but its use still depends on the consent of the male partner. Therefore, the development of safe and effective vaginal microbicides or virucides, which would locally inactivate HIV, and ideally not be noticeable by the partner, has become a priority for several international agencies [166-168].

Routine provision of anti-retroviral post-exposure prophylaxis for sexual and needle-sharing HIV exposures is still an issue of debate [52,169]. Even with potent anti-retroviral therapy and prolonged suppression of HIV-1 RNA in plasma, the virus may still be present in semen [170].

5. Remaining Issues and Principles and Practice of Anti-retroviral Therapy

The most important principle of anti-retroviral therapy is that one should aim to reduce viral replication to minimal levels to avoid development of resistance [52, 131,171], which, given the large degree of cross-resistance among anti-retrovirals, will not only affect the current drug regimen, but also compromise future therapeutic options [38].

It has been shown that triple-drug therapy does not represent the zenith of antiviral potency, and that treatment with five drugs leads to a more rapid decline of plasma HIV-1 RNA levels than standard three-drug therapy [172]. Moreover, even with prolonged therapy with potent anti-retroviral drug regimens and prolonged suppression of plasma HIV-1 RNA to undetectable levels, there is discrete evidence of ongoing viral replication [173-177]. This all indicates that antiviral potency still is an issue of primary concern in the treatment of HIV-1 infection and that to achieve lasting success, our pharmacological interventions need to become more sophisticated.

Apart from insufficient antiviral potency, a major stumbling block for attempts at pharmacological eradication of HIV, is the existence of a long-lived latent reservoir of infected cells [178-182]. Attempts to 'purge' this reservoir by means of immune activation have started [173,183], yet it remains to be seen whether this is possible in the absence of complete suppression of viral replication. Besides, there may be additional sanctuaries for HIV, that are unresponsive to this type of approach. Related to the potency issue and the difficulty of HIV eradication, is that strategies of induction-maintenance therapy, so successful in the treatment of tuberculosis, have thus far failed for HIV [102,184,185].

In seeming contrast with the above section on the overriding importance of antiviral potency are the findings of sustained $CD4^+$ cell responses, and decreased $CD4^+$ cell turnover as compared to untreated controls, in people displaying virological failure during treatment with protease-inhibitor-based regimens [92,186,187]. This has been ascribed to lack of fitness of protease-inhibitor-resistant

HIV-1 [188]. It is too early, however, to adapt continuation of drug regimens that fail to suppress HIV-1 plasma levels to below the limits of detection as the strategy of choice in patients who still have other anti-retroviral drug options. It can be predicted that, eventually, with ongoing high-level replication the virus will acquire compensatory mutations that will restore fitness, with consequences that can also be predicted.

There is a great need for simplification of therapy. Many multidrug regimens require drugs to be taken several times a day with specific food instructions. This is a great barrier to adherence with therapy of particular patients and populations. The ideal would be to have a regimen that can be given once a day (or even less frequent), without regard to food intake. This would allow for observed therapy. Fortunately, attempts to achieve the goal of once daily therapy are underway [47,189].

The whole issue of pediatric therapy has not been discussed in this chapter. Principles are the same as for adults [190], but the practice is even more difficult. Pharmacokinetics require specific attention, as these are often age-dependent. In addition, the propensity of high viral loads in pediatric HIV infection [191] asks for specific attention regarding antiviral potency.

One of the most contentious issues in the anti-retroviral therapy field from the very beginning has been when to start therapy. The theoretical advantages of intervening as soon as possible in the destructive process driven by viral replication and, by doing so, preventing further immunologic damage, are indisputable [192-194]. Starting early may, however, be offset by the difficulty of long-term compliance with therapy and the spectre of long-term side effects. This is clearly a decision that should be taken very carefully on an individual basis. An important guiding principle here is that the patient should be 'ready' to start therapy [38].

A major challenge now is to find therapeutic strategies that are valid in patients who have 'seen and failed' most available drugs. One approach to such salvage therapy is to temporarily interrupt anti-retroviral treatment, which will lead to reversal of the dominant HIV quasispecies in that particular patient from multidrugresistant mutant to wild-type [195]. Another or concomitant approach is to use combinations of more than six drugs [195,196]. Long-term suppression of viral replication will, however, only be achieved in a small minority of these patients [195,196].

Even when there were only three nRTIs available, the experience of primary care physicians in the management of AIDS was significantly associated with survival of their patients [197]. The complexity of current drug regimens, with each drug having distinct activities, toxicities, pharmacological profiles and patterns of drug resistance, calls for the administration of HIV chemotherapy by specialists, analogous to the practice of oncology [198].

6. Conclusion

The accomplishments of anti-retroviral therapy have been great. Slightly more than a decade after the discovery of HIV as the causative agent of AIDS, effective therapies against this virus have become available. Because of a large degree of cross-resistance among anti-retroviral agents in the same class (nRTIs, NNRTIs, or

PIs), once a therapy has failed virologically, future therapeutic options with currently available drugs are limited, however. Poor tolerability and toxicity may also compromise the long-term prospects of anti-retroviral therapy. Current anti-retroviral treatment strategies are too complex and far too costly for most developing country settings where the major burden of HIV disease occurs.

7. References

1. Fischl, M.A., Richman, D.D., Grieco, M.H., *et al.*: The efficacy of azidothymidine (AZT) in the treatment of patients with AIDS and AIDS-related complex: a double-blind, placebo-controlled trial, *New Engl. J. Med.* **317** (1987), 185-191.
2. De Wolf, F., Lange, J.M.A., Goudsmit, J., *et al.*: Effect of zidovudine on serum human immunodeficiency virus antigen levels in symptom-free subjects, *Lancet* **i** (1988), 373-376.
3. Fischl, M.A., Richman, D.D., Hansen, N., *et al.*: The safety and efficacy of zidovudine (AZT) in the treatment of subjects with mildly symptomatic human immunodeficiency type 1 (HIV) infection: a double-blind, placebo-controlled trial, *Ann. Intern. Med.* **112** (1990), 437-443.
4. Volberding, P.A., Lagakos, S.W., Koch, M.A., *et al.*: Zidovudine in asymptomatic human immunodeficiency virus infection: a controlled trial in persons with fewer than 500 CD4-positive cells per cubic millimeter, *New Engl. J. Med.* **322** (1990), 941-949.
5. Hamilton, J.D., Hartigan, P.M., Simberkoff, M.S., *et al.*: A controlled trial of early *versus* late treatment with zidovudine in symptomatic human immunodeficiency virus infection: results of the Veterans Affairs Cooperative study, *New Engl. J. Med.* **326** (1992), 437-443.
6. Concorde Coordinating Committee. Concorde: MRC/ANRS randomised double-blind control-led trial of immediate and deferred zidovudine in symptom-free HIV infection, Lancet **343** (1994), 871-881.
7. Yarchoan, R., Perno, C.F., Thomas, R.V., *et al.*: Phase 1 studies of 2',3'-dideoxycytidine in severe human immunodeficiency virus infection as a single agent and alternating with zidovudine (AZT), *Lancet* **i** (1988), 76-81.
8. Yarchoan, R., Mitsuya, H., Thomas, R., *et al.*: *In vivo* activity against HIV and favorable toxicity profile of 2',3'-dideoxyinosine, *Science* **245** (1989), 412-415.
9. Lambert, J.S., Seidlin, M., Reichman, R.C., *et al.*: 2',3'-dideoxyinosine (ddI) in patients with the acquired immunodeficiency syndrome or AIDS-related complex: a phase I trial, *New Engl. J. Med.* **322** (1990), 1333-1340.
10. Cooley, T.P., Kunches, L.M., Saunders, C.A., *et al.*: Once-daily administration of 2',3'-dideoxyinosine (ddI) in patients with the acquired immunodeficiency syndrome or AIDS-related complex: results of a phase I trial, *New Engl. J. Med.* **322** (1990), 1340-1345.
11. Fischl, M.A., Olson, R.M., Follansbee, S.E., *et al.*: Zalcitabine compared with zidovudine in patients with advanced HIV-1 infection who received previous zidovudine therapy, *Ann. Intern. Med.* **118** (1993), 762-769.
12. Alpha International Coordinating Committee. The Alpha trial: European/Australian randomized double-blind trial of two doses of didanosine in zidovudine-intolerant patients with symptomatic HIV disease, *AIDS* **10** (1996), 867-880.
13. Pauwels, R., Andries, K., Desmyter, D., *et al.*: Potent and selective inhibition of HIV-1 replication *in vitro* by a novel series of TIBO derivatives, *Nature* **343** (1990), 470-474.
14. Goldman, M.E., Nunberg, J.H., O'Brien, J.A., et al.: Pyridinone derivatives: specific human immunodeficiency virus type 1 reverse transcriptase inhibitors with antiviral activity, Proc. Natl. Acad. Sci. USA 88 (1991), 6863-6867.
15. Nunberg, J.H., Schleif, W.A., Boots, E.J., *et al.*: Viral resistance to human immunodeficiency virus type 1-specific pyridinone reverse transcriptase inhibitors, *J. Virol.***65** (1991), 4887-4892.
16. Richman, D.D., Shih,C.-K., Lowy, I., *et al.*: Human immunodeficiency virus type 1 mutants resistant to nonnucleoside inhibitors of reverse transcriptase arise in tissue culture, *Proc. Natl .Acad. Sci. USA* **88** (1991), 11241-11245.
17. Richman, D.D., Havlir, D., Corbeil, J., *et al.*: Nevirapine resistance mutations of human immunodeficiency virus type 1 selected during therapy, *J. Virol.* **68** (1994), 1660-1666.

18. De Jong, M.D., Loewenthal, M., Boucher, C.A.B., *et al.*: Alternating nevirapine and zidovudine treatment of human immunodeficiency virus type-1-infected persons does not prolong nevirapine activity, *J. Infect. Dis.* **169** (1994), 1346-1350.
19. Meng, T.-C., Fischl, M.A., Boota, A.M., *et al.*: Combination therapy with zidovudine and dideoxycytidine in patients with advanced human immunodeficiency virus infection, *Ann. Intern. Med.* **116** (1992), 13-20.
20. Yarchoan, R., Lietzau, J.A., Nguyen, B.-Y., *et al.*: A randomized pilot study of alternating or simulateneous zidovudine and didanosine therapy in patients with symptomatic human immunodeficiency virus infection, *J. Infect. Dis.* **169** (1994), 9-17.
21. Schooley, R.T., Ramirez-Ronda, C., Lange, J.M.A., Cooper, D.A., Lavelle, J., Lefkowitz, L., Moore, M., Larder, B.A., St. Clair, M., Mulder, J.W., McKinnis, R., Pennington, K., Harrigan, P.R., Kinghorn, I., Steel, H., Rooney, J.F., and the Wellcome Resistance Study Collaborative Group: Virologic and immunologic benefits of initial combination therapy with zidovudine and zalcitabine or didanosine compared with zidovudine monotherapy, *J. Infect. Dis.* **173** (1996), 1354-1366.
22. Delta Coordinating Committee. Delta: a randomised double-blind controlled trial comparing combinations of zidovudine plus didanosine or zalcitabine with zidovudine alone in HIV-infected individuals, *Lancet* **348** (1996), 283-291.
23. Hammer, S.M., Katzenstein, D.A., Hughes, M.D., *et al.*: A trial comparing nucleoside monotherapy with combination therapy in HIV-infected adults with CD4 cell counts from 200 to 500 per cubic millimeter, *New Engl. J. Med.* **335** (1996), 1081-1090.
24. Van Leeuwen, R., Lange, J.M.A., Hussey, E.K., *et al.*: The safety and pharmacokinetics of a reverse transcriptase inhibitor, 3TC, in patients with HIV infection: a phase I study, *AIDS* **6** (1992), 1471-1475.
25. Katlama, C., Ingrand, D., Loveday, C., *et al.*: Safety and efficacy of lamivudine-zidovudine combination therapy in antiretroviral naive patients: a randomized controlled comparison with zidovudine monotherapy, *JAMA* **276** (1996), 118-125.
26. Eron, J.J., Benoit, S.L., Jemsek, J., *et al.*: Treatment with lamivudine, zidovudine or both in HIV-positive patients with 200 to 500 $CD4^+$ cells per cubic millimeter, *New Engl. J. Med.* **333** (1995), 1662-1669.
27. Staszewski, S., Loveday, C., Picazo, J.J., *et al.*: Safety and efficacy of lamivudine-zidovudine combination therapy in zidovudine-experienced patients: a randomized controlled comparison with zidovudine monotherapy, *JAMA* **276** (1996), 111-117.
28. Bartlett, J.A., Benoit, S.L., Johnson, V.A., *et al.*: Lamivudine plus zidovudine compared with zalcitabine plus zidovudine in patients with HIV infection, *Ann. Intern. Med.* **125** (1996), 161-172.
29. Browne, M.J., Mayer, K.H., Chafee, S.B.D., *et al.*: 2',3'-didehydro-3'-deoxythymidine (d4T) in patients with AIDS or AIDS-related complex, *J. Infect. Dis.* **167** (1993), 21-29.
30. Spruance, S.L., Pavia, A.T., Mellors, J.W., *et al.*: Clinical efficacy of monotherapy with stavudine compared with zidovudine in HIV-infected, zidovudine-experienced patients: a randomized, double-blind, controlled trial, *Ann. Intern. Med.* **126** (1997), 355-363.
31. Saag, M.S., Holodniy, M., Kuritzkes, D.R., *et al.*: HIV viral load markers in clinical practice, *Nature Med.* **2** (1996), 625-629.
32. Wei, X., Ghosh, S.K., Taylor, M.E., *et al.*: Viral dynamics in human immunodeficiency virus type 1 infection, *Nature* **373** (1995), 117-122.
33. Ho, D.D., Neumann, A.U., Perelson, A.S., Chen, W., Leonard, J.M. and Markowitz, M.: Rapid turnover of plasma virions and CD4 lymphocytes in HIV-1 infection, *Nature* **373** (1995), 123-126.
34. Perelson, A.S., Neumann, A.U., Markowitz, M., Leonard, J.M. and Ho, D.H.: HIV-1 dynamics *in vivo*: virion clearance rate, infected cell life-span, and viral generation time, *Science* **271** (1996), 1582-1586.
35. Perelson, A.S., Essunger, P., Cao, Y., *et al.*: Decay characteristics of HIV-1- infected compartments during combination therapy, *Nature* **387** (1997), 188-191.
36. Mansky, L.M. and Temin, H.M.: Lower *in-vivo* mutation rate of human immunodeficiency virus type 1 than that predicted from the fidelity of purified reverse transcriptase, *J. Virol.* **69** (1995), 5087-5094.
37. Coffin, J.M.: HIV population dynamics *in vivo*: implications for genetic variation, pathogenesis, and therapy, *Science* **267** (1995), 483-489.

38. Lange, J.M.A.: Current problems and the future of antiretroviral drug trials, *Science* **276** (1997), 548-550.
39. Danner, S.A., Carr, A., Leonard, J.M., *et al.*: A short-term study of the safety, pharmacokinetics and efficacy of ritonavir, an inhibitor of HIV-1 protease, *New Engl. J. Med.* **333** (1995), 1528-1533.
40. Markowitz, M., Saag, M., Powderly, W.G., *et al.*: A preliminary study of ritonavir, an inhibitor of HIV-1 protease, to treat HIV-1 infection., *New Engl. J. Med.* **333** (1995), 1534-1539.
41. Gulick, R.M., Mellors, J.W., Havlir, D., *et al.*: Treatment with indinavir, zidovudine, and lamivudine in adults with human immunodeficiency virus infection and prior antiretroviral therapy, *New Engl. J. Med.* **337** (1997), 734-739.
42. Hammer, S.M., Squires, K.E., Hughes, M.D., *et al.*: A controlled trial of two nucleoside analogues plus indinavir in persons with human immunodeficiency virus infection and CD4 cell counts of 200 per cubic millimeter or less, *New Engl. J. Med.* **337** (1997), 725-733.
43. Mouton, Y., Alfandari, S., Valette, M., *et al.*: Impact of protease inhibitors on AIDS-defining events and hospitalizations in 10 French AIDS reference centres, *AIDS* **11** (1997), F101-F105.
44. Palella, F.J., Delaney, K.M., Moorman, A.C., *et al.*: Declining morbidity and mortality among patients with advanced human immunodeficiency virus infection, *New Engl. J. Med.* **338** (1998), 853-860.
45. Hogg, R.S., Yip, B., Kully, C., *et al.*: Improved survival among HIV-infected patients after initiation of triple-drug antiretroviral regimens, *Canad. Med. Assoc. J.* **160** (1999), 659-665.
46. Montaner, J.S.G., Reiss, P., Cooper, D., *et al.*: A randomized, double-blinded trial comparing combinations of nevirapine, didanosine and zidovudine for HIV-infected patients - the Incas trial, *JAMA* **279** (1998), 930-937.
47. Katlama, C., Murphy, R., Johnson, V., *et al.*: The Atlantic study: a randomised open-label study comparing two protease inhibitors (PI)-sparing strategies *versus* a standard PI containing regimen. In: *6th Conference on Retroviruses and Opportunistic Infections*, Chicago, January 31-February 1999 [abstract 18].
48. Tashima, K., Staszewski, S., Stryker, R., *et al.*: A phase III, multicenter, randomized, open-label study to compare the antiretroviral activity and tolerability of efavirenz (EFV) + indinavir (IDV), versus EFV + zidovudine (ZDV) + lamivudine (3TC), versus IDV + ZDV + 3TC at 48 weeks (Study DMP 266-006). In: *6th Conference on Retroviruses and Opportunistic Infections*, Chicago, January 31-February 4, 1999 [abstract LB16].
49. Staszewski, S., Keiser, P., Gathe, J., *et al.*: Ziagen/Combivir is equivalent to indinavir/Combivir in antiretroviral therapy naive adults at 24 weeks. In: *6th Conference on Retroviruses and Opportunistic Infections*, Chicago, January 31-February 4, 1999 [abstract 20].
50. Fischl, M., Greenberg, S., Clumeck, N., *et al.*: Ziagen (abacavir,ABC, 1592) combined with 3TC & ZDV is highly effective and durable through 48 weeks in HIV-1-infected antiretroviral therapy-naive subjects (CNAA3003). In: *6th Conference on Retroviruses and Opportunistic Infections*, Chicago, January 31-February 4, 1999 [abstract 19].
51. Richman, D.D.: Nailing down another HIV target, *Nature Med.* **4** (1998), 1232-1233.
52. Carpenter, C.C.J., Fischl, M.A., Hammer, S.M., *et al.*: Antiretroviral therapy for HIV infection in 1998: updated recommendations of the International AIDS Society-USA Panel, *JAMA* **280** (1998), 78-86.
53. Mitsuya, H., Weinhold, K.J., Furman, P.A., *et al.*: 3'-azido-3'-deoxythymidine (BW A509U): an antiviral agent that inhibits the infectivity and cytopathic effect of human T-lymphotropic virus type III/lymphadenopathy-associated virus *in vitro*, *Proc. Natl. Acad. Sci. USA* **82** (1985), 7096-7100.
54. Shirasaka, T., Chokekijchai, S., Yamada, A., Gosselin, G., Imbach, J.-L. and Mitsuya, H.: Comparative analysis of anti-human immunodeficiency virus type 1 activities of dideoxynucleoside analogs in resting and activiated peripheral blood mononuclear cells, *Antimicrob. Agents Chemother.* **39** (1995), 2555-2559.
55. Van 't Wout, A.B., Ran, L.J., de Jong, M.D., Bakker, M., van Leeuwen, R., Notermans, D.W., Loeliger, A.E., de Wolf, F., Danner, S.A., Reiss, P., Boucher, C.A.B., Lange, J.M.A. and Schuitemaker, H.: Selective inhibition of syncytium-inducing and non-syncytium inducing HIV-1 variants in individuals receiving didanosine or zidovudine, respectively, combination therapy, *J. Clin. Invest.* **100** (1997), 2325-2332.

56. Merrill, D.P., Moonis, M., Chou,T.-C. and Hirsch, M.S.: Lamivudine or stavudine in two- and three-drug combinations against human immunodeficiency virus type 1 replication *in vitro*, *J. Infect. Dis.* **173** (1996), 355-364.

57. Havlir, D.V., Friedland, G., Pollard, R., *et al.*: Combination zidovudine and stavudine therapy *versus* other nucleosides: report of two randomized trial (ACTG 290 and 298). In: *5th Conference on Retroviruses and Opportunistic Infections*, Chicago, February 1-5, 1998 [abstract 2].

58. Foudraine, N.A., Hoetelmans, R.M.W., Lange, J.M.A., *et al.*: Cerebrospinal fluid (CSF) HIV-RNA and drug concentrations during treatment with lamivudine (3TC) in combination with zidovudine (AZT) or stavudine (d4T), *Lancet* **351** (1998), 1547-1551.

59. Foudraine, N.A., de Jong, J.J., Weverling, G.J., *et al.*: An open randomised controlled trial of zidovudine (AZT) + lamivudine (3TC) *versus* stavudine (d4T) + lamivudine in antiretroviral therapy-naive patients infected with human immunodeficiency virus type 1, *AIDS* **12** (1998), 1513-1519.

60. Kuritzkes, D.R., Marschner, I.C., Johnson, V.A., *et al.*: Lamivudine in combination with zidovudine, stavudine, or didanosine in patients with HIV-1 infection: a randomized, double-blind, placebo-controlled trial, *AIDS* **13** (1999), 685-694.

61. Richman, D.D., Fischl, M.A., Grieco, M.H., *et al.*: The toxicity of azidothymidine (AZT) in the treatment of patients with AIDS and AIDS-related complex: a double-blind, placebo-controlled trial, *New Engl. J. Med.* **317** (1987), 192-197.

62. Caesar Coordinationg Committee. Randomised trial of addition of lamivudine or lamivudine plus loviride to zidovudine-containing regimens for patients with HIV-1 infection, *Lancet* **349** (1997), 1413-1421.

63. Larder, B.A., Kemp, S.D. and Harrigan, P.R.: Potential mechanism for sustained antiretroviral efficacy of AZT-3TC combination therapy, *Science* **269** (1995), 696-699.

64. Gulick, R.M., Mellors, J.W., Havlir, D., *et al.*: Simultaneous *vs* sequential initiation of therapy with indinavir, zidovudine, and lamivudine for HIV-1 infection: 100 week follow-up, *JAMA* **280** (1998), 35-41.

65. Pollard, R., Peterson, D., Hardy, D., *et al.*: Antiviral effect and safety of stavudine (d4T) and didanosine (ddI) combination therapy in HIV-infected subjects in an ongoing pilot randomized double-blinded trial.). In: *3rd Conference on Retroviruses and Opportunistic Infections*, Washington DC, January 28 - February 1, 1996 [abstract 197].

66. Wakeford, C., Hulett, L., Quinn, J., *et al.*: A phase I/II randomized, controlled study of FTC *versus* 3TC in HIV-infected patients. In: *6th Conference on Retroviruses and Opportunistic Infections*, Chicago, January 31 - February 4, 1999 [abstract 16].

67. Staszewski, S., Katlama, C., Harrer, T., *et al.*: A dose-ranging study to evaluate the safety and efficacy of abacavir alone or in combination with zidovudine and lamivudine in antiretroviral treatment-naive subjects, *AIDS* **12** (1998), F197-F202.

68. Saag, M.S., Sonnerborg, A., Torres, R.A., *et al.*: Antiretroviral effect and safety of abacavir alone and in combination with zidovudine in HIV-infected adults, *AIDS* **12** (1998), F203-F209.

69. Hammer, S., Squires, K., DeGruttola, V., *et al.*: Randomized trial of abacavir (ABC) & nelfinavir (NFV) in combination with efavirenz (EFV) + adefovir dipivoxil (ADV) as salvage therapy in patients with virologic failure receiving indinavir (IDV).). In: *6th Conference on Retroviruses and Opportunistic Infections*, Chicago, January 31-February 4, 1999 [abstract 490].

70. Brinkman, K., ter Hofstede, H.J., Burger, D.M., Smeitink, J.A. and Koopmans, P.P.: Adverse effects of reverse transcriptase inhibitors: mitochondrial toxicity as common pathway, *AIDS* **12** (1998), 1735-1744.

71. McKenzie, R., Fried, M.W., Sallie, R., *et al.*: Hepatic failure and lactic acidosis due to fialuridine (FIAU), an investigational nucleoside analogue for chronic hepatitis B, *New Engl. J. Med.* **333** (1995), 1099-105.

72. Deeks, S.G., Collier, A., Lalezari, J., *et al.* The safety and efficacy of adefovir dipivoxil, a novel anti-human immunodeficiency virus (HIV) therapy, in HIV-infected adults: a randomized, double-blind placebo-controlled trial, *J. Infect. Dis.* **176** (1997), 1517-1523.

73. Nuessle, S.J., Barriere, S.L., Rooney, J.F., *et al.* The Preveon Expanded Access Program: safety of adefovir dipivoxil in antiretroviral treatment experienced patients with advanced HIV dis-ease. In: *6th Conference on Retroviruses and Opportunistic Infections*, Chicago, January 31-February 4, 1999 [abstract 379].

74. Tsai, C.-C., Follis, K.E., Sabo, A., *et al.* Prevention of SIV infection in macaques by (R)-9-(2-phosphonylmethoxypropyl)adenine, *Science* **270** (1995), 1179-1199.

75. Deeks, S.G., Barditch-Crovo, P., Lietman, P.S., *et al.* The safety and efficacy of PMPA prodrug monotherapy: preliminary results of a phase I/II dose-escalation study. In: *5th Conference on Retroviruses and Opportunistic Infections*, Chicago, February 1-5, 1999 [abstract LB8].

76. Spence, R.A., Kati, W.M., Anderson, K.S. and Johnson, K.A.: Mechanism of inhibition of HIV-1 reverse transcriptase by nonnucleoside inhibitors, *Science* **267** (1995), 988-993.

77. Emini, E.A.: Non-nucleoside reverse transcriptase inhibitors - mechanisms. In: *Antiviral Drug Resistance* (D.D. Richman, ed.), Chichester: John Wiley & Sons, 1996, pp. 225-240.

78. Freimuth, W.W.: Delavirdine mesylate, a potent non-nucleoside reverse transcriptase inhibitor, *Adv. Exp. Med. Biol.* **394** (1996), 279-289.

79. Para, M.F., Meehan, P., Holden-Wiltze, J. *et al.*: ACTG 260: a randomized, phase I-II, dose-ranging trial of the anti-human immunodeficiency virus activity of delavirdine monotherapy, *Antimicrob. Agents Chemother.* **43** (1999), 1373-1378.

80. Adkins, J.C. and Noble, S.: Efavirenz, *Drugs* **56** (1998), 1055-1064.

81. McCreedy, B., Borroto-Esoda, K., Harris, J., Klish, C., Fang, L. and Miralles, D.: Genotypic and phenotypic analysis of HIV-1 from patients receiving combination therapy containing two nucleoside reverse transcriptase inhibitors (NRTIs) and the non-NRTI, emivirine (MKC-442), *Antiviral Ther.* **4 (suppl. 1)** (1999) (Abstract 13), 9.

82. Potts, K.E., Fujiwara, T., Sato, A., *et al.*: Antiviral activity and resistance profile of AG1549, a novel nonnucleoside reverse transcriptase inhibitor. In: *6th Conference on Retroviruses and Opportunistic Infections*, Chicago, January 31-February 4, 1999 [abstract 12].

83. Erickson-Viitanen, S., Corbett, J., Ko, S., *et al.*: DMP 961 and DMP 963: 2nd generation non-nucleoside reverse transcriptase inhibitors active against the RT K103N mutant. In: *6th Conference on Retroviruses and Opportunistic Infections*, Chicago, January 31-February 4, 1999 [abstract 13].

84. Carr, A., Samaras, K., Burton, S., *et al.*: A syndrome of peripheral lipodystrophy, hyperlipidaemia and insulin resistance in patients receiving HIV protease inhibitors, *AIDS* **12** (1998), F51-F58.

85. Gervasoni, C., Ridolfo, A.L., Trifiro, G., *et al.*: Redistribution of body fat in HIV-infected women undergoing combined antiretroviral therapy, *AIDS* **13** (1999), 465-471.

86. Flexner, C.: HIV-protease inhibitors, *New Engl. J. Med.* **338** (1998), 1281-1291.

87. Kitchen, V.S., Skinner, C., Ariyoshi, K., *et al.*: Safety and activity of saquinavir in HIV infection, *Lancet* **345** (1995), 952-955.

88. Vella, S., Lazzarin, A., Carosi, C., *et al.*: A randomized controlled trial of a protease inhibitor (saquinavir) in combination with zidovudine in previously untreated patients with advanced HIV infection, *Antiviral. Ther.* **1** (1996), 129-140.

89. Flexner, C.: Pharmacokinetics and pharmacodynamics of HIV protease inhibitors, *Infect. Med.* **13 (suppl. F)** (1996), 16-23.

90. Cohen Stuart, J.W.T., Schuurman, R., Burger, D.M., *et al.*: Randomized trial comparing saquinavir soft gelatin capsules *versus* indinavir as part of triple therapy (Cheese study), *AIDS* **13** (1999), F53-F58.

91. Cameron, D.W., Heath-Chiozzi, M.H., Danner, S.A., Cohen, C., Kravcik, S., Maurath, C., Sun, E., Henry, D., Rode, R., Potthof, A. and Leonard, J. for the Advanced HIV Disease Ritonavir Study Group, *Lancet* **351** (1998), 543-549.

92. Wit, F.W.N.M., van Leeuwen, R., Weverling, G.J., Jurriaans, S., Nauta, K., Steingrover, R., Schuijtemaker, J., Eyssen, X., Fortuin, D., Weeda, M., de Wolf, F., Reiss, P., Danner, S.A. and Lange, J.M.A.: Outcome and predictors of failure of highly active antiretroviral therapy: one year follow-up of an unselected cohort of HIV-1 infected individuals, *J. Infect. Dis.* **179** (1999) 790-798.

93. Kempf, D., Marsh, K., Kumar, G., *et al.*: Pharmacokinetic enhancement of inhibitors of the human immunodeficiency virus potease by coadministration with ritonavir, *Antimicrob. Agents Chemother.* **41** (1997), 654-660.

94. Merry, C., Barry, M.G., Mulcahy, F., *et al.*: Saquinavir pharmacokinetics alone and in combination with ritonavir in HIV-infected patients, *AIDS* **11** (1997), F29-F33.

95. Dieleman, J.P., Gyssens, I.C., van der Ende, M.E., de Marie, S. and Burger, D.M.: Urological complaints in relation to indinavir plasma concentrations in HIV-infected patients, *AIDS* **13** (1999), 473-478.

96. Burger, D.M., Hugen, P.W.H., Prins, J.M., van de Ende, M.E., Reiss, P. and Lange, J.M.A.: Pharmacokinetics of an indinavir/ritonavir 800/100 mg BID regimen. In: *6th Conference on*

Retroviruses and Opportunistic Infections, Chicago, January 31-February 4, 1999 [abstract 363].

97. Van Heeswijk, R.P.G., Veldkamp, A., Hoetelmans, R.M.W., *et al.*: The steady state pharmacokinetics of indinavir alone in combination with a low dose of ritonavir in twice daily dosing regimens in HIV-1-infected individuals, *AIDS* 1999, in press.

98. Jarvis, B. and Faulds, D.: Nelfinavir, a review of its therapeutic efficacy in HIV infection, *Drugs* **56** (1998), 147-167.

99. Myers, R.E., Snowden, W., Randall, S. and Tisdale, M.: Unique resistance profile of the protease inhibitor amprenavir (141W94) observed *in vitro* and in the clinic. In: *2nd International Workshop on HIV Drug Resistance & Treatment Strategies*, Lake Maggiore, June 24-27, 1998, *Antiviral. Ther.* **3 (suppl. 1)** (1998), 59-60 [abstract 86].

100. Cameron, D.W., Japour, A.J., Xu, Y., *et al.*: Ritonavir and saquinavir combination therapy for the treatment of HIV infection, *AIDS* **13** (1999), 213-224.

101. Kirk, O., Katzenstein, T.L., Gerstoft, J., *et al.*: Combination therapy containing ritonavir plus saquinavir has superior short-term antiretroviral efficacy: a randomized trial, *AIDS* **13** (1999), F9-F16.

102. Reijers, M.H.E., Weverling, G.J., Jurriaans, S., *et al.*: Maintenance therapy after quadruple induction therapy in HIV-1-infected individuals: Amsterdam duration of antiretroviral medica-tion (ADAM) study, *Lancet* **352** (1998), 185-190.

103. Sham, H.L., Kempf, D.J., Molla, A., *et al.*: ABT378, a high potent inhibitor of the human immunodeficiency virus protease, *Antimicrob. Agents Chemother.* **42** (1998), 3218-3224.

104. Wang, Y., Freimuth, W.W., Daenzer, C.L., *et al.*: Safety and efficacy of PNU-140690, a new non-peptidic HIV protease inhibitor, and HIV genotypic changes in patients in a phase II study. In: *2nd International Workshop on HIV Drug Resistance & Treatment Strategies*, Lake Maggiore, June 24-27, 1998, *Antiviral. Ther.* **3 (suppl. 1)** (1998), 5 [abstract 5].

105. Potts, K.E., Fujiwara, T., Sato, A., *et al.*: Antiviral activity and resistance profile of AG1776, a novel inhibitor of HIV-1 protease. In: *6th Conference on Retroviruses and Opportunistic Infections*, Chicago, January 31-February 4, 1999 [abstract 11].

106. Gong, Y., Robinson, B., Rose, R., *et al.*: Resistance profile and drug combination studies of anHIV-1 protease inhibitor BMS-232632. In: *6th Conference on Retroviruses and Opportunistic Infections*, Chicago, January 31-February 4, 1999 [abstract 603].

107. Lori, F., Malykh, A., Cara, A., *et al.*: Hydroxyurea as an inhibitor of human immunodeficiency virus-type 1 replication, *Science* **266** (1994), 801-805.

108. Clotet, B., Ruiz, L., Cabrera, C., *et al.*: Short-term anti-HIV activity of the combination of didanosine and hydroxyurea, *Antiviral.Ther.* **1** (1996), 189-193.

109. Foli, A., Maserati, R., Tinelli, C., Minoli, L. and Lisziewicz, J.: Hydroxyurea and didanosine is a more potent combination than hydroxyurea and zidovudine, *Antiviral. Ther.* **2** (1997), 31-38.

110. Lori, F., Malykh, A.G., Foli, A., *et al.*: Combination of a drug targeting the cell with a drug targetting the virus controls human immunodeficiency virus type 1 resistance, *AIDS Res. Human Retrovir.* **13** (1997), 1403-1409.

111. Montaner, J.S., Zala, C., Conway, B., *et al.*:A pilot study of hydroxyurea among patients with advanced human immunodeficiency virus (HIV) disease receiving didanosine therapy: Canadian HIV Trials Network Protocol 080, *J. Infect. Dis.* **175** (1997), 801-806.

112. Rutschmann, O.T., Opravil, M., Iten, A., *et al.*: A placebo-controlled trial of didanosine plus stavudine, with and without hydroxyurea, for HIV infection, *AIDS* **12** (1998), F71-F77.

113. De Boer, R.J., Boucher, C.A.B. and Perelson, A.S.: Target cell availability and the successful suppression of HIV by hydroxyurea and didanosine, *AIDS* **12** (1998), 1567-1570.

114. Boelaert, J.R. and Sperber, K.: Antiretroviral treatment, *Lancet* **352** (1998), 1224-1225.

115. Kovacs, J.A., Baseler, M., Dewar, R.J., *et al.*: Increases in CD4 T lymphocytes with intermittent courses of interleukin-2 in patients with human immunodeficiency virus infection: a preliminary study, *New Engl. J. Med.* **332** (1995), 567-575.

116. Kovacs, J.A., Vogel, S., Albert, J.M., *et al.*: Controlled trial of interleukin-2 infusions in patients infected with the human immunodeficiency virus, *New Engl. J. Med.* **335** (1996), 1350-1356.

117. Carr, A., Emery, S., Lloyd, A., *et al.*: Outpatient continuous intravenous interleukin-2 or subcutaneous, polyethylene glycol modified interleukin-2 in human immuno-deficiency virus-infected patients: a randomised, controlled, multicenter study, *J. In-fect. Dis.* **178** (1998), 992-999.

118. Hoff, R. and McNamara, J.: Therapeutic vaccines for preventing AIDS: their use with HAART, *Lancet* **353** (1999), 1723-1724.
119. Veenstra, J., Williams, I.G., Colebunders, R., Dorrell, L., Tchamouroff, S.E., Patou, G., Lange, J.M.A., Weller, I.V.D., Goeman, J., Uthayakumar, S., Gow, I.R., Weber, J.N. and Coutinho, R.A.: Immunization with recombinant p17/p24:Ty virus-like particles in human immunodeficiency virus-infected persons, *J. Infect. Dis.* **174** (1996), 862-866.
120. Valentine, F.T., DeGruttola, V., Kaplan, M., *et al.*: Effects of HAART compared to HAART plus an inactivated HIV immunogen on lymphocyte proliferative responses (LPR) to HIV antigens. In: *12th World AIDS Conference*, Geneva, June 28-July 3, 1998 [abstract LB31227].
121. Ratto-Kim, S., Sitz, K.V., Garner, R.P., *et al.*: Repeated immunization with recombinant gp160 human immunodeficiency virus (HIV) envelope protein in early HIV-1 infection: evaluation of cell proliferative response, *J. Infect. Dis.* **179** (1999), 337-344.
122. Sandström, E., Wahren, B. and Nordic VAC-04 Study Group.: Therapeutic immunisation with recombinant gp160 in HIV-1 infection: a randomised double-blind placebo-controlled trial, *Lancet* **353** (1999), 1735-1742.
123. Rosenberg, E.S., Billingsley, J.M., Caliendo, A.M., *et al.*: Vigorous HIV-1-specific CD4+ T-cell responses associated with control of viremia, *Science* **278** (1997), 1447-1450.
124. Hazuda, D.J.: Inhibitors of HIV integrase: antiviral activity and mechanism, *Antiviral Ther.* **4 (suppl. 1)** (1999) (Abstract 1), 3.
125. Kilby, J.M., Hopkins, S., Venetta, T.M. *et al.*: Potent suppression of HIV-1 replication in humans by T-20, a peptide inhibitor of gp14-mediated virus entry, *Nature Med.* **4** (1998), 1302-1307.
126. Rice, W.G., Supko, J.G., Malspeis, L., *et a*l.: Inhibitors of HIV nucleocapside protein zinc fingers as candidates for the treatment of AIDS, *Science* **270** (1995), 1194-1197.
127. Mei, H.-Y., Cui, M., Heldsinger, A., *et al.*: Inhibitors of protein-RNA complexation that target the RNA: specific recognition of human immunodeficiency virus type 1 TAR RNA by small organic molecules, *Biochemistry* **37** (1998), 14204-14212.
128. Burger, D.M., Hoetelmans, R.M.W., Koopmans, P.P., Meenhorst, P.L., Mulder, J.W., Hekster, Y.A. and Beijnen, J.H.: Clinically relevant drug interactions with antiretroviral agents, *Antiviral. Ther.* **2** (1997), 149-165.
129. Murphy, R.L. and Montaner, J.: Nevirapine: a review of its dvelopment, pharmacological profile and potential for clinical use, *Expert Opinion in Investigational Drugs* **5** (1996), 1183-1199.
130. Cheng, C.-L., Smith, D.E., Cox, S.R., *et al.*: Steady-state (SS) pharmacokinetics (PK) of delavirdine (DLV) in HIV+ patients: *in-vivo* effect of DLV on the erythromycin breath test (ERMBT). In: *36th Interscience Conference on Antimicrobial Agents and Chemotherapy*, New Orleans, September , 1996 [abstract A056].
131. Hirsch, M.S., Conway, B., DíAquila, R.T., *et al.*: Antiretroviral drug resistance testing in adults with HIV infection: implications for clinical management, *JAMA* **279** (1998), 1984-1991.
132. Pauwels, R., Hertogs, K., Kemp, S., *et al.*: Comprehensive HIV drug resistance monitoring using rapid, high-throughput phenotypic and genotypic assays with correlative data analysis, *Antiviral. Ther.* **3 (suppl. 1)** (1998), 35-36 [abstract 51].
133. Baxter, J.D., Mayers, D.L., Wentworth, D.N., *et al.*: A pilot study of the short-term effects of antiretroviral management based on plasma genotypic antiretroviral resistance testing (GART) in patients failing antiretroviral therapy. In: *6th Conference on Retroviruses and Opportunistic Infections,* Chicago, January 31-February 4, 1999 [abstract LB8].
134. Durant, J., Clevenbergh, P., Halfon, P., Delgiudice, P., Porsin, S., Simonet, P., Montagne, N., Boucher, C.A.B., Schapiro, J.M. and Dellamonica, P.: Drug-resistance genotyping in HIV-1 therapy: the VIRADAPT randomised controlled trial, *Lancet* **353** (9171) (1999), 2195-2199.
135. Havlir,D.V., Gamst, A., Eastman, S. and Richman, D.D.: Nevirapine-resistant human immuno-deficiency virus: kinetics of replication and estimated prevalence in untreated patients, *J. Virol.* **70** (1996), 7894-7899.
136. Schuurman, R., Nijhuis, M., van Leeuwen ,R., *et al.*: Rapid changes in human immuno-deficiency virus type 1 RNA load and appearance of drug-resistant virus populations in persons treated with lamivudine (3TC), *J. Infect. Dis.* **171** (1995), 1411-1419.
137. Larder, B.A. and Kemp, S.D.: Multiple mutations in HIV-1 reverse transcriptase confer high-level resistance to zidovudine, *Science* **246** (1989), 1155-1158.

138. Boucher, C.A.B., O'Sullivan, E., Mulder, J.W., *et al.*: Ordered appearance of zidovudine resistance mutations during treatment of 18 human immunodeficiency virus-positive subjects, *J. Infect. Dis.* **165** (1992), 105-110.
139. Condra, J.H., Schleif, W.A., Blahey, O.M., *et al.*: *In-vivo* emergence of HIV-1 variants resistant to multiple protease inhibitors, *Nature* **374** (1995), 569-571.
140. Molla, A., Korneyeva, M., Gao, Q., *et al.*: Ordered accumulation of mutations in HIV protease confer resistance to ritonavir, *Nature Med.* **2** (1996), 760-766.
141. Larder, B.A., Darby, G. and Richman, D.D.: HIV with reduced sensitivity to zidovudine (AZT) isolated during prolonged therapy, *Science* **243** (1989), 1731-1734.
142. Richman, D.D., Grimes, J.M. and Lagakos, S.W.: Effect of stage of disease and drug dose on zidovudine susceptibilities of isolates of human immunodeficiency virus, *J. Ac. Immune Def. Syndr. Human Retrovir.* **3** (1990), 743-746.
143. De Jong, M.D., Veenstra, J., Stilianakis, N.I., *et al.*: Host-parasite dynamics and outgrowth of virus containing a single K70R amino acid change in reverse transcriptase are responsible for the loss of HIV-1 RNA load suppression by zidovudine, *Proc. Natl. Acad. Sci. USA* **93** (1996), 5501-5506.
144. Lacey, S.F. and Larder, B.A.: Novel mutation (V75T) in human immunodeficiency virus type 1 reverse transcriptase confers resistance to 2',3'-didehydro-2'3'-dideoxythymidine in cell culture, *Antimicrob. Agents Chemother.* **38** (1994), 1428-1432.
145. Dueweke, T.J., Pushkarskaya, T., Poppe, S.M., *et al.*: A mutation in reverse transcriptase of (bis)piperazine-resistant human immunodeficiency virus type 1 that confers increased sensitivity to other nonnucleoside inhibitors, *Proc. Natl. Acad. Sci. USA* **90** (1993), 4713-4717.
146. Bloor, S., Hertogs, K., Desmet, R.L., Pauwels, R. and Larder, B.A.: Virologic basis for HIV-1 resistance to stavudine investigated by analysis of clinical samples. In: *2nd International Work-shop on HIV Drug Resistance & Treatment Strategies*, Lake Maggiore, June 24-27, 1998, *Antiviral. Ther.* **3 (suppl. 1)** (1998), 13-14 [abstract 15].
147. Whitcomb, J.M., Limoli, K., Smith, D., *et al.*: Phenotypic and genotypic analyisis of stavudine-resistant isolates of HIV-1. In: *2nd International Workshop on HIV Drug Resistance & Treatment Strategies*, Lake Maggiore, June 24-27, 1998, *Antiviral. Ther.* **3 (suppl. 1)** (1998), 14-15 [abstract 17].
148. Shafer, R.W., Kozal, M.J., Winters, M.A., *et al.*: Combination therapy with zidovudine and didanosine selects for drug-resistant human immunodeficiency virus type 1 strains with unique patterns of pol gene mutations, *J. Infect .Dis.* **169** (1994), 722-729.
149. Winters, M.A., Coolley, K.L., Girard, Y.A., *et al.*: Phenotypic and molecular analysis of HIV-1 isolates possessing 6 bp inserts in the reverse transcriptase gene that confer resistance to nucleoside analogues. In: *2nd International Workshop on HIV Drug Resistance & Treatment Strategies*, Lake Maggiore, June 24-27, 1998, *Antiviral. Ther.* **3 (suppl. 1)** (1998), 14 [abstract 16].
150. De Jong, J.J., Jurriaans, S., Goudsmit ,J., *et al.*: Insertion of two amino acids in reverse transcriptase (RT) during antiretroviral combination therapy: implications for resistance against nucleoside RT inhibiotors. In: *2nd International Workshop on HIV Drug Resistance & Treatment Strategies*, Lake Maggiore, June 24-27, 1998, *Antiviral.Ther.* **3 (suppl. 1)** (1998), 15 [abstract 18].
151. St. Clair, M.H., Martin, J.L., Tudor-Williams, G., *et al.*: Resistance to ddI and sensitivity to AZT induced by a mutation in HIV-1 reverse transcriptase, *Science* **253** (1991), 1557-1559.
152. Miller, M.D., Anto, K.E., Mulato, A.S., Lamy, P.D., Margot, N.A. and Cherrington, J.M.: HIV-1 expressing the lamivudine-associated M184V mutation in reverse transcriptase (RT) shows increased susceptibility to adefovir and PMPA as well as decreased replication capacity *in vitro*. In: *2nd International Workshop on HIV Drug Resistance & Treatment Strategies*, Lake Maggiore, June 24-27, 1998, *Antiviral. Ther.* **3 (suppl. 1)** (1998), 24 [abstract 34].
153. Hecht, F.M., Grant, R.M. and Petropoulos, C.J.: Sexual transmission of an HIV-1 variant resistant to multiple reverse-transcriptase and protease inhibitors, *New Engl. J. Med.* **339** (1998), 307-311.
154. Connor, E.M., Sperling, R.S., Gelber, R., *et al.*: Reduction of maternal-infant transmission of human immunodeficiency virus type 1 with zidovudine treatment, *New Engl. J. Med .* **331** (1994), 1173-1180.
155. Shaffer, N., Chuachoowong , R., Mock, P.A., *et al.*: Short-course zidovudine for perinatal HIV-1 transmission in Bangkok, Thailand: a randomised controlled trial, *Lancet* **353** (1999), 773-780.

156. Wiktor, S.Z., Ekpini, E., Karon, J.M., *et al.*: Short-course oral zidovudine for prvention of mother-to-child transmission of HIV-1 in Abidjan, Cote d'Ivoire: a randomised trial, *Lancet* **353** (1999), 781-785.

157. Dabis, F., Msellati, P., Meda, N., *et al.*: 6-month efficacy, tolerance, and acceptability of a short regimen of oral zidovudine to reduce vertical transmission of HIV in breastfed children in Cote d'Ivoire and Burkina Fasso: a double-blind placebo-controlled multicentre trial, *Lancet* **353** (1999), 786-792.

158. Saba, J.: The results of the PETRA intervention trial to prevent perinatal transmission in SubSaharan Africa. In: *6th Conference on Retroviruses and Opportunistic Infections*, Chicago, January 31-February 4, 1999 [abstract S7].

159. Mandelbrot, L., Le Chenedac, J., Berrebi, A., *et al.*: Perinatal HIV-1 transmission: interaction between zidovudine prophylaxis and mode of delivery in the French perinatal cohort, *JAMA* **280** (1998), 55-60.

160. The European Mode of Delivery Collaboration: Elective caesarian-section versus vaginal delivery in prevention of vertical HIV-1 transmission, *Lancet* **353** (1999), 1035-1039.

161. The International Perinatal HIV Group: The mode of delivery and the risk of vertical transmission of human immunodeficiency virus type 1: a meta-analysis of 15 prospective cohort studies, *New Engl. J. Med.* **340** (1999), 977-987.

162. Cardo, D.M., Culver, D.H., Ciesielski, C.A., *et al.*: A case-control study of HIV sero-conversion in health care workers after percutaneous exposure, *New Engl. J. Med.* **337** (1997), 1485-1490.

163. Centers for Disease Control and Prevention: Public Health Service guidelines for the management of health-care worker exposures to HIV and recommendations for postexposure prophylaxis, *Morbid Mortal Wkly Rep.* **47 (RR-7)** 1998), 1-33.

164. Stein, Z.A.: HIV prevention: the need for methods women can use, *Am. J. Public Health* **80** (1990), 460-462.

165. Fontanet, A.L., Saba, J., Chandelying, V., Sakondhavat, C., Bhiraleus, P., Rugpao, S., Chongsomchai, C., Kiriwat, O., Tovanabutra, S., Dally, L., Lange, J.M.A. and Rojanapithayakorn, W.: Protection against sexually transmitted diseases by granting sex workers in Thailand the choice of using the male or female condom: results from a randomized controlled trial, *AIDS* **12 (14)** (1998), 1851-1859.

166. Elias, C.J. and Heise, L.L.: Challenges for the development of female-controlled vaginal micro-bicides, *AIDS* **8** (1994), 1-9.

167. Stone, A.B. and Hitchcock, P.: Vaginal microbicides for preventing the sexual transmission of HIV, *AIDS* **8 (suppl. 1)** (1994), S285-S293.

168. Global Programme on AIDS, World Health Organization: Report on a meeting on the development of vaginal microbicides for the prevention of heterosexual transmission of HIV [Document WHO/GPA/RID/CRD/94.1; 11-13 November 1993]. Geneva: GPA/WHO;1993.

169. Katz, M.H. and Gerberding, J.L.: The care of persons with recent sexual exposure to HIV, *Ann. Intern. Med.* **128** (1998), 306-312.

170. Zhang, H., Dornadula, G., Beumont, M., *et al.*: Human immunodeficiency virus type 1 in the semen of men receiving highly active antiretroviral therapy, *New Engl. J. Med.* **339** (1998), 1803-1809.

171. Raboud, J.M., Montaner, J.S.G., Conway, B., *et al.*: Suppression of plasma viral load below 20 copies/ml is required to achieve a long-term response to therapy, *AIDS* **12** (1998), 1619-1624.

172. Weverling, G.J., Lange, J.M.A., Jurriaans, S., *et al.*: Alternative multidrug regimen provides improved suppression of HIV-1 replication over triple therapy, *AIDS* **12** (1998), F117-F122.

173. Prins, J., Jurriaans, S., van Praag, R., *et al.*: OKT3 and rhIL-2 in HIV-1 patients with prolonged suppression of plasma viremia. In: *6th Conference on Retroviruses and Opportunistic Infections*, Chicago, January 31-February 4, 1999 [abstract LB6].

174. Wong, J., Günthard, H., Fiscus, S., *et al.*: Residual HIV RNA and DNA in lymph node and HIV RNA in genital secretions and in CSF after two years of suppression of viremia in the Merck 035 cohort. In: *6th Conference on Retroviruses and Opportunistic Infections*, Chicago, January 31-February 4, 1999 [abstract 6].

175. Natarajan ,V., Bosche, M., Metcalf, J.A., Ward, D.J., Lane, H.C. and Kovacs, J.A.: HIV-1 replication in patients with undetectable plasma virus receiving HAART, *Lancet* **353** (1999), 119-120.

176. Zhang, L., Ramratnam, B., Tenner-Racz, K., *et al.*: Quantifying residual HIV-1 replication in patients receiving combination antiretroviral therapy, *New Engl. J. Med.* **340** (1999), 1605-1613.
177. Furtado, M.R., Callaway, D.S., Phair, J.P., *et al.*: Persistence of HIV-1 transcription in peripheral-blood mononuclear cells in patients receiving potent antiretroviral therapy, *New Engl. J. Med.* **340** (1999), 1614-1622.
178. Chun, T.-W., Carruth, L., Finzi, D., *et al.*: Quantification of latent tissue reservoirs and total body viral load in HIV-1 infection, *Nature* **387** (1997), 183-188.
179. Wong, J.K., Hezareh, M., Günthard, H.F., *et al.*: Recovery of replication-competent HIV despite prolonged suppression of plasma viremia, *Science* **278** (1997), 1291-1295.
180. Finzi, D., Hermankova, M., Pierson, T., *et al.*: Identification of a reservoir for HIV-1 in patients on highly active antiretroviral therapy, *Science* **278** (1997), 1295-1300.
181. Chun, T.-W., Stuyver, L., Mizell, S.B., *et al.*: Presence of an inducible HIV-1 latent reservoir during highly active antiretroviral therapy, *Proc. Natl. Acad. Sci. USA* **94** (1997), 13193-13197.
182. Finzi, D., Blankson, J., Siliciano, J.D., *et al.*: Latent infection of $CD4^+$ cells provides a mechanism for lifelong persistence of HIV-1, even in patients on effective combination therapy, *Nature Med.* **5** (1999), 512-517.
183. Chun, T.W., Engel, D., Mizell, S.B., Ehler, L.A. and Fauci, A.S.: Induction of HIV-1 replication in latently infected $CD4^+$ T cells using a combination of cytokines, *J. Exp. Med.* **188** (1998), 83-91.
184. Havlir, D.V., Marschner, I.C., Hirsch, M.S., *et al.*: Maintenance antiretroviral therapies in HIV-infected subjects with undetectable plasma HIV RNA after triple drug therapy, *New Engl. J. Med.* **339** (1998), 1261-1268.
185. Palioux, G., Raffi, F., Brun-Vezinet, F., *et al.*: A randomized trial of three maintenance regimens given after three months of induction therapy with zidovudine, lamivudine, and indinavir in previously untreated HIV-1-infected patients, *New Engl. J. Med.* **339** (1998), 1269-1276.
186. Kaufmann, D., Pantaleo, G., Sudre, P. and Telenti, A., for the Swiss HIV Cohort Study: CD4-cell count in HIV-1-infected individuals remaining viraemic with highly active antiretroviral therapy (HAART), *Lancet* **351** (1998), 723-724.
187. Deeks, S., Hoh, R., Hanley, M.B., *et al.*: T-cell turnover kinetics in patients with a sustained CD4 cell response after experiencing virologic failure of a protease inhibitor-based regimen. In: *6th Conference on Retroviruses and Opportunistic Infections*, Chicago, January 31-February 4, 1999 [abstract LB2].
188. Stoddart, C., Mammano, F., Moreno, M., *et al.*: Lack of fitness of protease inhibitor-resistant HIV-1 *in vivo*. In: *6th Conference on Retroviruses and Opportunistic Infections*, Chicago, January 31-February 4, 1999 [abstract 4].
189. Hoetelmans, R.M.W., van Heeswijk, R.P.G., Profijt, M., Mulder, J.W., Meenhorst, P.L., Lange, J.M.A., Reiss, P. and Beijnen, J.H.: Comparison of the plasma pharmacokinetics and renal clearance of didanosine during once and twice daily dosing in HIV-1-infected individuals, *AIDS* **12** (1998), F211-F216.
190. Centers for Disease Control and Prevention: Guidelines for the use of antiretroviral agents in pediatric HIV infection, *Morbid Mortal Wkly Rep.* **47** (1998), RR-4.
191. Palumbo, P.E., Raskino, C., Fiscus, S., *et al.*: Disease progression in HIV-infected infants and children: predictive value of quantitative plasma HIV RNA and CD4 lymphocyte count, *JAMA* **279** (1998), 756-761.
192. Havlir, D. and Richman, D.D.: Zidovudine should be given before HIV-positive individuals develop symtoms, *Rev. Med. Virol.* **4** (1994), 75-80.
193. Ho, D.D.: Time to hit HIV early and hard, *New Engl. J. Med.* **333** (1995), 450-451.
194. Walker, B.D.: HIV infection: the body fights back. In: *6th Conference on Retroviruses and Opportunistic Infections*, Chicago, January 31-February 4, 1999 [abstract L4].
195. Miller, V., Rottmann, C., Hertogs, K., *et al.*: Mega-HAART, resistance and drug holidays. In: *2nd International Workshop on Salvage Therapy for HIV Infection*, Toronto, May 19-21, 1999, *Antiviral Ther.* **4 (suppl. 1)** (1999), 27-28 [abstract 030].
196. Montaner, J.S.G., Harrigan, P.R., Jahnke, N.A., *et al.*: Multidrug rescue therapy for HIV-infected individuals with prior virologic failure to multiple regimens: results from an initial cohort. In: *2nd International Workshop on Salvage Therapy for HIV Infection*, Toronto, May 19-21, 1999, *Antiviral Ther.* **4 (suppl. 1)** (1999), 17 [abstract 015].

197. Kitahata, M.M., Koepsell, T.D. Deyo, R.A., Maxwell, C.L., Dodge, W.T. and Wagner, E.H.: Physician's experience with the acquired immunodeficiency syndrome as a factor in patient's survival, *New Engl. J. Med.* **334** (1996), 701-706.

198. Richman, D.D. and Lange, J.M.A.: Playing with evolution requires planning, *Antiviral Ther.* **1** (1996), 208-209.

CHAPTER 13

PROGNOSTIC MARKERS AND IMMUNOLOGICAL RECONSTITUTION DURING HIV-1 INFECTION

MARIJKE TH.L. ROOS[1], NADINE G. PAKKER[1] AND PETER TH.A. SCHELLEKENS[1,2]

[1]Department of Clinical Viro-Immunology, CLB, Sanquin Blood Supply Foundation, Laboratory for Experimental and Clinical Immunology of the University of Amsterdam, and [2]Department of Internal Medicine, Academic Medical Center; Amsterdam, The Netherlands.

1. Introduction

Although anti-retroviral treatment results in improvement of the clinical condition of AIDS patients and interferes with progression to disease in asymptomatic HIV-1 infected persons, evaluation of efficacy of anti-retroviral therapies using clinical end-points alone is rather difficult. Shift of treatment to earlier asymptomatic stages of HIV-1 infection, especially with the modern potent anti-retroviral drugs and prophylactic treatment for opportunistic infections results in exceedingly long follow-up periods to clinical endpoints. Because of this, markers of progression of HIV-1 infection have to be used as surrogate markers for clinical efficacy. Ideally, these surrogate markers should be both predictive for disease progression in all stages of asymptomatic infection and validated in therapy protocols that use true clinical endpoints in parallel.

For predictive markers to be used for evaluation of anti-retroviral therapy one should primarily focus on markers that are pathogenetically involved in HIV-1 disease induction. Changes in these markers are expected to be of direct clinical relevance since they are causally related with HIV-1-induced immune deficiency or HIV-1-induced pathology. Identification of this type of markers and their prognostic relevance has come from large prospective cohort studies aimed at elucidating the sequence of pathogenetically critical events in HIV-1 infection in progression to AIDS. First, we will briefly summarize studies with respect to prognostic markers for progression to AIDS, and the relevance of some of these markers to determine the extent of immunological reconstitution during anti-retroviral treatment.

H. Schuitemaker and F. Miedema (ed.), Aids Pathogenesis, 243–259.

2. Progression to AIDS and Prognostic Markers

2.1 HIV-1 VARIANTS

A feature of HIV-1 is its great variability with respect to biological properties, such as cytotropism, replication rate and syncytium-inducing (SI) capacity [1]. The HIV-1 biological phenotype is an important determinant in the variable clinical course of the infection. Usually, HIV-1 infection starts with slow replicating macrophage-tropic non-syncytium-inducing (NSI) viruses [2-4]. With progression towards disease, as viral burden increases, the virus population shifts towards more rapid replicating, preferentially T-cell tropic viruses. In 50% of the cases this shift is associated with the emergence of SI HIV-1 variants [5], which can easily be detected employing the MT-2 indicator cells [6].

2.2 VIRAL LOAD

Viral RNA can be detected in plasma or serum and reflects the total number of viral particles, both infectious and non-infectious (defective or neutralised) viruses. Infectious viral load, both the amount of virus RNA in plasma and the number of productively infected cells seem to increase substantially with progression to disease [2,7-11]. Viral load is a strong progression marker for clinical disease especially early in infection [12-15].

2.3 $CD4^+$ T-CELL DEPLETION

The most widely used marker for staging of asymptomatic HIV-1 infection is the $CD4^+$ T-cell count. The predictive value of $CD4^+$ T-cell count for the development of AIDS in HIV-1-infected individuals has been confirmed in numerous studies [16]. However, persons with similar $CD4^+$ T-cell counts may differ in their future rate of $CD4^+$ T-cell decline and risk for clinical progression [17,18]. Evidence has been obtained for differential rates of $CD4^+$ T-cell decline associated with presence of certain HIV-1 biological variants [19-21]. Thus, these findings limit the usefulness of $CD4^+$ T-cell count as a sole criterion to predict both clinical outcome and effect of anti-retroviral therapy. Therefore, application of additional predictive markers is of major importance. Especially after the introduction of HAART, a discussion started on mechanisms to explain this decline in $CD4^+$ T cells.

2.4 LOSS OF NAIVE AND INCREASE OF ACTIVATED $CD8^+$ T CELLS

A consistent feature of HIV-1 infection, besides the progressive loss of $CD4^+$ T lymphocytes, is the increase in $CD8^+$ T lymphocytes. It has been proposed that this is related to a form of blind T-cell homeostasis that is maintained for long periods after

HIV-1 infection, but is lost in the last two years preceding AIDS [22]. T cells can be subdivided in naive cells, *i.e.* cells that never encountered antigen, and memory or effector T cells that did so. Isoforms of CD45 have been used to separate T cells in naive and memory/effector cells. However, these CD45 isoforms (CD45RA and CD45RO) clearly do not have the required specificity to distinguish between naive and memory T cells. One problem is that T cells expressing CD45RO, usually seen as the characteristic of memory/effector cells, can reconvert to "naive" cells expressing CD45RA [23-25] when no longer triggered by antigen. Roederer *et al.* [26] reported that by performing three-color flow cytometry with the addition of CD62L (L-selectin), an adhesion marker on the cell surface, identification of naive and memory cells is possible. $CD45RA^+$ T cells that express CD62L are regarded as truly naive T cells, whereas T cells expressing CD45RO, and T cells that express CD45RA but lack CD62L are regarded as memory lymphocytes. Recent detailed phenotypic and functional studies investigating CD45RA expression in combination with the expression of the co-stimulatory molecule CD27 showed that naive T cells are $CD45RA^+CD27^+$, while memory and effector T cells lack either CD45RA and/or CD27 expression [27,28]. A selective loss of naive $CD8^+$ T cells during progression of HIV-1 infection has been reported [26], implying a concomitant loss in the ability to mount immune responses to new antigens, including the continually mutating HIV-1 itself.

In HIV-1-infected individuals, many signs of chronic immune activation can be recognized [29]. Expansion of CD8 populations is thought to result from immune activation, although this does not seem effective in preventing disease progression. Especially the expression of the activation marker CD38 on $CD8^+$ cells has been studied. The percentage $CD8^+$ $D38^+$ cells predicts progression to disease independently of the $CD4^+$ T-cell counts [30]. Giorgi and coworkers compared percentage and relative fluorescence intensity measurements of CD38 antigen expression on $CD8^+$ lymphocytes. The prognostic value of elevated mean fluorescence intensity of CD38 expression on $CD8^+$ cells is almost identical to the prognostic value of the percentage of $CD8^+$ cells that are positive for expression of CD38 [31,32].

Besides increased numbers of $CD8^+$ T cells expressing activation markers, also $CD8^+$ T cells that lack CD28 expression have been described [33-36]. Gruters *et al.* [33] demonstrated in persons studied from seroconversion to AIDS, that the appearance of $CD8^+CD28^-$ T cells parallels the emergence of cells expressing the CD38 marker during HIV-1 infection.

2.5 OTHER ACTIVATION MARKERS

Activation of monocytes and elevated serum levels of neopterin, β_2-microglobulin (β_2-M), soluble IL-2, CD8 and tumor-necrosis factor (TNF) receptors are also indicative for an activated immune system [37-43]. This might reflect the continuous attempt of the immune system to fight the viral infection. The persistence of HIV-1 and HIV-1 replication throughout the infection might result in the maintenance of this state of immune activation. These markers therefore may be related to HIV-1 persistence and not to progression and do not measure residual immune capacity as such. They were shown to have strong predictive value in many untreated study populations independent of $CD4^+$ T-cell counts [38,43-45], but not in a multivariate model with $CD4^+$ T-cell counts, HIV-1 RNA load and p24 antigen as covariates [46].

2.6 T-CELL PROLIFERATIVE CAPACITY

Already before the dramatic decline of $CD4^+$ T-helper cells, immunological abnormalities, including disturbed T-cell, B-cell and antigen-presenting cell (APC) functions, are observed. The presence of predominantly macrophage-tropic HIV-1 variants in asymptomatic infection is compatible with the hypothesis that changes in macrophage/ APC function might underlie the early immunological abnormalities in HIV-1 infection.

Once it was shown that loss of *in-vitro* T-cell reactivity may precede $CD4^+$ T-cell decline, this additional parameter for immune status was analyzed as a predictor for progression. Hofmann *et al.* [47] showed that, next to decreased $CD4^+$ T-cell count, low responsiveness to pokeweed mitogen (PWM), a polyclonal T-cell activator, has predictive value for progression to AIDS. Shearer and collaborators showed in HIV-1-infected asymptomatic men, in contrast to normal responsiveness to allo-antigens (ALLO), a selective loss of T-cell reactivity with respect to proliferation, IL-2 production and CTL generation in response to recall antigens presented by autologous MHC (self-MHC) molecules [48-50]. They have obtained evidence that there is a sequential loss of responses to recall antigen, allo-antigen and PHA, respectively, in the course of HIV-1 infection [51].

Antigens are presented to the T-cell receptor (TCR) in the context of molecules of the HLA system. The signal received by the TCR is transduced intracellularly via the CD3 complex and will eventually lead to cell proliferation and differentiation. Monoclonal antibodies (mAb) directed to CD3 are mitogenic for T cells and therefore are believed to mimic antigenic stimulation. The activation of T cells via TCR is often called 'signal one'. For optimal activation of T cells, secondary signals are needed as well. These signals also can be provided by molecules present on the membrane of APC [52]. An important molecule capable of transducing these signals to T cells is CD28. Ligation of CD28 by its ligand on APC generates a potent costimulatory signal for IL-2 production and proliferation. Therefore, not only stimulation by CD3 mAb but also by CD3 plus CD28 mAb was investigated.

These measurements required laborious cell-separation techniques for purification of mononuclear cells from peripheral blood (PBMC), unpractical for routine testing of large numbers of samples. The development of a whole-blood culture system for T-cell proliferative capacity enabled us to perform T-cell reactivity tests in large groups routinely. Using this system it was shown that, shortly after HIV-1 seroconversion, T-cell reactivity to CD3 mAb, expressed as reactivity per 10^3 $CD3^+$ T lymphocytes, is decreased to 20% of the normal values, returning to 60% within three months [53]. Independently from low $CD4^+$ T-cell numbers and the presence of SI HIV-1 variants, low T-cell responses to CD3 mAb turns out to be a strong prognostic marker for development of AIDS [53,54]. Because CD28 mAb co-stimulation considerably enhances CD3 mAb-induced T-cell reactivity and is accompanied by notably reduced coefficients of variation [55], the predictive value for development of AIDS of the T-cell proliferative capacity to the combination of these mAb was measured in subsequent studies. Both at baseline and in a time-dependent proportional hazards analysis, T-cell reactivity to CD3 mAb in the presence of CD28 mAb is a stronger marker for progression to AIDS than T-cell reactivity to CD3 mAb alone [55,56].

2.7 COMBINED VIROLOGICAL AND IMMUNOLOGICAL APPROACHES

Serologic assays that detect p24 antibodies or the core antigen p24 itself can define the risk for progression to serious clinical disease, but in a multivariate analysis these markers do not add independently to the information provided by the $CD4^+$ T-lymphocyte count [16,57].

Previous studies reported combined testing of immunological and virological markers and showed that early in infection a high HIV-1 RNA level in serum is the only predictive marker of disease progression [15,58]. Later in infection, low $CD4^+$ T-cell counts and subsequently decreased T-cell function gain predictive importance, whereas a high HIV-1 RNA level becomes less important [15,58]. In recent studies, T-cell responses were measured to the combination of CD3 and CD28 mAb. Combined testing of T cell reactivity with $CD4^+$ T-cell numbers, HIV-1 phenotype and viral load in a proportional hazards analysis, both in a baseline and longitudinal model, demonstrated that all these markers are independently predictive for progression at baseline. In a time-dependent analysis, $CD4^+$ T cells and T-cell reactivity were still predictive, but viral load had then lost its predictive value [56]. This reciprocal behaviour of immunological and viral prognostic markers over time is consistent with ongoing viral replication leading to $CD4^+$ T-cell depletion and functional impairment of T cells, which finally renders the patient vulnerable to opportunistic infections and HIV-1-related malignancies.

3. Immunological Reconstitution During Anti-retroviral Therapy

During the last few years new strategies to treat early- and late-stage HIV-1-infected persons are predominated by different combinations of (non)-nucleoside and protease inhibitors. Since this highly active anti-retroviral therapy (HAART) became available, it has given clinicians the opportunity to treat patients more effectively than could be done before with treatment based on a single drug [59-63].

In parallel with these new treatment strategies, it has become of even more interest to evaluate optimal surrogate markers. These markers will be needed for monitoring patients during treatment to predict their (clinical) outcome, the degree of immune reconstitution and to assess whether the immune system is sufficiently restored to end anti-retroviral treatment or to switch to maintenance therapy [64,65]. The main and most obvious responses on potent therapy, and therefore the most likely candidates as parameters for monitoring patients, are a declining viral load and an increase in the number of $CD4^+$ T cells [66,67].

3.1 VIRAL LOAD

The decrease in viral load as caused by monotherapy generally appears to be transient but can be prolonged in case different combinations of anti-retrovirals with or without protease inhibitors are being used [68-72]. Potent therapy even made it possible to reduce the number of viral RNA copies in the blood to 'unquantifiable' levels, although current achievements to improve virologcal assays have lowered the quantification limit from 400 to 20 copies/ml and extended the debate on the definition of 'undetectable HIV RNA levels'. Besides the blood compartment, reduction of HIV RNA levels has also been described in other compartments, like lymphoid tissues and cerebrospinal fluid [71,73-75]. However, several reports showed that even when plasma HIV RNA could not be detected using an ultrasensitive assay, HIV could still be detected in lymphoid tissue or cultured from PBMC [76-79]. Also, proviral DNA remained detectable in patients with significantly reduced viral RNA load [74,80]. Therefore, it has to be concluded that thusfar total eradication of the HIV-1 virus has not yet been realised.

Possibly, because of highly unknown factors associated with the compartmentalisation of the virus, and because discrimination is impossible as soon as viral load reaches unquantifiable levels, HIV RNA is useful but lacks the ability to function as the ideal surrogate marker to monitor disease progression and assess the effect of anti-retroviral therapy [81-84].

3.2 $CD4^+$ T CELLS AND T-CELL REACTIVITY

Based on observations in patients treated with a protease inhibitor, in 1995 two studies [85,86] provided evidence for a high turnover of $CD4^+$ T cells during HIV-1 infection driven by large amounts of virus that are produced daily, suggesting physical exhaustion of $CD4^+$ T-cell renewal as a cause of $CD4^+$ T-cell depletion. If so, antiviral treatment, despite very efficient control of virus replication, could have it limits as to the restoration of the immune system. The hypothesis of a high daily $CD4^+$ T-cell turnover is based on the assumption that the increase in $CD4^+$ T cells as seen directly after therapy is caused by peripheral expansion of T cells. This daily T-cell production rate is believed to reflect a rate similar as before treatment, however, since the viral pressure has been diminished lymphocytes are now rescued from virus induced death. However, the number of $CD4^+$ T cells as measured in the peripheral blood represents only a fraction of total body $CD4^+$ T cells, estimated to be on the order of two percent. Hence, the origin of $CD4^+$ T cells reappearing in the blood as part of immune reconstitution during potent anti-retroviral therapy is still controversial. We have shown that the initial $CD4^+$ T-cell recovery may result from redistribution of cells from the lymphoid tissue to the peripheral blood, rather than generation of T cells via a thymic or extra thymic pathway [87-89].

It has been reported that the response in $CD4^+$ T cells does not always provide optimal estimates for therapeutic benefit [82-84]. Other variables, such as T-cell function and the degree of activation of the immune system, might also be used as surrogate markers. One of the first detailed descriptions of immunological

changes during potent therapy was provided by Kelleher *et al.*[90], who studied alterations in T-lymphocyte subsets and proliferative responses to antigens *in vitro*. They studied 21 patients, who were treated with the protease inhibitor ritonavir for 16 weeks and found an immediate increase in $CD4^+CD45RO^+$ cells and a delayed increase in $CD4^+CD45RA^+$ cells. The number of $CD4^+$ and $CD8^+$ T cells expressing CD38 declined and the proliferative capacity to PHA and recall antigens raised. The latter was also seen in a study describing the pattern of T-cell repopulation, viral load reduction and T-cell function in patients treated with different monotherapies [91]. In general, it appeared that improvement of T-cell function co-incided with a rise of $CD4^+$ T cells after treatment. However, in the study group treated with the protease inhibitor ritonavir, the rise in T-cell function as measured by proliferation after stimulation with CD3 mAb was only temporal and was inversely correlated with the reduction in viral load.

3.3 T-CELL SUBPOPULATIONS

One way to address the question about the origin of reappearing T cells after therapy is to see whether these T cells have been challenged with antigens before, *i.e.* whether these cells are memory or naive cells. Adapting the already described method for subtyping $CD4^+$ and $CD8^+$ T-cell populations in naive and memory cells, it was demonstrated by Autran *et al.* [92], by Pakker *et al.* [89] and more recently by Lederman *et al.* [93] that during triple combination treatment repopulation of $CD4^+$ T cells has a biphasic pattern, and is distinct for memory and naive cells. The immediate rapid but temporary increase in memory $CD4^+$ and $CD8^+$ T cells in the first 3-6 weeks was calculated to be on the order of 10^9 cells/day. In addition, $CD4^+$ and $CD8^+$ naive T cells increase continuously but slowly at a rate on the order of 10^8 cells per day, similar to what has been reported in other groups of patients with severe immunodeficiency [94-97]. Mathematical modelling showed that the rapid $CD4^+$ T-cell increase can be explained by either expansion or redistribution of memory cells. Redistribution, however, could explain the striking correlation between the initial $CD4^+$ and $CD8^+$ memory T-cell repopulation in the blood since it is biologically plausible that as soon as the antigenic challenge diminishes both the trapped $CD4^+$ and $CD8^+$ T cells flow into the periphery. The increase in T cells following antiviral therapy, thus, most likely is a composite of an initial fast redistribution of memory cells to the blood, accompanied by a continuous slow repopulation with newly produced naive T cells.

Reported $CD8^+$ T-cell responses to different drug regimens seem to be controversial. Until now, it remains unclear whether the observed pattern in $CD8^+$ T-cell numbers after administration of anti-retroviral therapy is dependent on the drug regimen used, the specific presence of ritonavir, the magnitude of antiviral effect, the degree of viral suppression, the base line level $CD8^+$ T cells or the stage of HIV infection. Clearly, lymphocyte activation markers, expressed on the surface of $CD8^+$ T cells, and TNF-α levels diminish during HAART [92,93,98,99].

Several studies have reported disturbance of haematopoiesis due to HIV infection of bone marrow precursor cells ($CD34^+$ cells) or because of indirect mechanisms affecting T-cell renewal [100-103]. A recent study, analysing T-cell development capacity in fetal thymus organ culture (FTOC), showed that those patients who progress to AIDS have an early and dramatic loss of T-cell development capacity compared to long-term asymptomatic patients, and improvement of T-cell development during HAART correlated with the increase in the number of naive $CD4^+$ T cells [104]. It has been described that dysfunction of the thymus progresses with age [96,105], and although disruption of thymopoieses might be reversible [106], the remaining thymic function at the start of treatment with HAART could determine naive T-cell repopulation. In HIV-infected infants and children treated with triple-combination therapy it was shown that potent therapy could induce significant immunological improvement, though the degree of immunological recovery was not related to age and was not more rapid than in adults. The number of naive $CD4^+$ and $CD8^+$ T cells at the start of study and baseline HIV RNA levels were predictive for naive T-lymphocyte repopulation, and a strong correlation was observed between naive $CD4^+$ and $CD8^+$ T-cell recovery rates (Scherpbier *et al.* submitted).

Walker *et al.* [107] argued that the recovery of naive cells during HAART does not reflect thymic-derived unprimed cells though mainly exists of reverted memory cells. This is based on the occurrence of reverted naive type (CD45RA) cells containing the neomycin phosphotransferase marker gene (NPT), after infusion of mainly memory (CD45RO) NPT-positive T cells. However, the repopulating cells during HAART that were called naive cells co-expressed either CD62L or CD27, and therefore possess not only phenotypical but also functional characteristics of genuinely naive cells [26,27].

3.4 T-CELL RECEPTOR REPERTOIRE

Immunological changes during the natural course of HIV infection do not only involve numeric and functional failure, but also distortion of the antigen-receptor repertoire, *i.e.* the variable region of the TCR ß-chain [108-110]. A combination of these factors probably gives rise to the manifestation of opportunistic infections and eventually causes clinical disease. The ultimate goal for therapy will be to restore all facets of the immune system. A study by Connors *et al.* [111] showed that we should not be too optimistic about the immune restorative effect of HAART. In combination with IL-2 , that invokes an even more pronounced increase in $CD4^+$ T cells than HAART alone, hardly any recovery from the distorted patterns of the TCR Vß family was observed. It might be argued that, due to the very slow increase of naive T cells alterations of Vß subfamilies can only be expected after a long-term of treat-ment. Data on the specific restoration of the T-cell repertoire in separate naive and memory $CD4^+$ subpopulations during HAART over a period of 40 weeks support this theory. In the first eight weeks, during the initial steep memory $CD4^+$ T-cell increase, an increase of the distorted Vß patterns was demonstrated. This could either be due to an oligoclonal expansion of these cells, which would argue against redistribution, or migration out of the lymphoid tissues of specific memory $CD4^+$ T cells with limited TCR diversity, which favours redistribution. At later time-points, it has been reported that an expansion of TCR diversity occurs [112,113]. Recently, in nine patients treated with a double or triple combination of reverse transcriptase inhibitors, who had maintained 'undetectable' viral RNA load levels (<20 copies/ml) for at least two years, long-term naive and memory $CD4^+$ and $CD8^+$ T-cell recovery was studied [72]. As in these patients naive T-cell recovery does not per definition occur, it will be of major importance to investigate their TCR diversity patterns.

3.5 DISCUSSION

Obviously the long-term and ultimate goal of HAART is complete eradication of HIV paralleled by reconstitution of all facets of the immune system. One of the short-term and more feasible goals of therapy is to minimise the risk of opportunistic infections for patients. Partial immune restoration during HAART has been observed and provides evidence that anti-retroviral treatment improves functional host defenses [89,90,92,93,114-116]. Sustained administration of protease-containing regimens has been associated with a diminished frequency of major opportunistic infections and death [117,118]. However, in some case reports reappearance of viral infections, like CMV, hepatitis and candidiasis during HAART has been described despite dramatic increases in $CD4^+$ T cells [119-121]. The question arises at what time during HAART prophylactic therapy for opportunistic infections can be safely discontinued, *i.e.* whether the magnitude of immune reconstitution is sufficient to ensure that regimens used for prophylaxis of opportunistic infections can safely be withdrawn as $CD4^+$ T-cell counts rise above levels commonly used to begin prophylaxis [122]. Evidence exists that during the natural course of HIV infection the occurrence of *Pneumocystis carinii* pneumonia (PCP) can be better predicted on both $CD4^+$ T-cell numbers and the level of T-cell function than on $CD4^+$ T-cell numbers alone (unpublished observation). It will be of extreme interest to investigate whether the decision to withdraw prophylaxis during HAART should be dependent on the degree of quantitative or qualitative T-cell regeneration, the ratio between naive and memory subsets or other lymphocyte subpopulations, changes in the Vß repertoire or a combination of restoration of these immunological factors. It might even be that the immune system maintains a certain redundancy and reserve so that subnormal recovery of the immune system provides enough protective immunity to-wards opportunistic events.

3.6 CONCLUSION

Based on the observations during the natural course of HIV-1 infection and the response on anti-retroviral therapy, monitoring should, besides frequent measurements of HIV-1 RNA levels and CD4$^+$ T-cell counts, include functional lymphocyte tests and naive CD4+ T cells counts to get a more detailed insight in immune restorattion .

4. References

1. Miedema, F., Tersmette, M. and Van Lier, R.A.W.: AIDS pathogenesis: a dynamic interaction between HIV and the immune system, *Immunol. Today* **11** (1990), 293-297.
2. Schuitemaker, H., Koot, M., Kootstra, N.A., Dercksen, M.W., De Goede, R. E.Y., Van Steenwijk, R.P., Lange, J.M.A., Eeftink Schattenkerk, J.K.M., Miedema, F. and Tersmette, M.: Biological phenotype of human immunodeficiency virus type 1 clones at different stages of infection: progression of disease is associated with a shift from monocytotropic to T-cell-tropic virus populations, *J. Virol.* **66** (1992), 1354-1360.
3. Zhu, T., Mo, H., Wang, N., Nam, D.S., Cao, Y., Koup, R.A. and Ho, D.D.: Genotypic and phenotypic characterization of HIV-1 in patients with primary infection, *Science* **261** (1993), 1179-1181.
4. Van 't Wout, A.B., Kootstra, N.A., Mulder-Kampinga, G.A., Albrecht-van Lent, N., Scherpbier, H.J., Veenstra, J., Boer, K., Coutinho, R.A., Miedema, F. and Schuitemaker, H.: Macrophage-tropic variants initiate human immunodeficiency virus type 1 infection after sexual, parenteral and vertical transmission, *J. Clin. Invest.* **94** (1994), 2060-2067.
5. Koot, M., Keet, I.P.M., Vos, A.H.V., De Goede, R.E.Y., Roos, M.Th.L., Coutinho, R.A., Miedema, F., Schellekens, P.Th.A. and Tersmette, M.: Prognostic value of human immunodeficiency virus type 1 biological phenotype for rate of CD4$^+$ cell depletion and progression to AIDS, *Ann. Intern. Med.* **118** (1993), 681-688.
6. Koot, M., Vos, A.H.V., Keet, I.P.M., De Goede, R.E.Y., Dercksen, W., Terpstra, F.G., Coutinho, R.A., Miedema, F. and Tersmette, M.: HIV-1 biological phenotype in long-term infected individuals, evaluated with an MT-2 cocultivation assay, *AIDS* **6** (1992), 49-54.
7. Ho, D.D., Moudgil, T. and Alam, M.: Quantitation of human immunodeficiency virus type 1 in the blood of infected persons, *New Engl. J. Med.* **321** (198), 1621-1625.
8. Connor, R.I., Mohri, H., Cao, Y. and Ho, D.D.: Increased viral burden and cytopathicity correlate temporally with CD4$^+$ T-lymphocyte decline and clinical progression in human immunodeficiency virus type-1-infected individuals, *J. Virol.* **67** (1993), 1772-1777.
9. Piatak Jr., M., Saag, M.S., Yang, L.C., Clark, S.J., Kappes, J.C., Luk, K.-C., Hahn, B.H., Shaw, G.M. and Lifson, J.D.: High levels of HIV-1 in plasma during all stages of infection determined by competitive PCR, *Science* **259** (1993), 1749-1754.
10. Jurriaans, S., van Gemen, B., Weverling, G.J., Van Strijp, D., Nara, P., Coutinho, R.A., Koot, M., Schuitemaker, H. and Goudsmit, J.: The natural history of HIV-1 infection: virus load and virus phenotype-independent determinants of clinical course?, *Virology* **204** (1994), 223-233.
11. Koot, M., Van 't Wout, A.B., Kootstra, N.A., De Goede, R.E.Y., Tersmette, M. and Schuitemaker, H.: Relation between changes in cellular load, evolution of viral phenotype, and the clonal composition of virus populations in the course of human immunodeficiency virus type 1 infection, *J. Infect. Dis.* **173** (1996), 349-354.
12. Mellors, J.W., Kingsley, L., Rinaldo Jr., C.R., Todd, J., Hoo, B.S., Kokka, R. P. and Gupta, P.: Quantitation of HIV-1 RNA in plasma predicts outcome after seroconversion, *Ann. Intern. Med.* **122** (1995), 573-579.
13. Henrard, D.R., Phillips, J.F., Muenz, L.R., Blattner, W.A., Wiesner, D., Eyster, E. and Goedert, J.J.: Natural history of HIV-1 cell-free viremia, *JAMA* **274** (1995), 554-558.
14. Mellors, J.W., Rinaldo Jr., C.R., Gupta, P., White, R.M., Todd, J.A. and Kingsley, L.A.: Prognosis in HIV-1 infection predicted by the quantity of virus in plasma, *Science* **272** (1996), 1167-1170.
15. De Wolf, F., Spijkerman, I., Schellekens, P.Th.A., Langendam, M., Kuiken, C.L., Bakker, M.,

Roos, M.Th.L., Coutinho, R.A., Miedema, F. and Goudsmit, J.: AIDS prognosis based on HIV-1 RNA, CD4$^+$ T-cell count and function: Markers with reciprocal predictive value over time after seroconversion, *AIDS* **11** (1997), 1799-1806.

16. Lange, J.M.A., De Wolf, F. and Goudsmit, J.: Markers for progression in HIV infection, *AIDS* **3 (suppl. 1)** (1990), 153-160.
17. Eyster, M.E., Gail, M.H., Ballard, J.O., Al-Mondiry, H. and Goedert, J.J.: Natural history of human immunodeficiency virus infections in hemophiliacs: Effects of T-cell subsets, platelet counts, and age, *Ann. Intern. Med.* **107** (1987), 1-6.
18. Kaplan, J.E., Spira, T.J., Fishbein, D.B., Bozeman, L.H., Pinsky, P.F. and Schonberger, L.B.: A six-year follow-up of HIV-infected homosexual men with lymphadenopathy, *JAMA* **260** (1988), 2694-2697.
19. Tersmette, M., Lange, J.M.A., De Goede, R.E.Y., De Wolf, F., Eeftink Schattenkerk, J.K.M., Schellekens, P.ThA., Coutinho, R.A., Huisman, J.G., Goudsmit, J. and Miedema, F.: Association between biological properties of human immunodeficiency virus variants and risk for AIDS and AIDS mortality, *Lancet* **i** (1989), 983-985.
20. Schellekens, P.Th.A., Tersmette, M., Roos, M.Th.L., Keet, I.P.M., De Wolf, F., Coutinho, R.A. and Miedema, F.: Biphasic rate of CD4$^+$ cell decline during progression to AIDS correlates with HIV-1 phenotype, *AIDS* **6** (1992), 665-669.
21. Koot, M., Schellekens, P.Th.A., Mulder, J.W., Lange, J.M.A., Roos, M.Th. L., Coutinho, R.A., Tersmette, M. and Miedema, F.: Viral phenotype and T-cell reactivity in human immunodeficiency virus type 1-infected asymptomatic men treated with zidovudine, *J. Infect. Dis.* **168** (1993), 733-736.
22. Margolick, J.B., Muñoz, A., Donnenberg, A.D., Park, L.P., Galai, N., Giorgi, J.V., O'Gorman, M.R.G., Ferbas, J. and for the Multicenter AIDS Cohort Study: Failure of T-cell homeostasis preceding AIDS in HIV-1 infection, *Nature Med.* **1** (1995), 674-680.
23. Bell, E.B. and Sparshott, S. M.: Interconversion of CD45R subsets of CD4 T cells *in vivo*, *Nature* **348** (1990), 163-166.
24. Bunce, C. and Bell, E.B.: CD45RC isoforms define two types of CD4 memory T cells, one of which depends on persisting antigen, *J. Exp. Med.* **185** (1997), 767-776.
25. Sparshott, S.M. and Bell, E.B.: Membrane CD45R isoform exchange on CD4 T cells is rapid, frequent and dynamic *in vivo*, *Eur. J. Immunol.* **24** (1994), 2573-2573.
26. Roederer, M., Gregson Dubs, J., Anderson, M.T., Raju, P.A., Herzenberg, L. A. and Herzenberg, L.: CD8 naive T cell counts decrease progressively in HIV-infected adults, *J. Clin. Invest.* **95** (1995), 2061-2066.
27. Hamann, D., Baars, P.A., Rep, M.H.G., Hooibrink, B., Kerkhof-Garde, S.R., Klein, M.R. and Van Lier, R.A.W.: Phenotypic and functional separation of memory and effector Human CD8pos T cells, *J. Exp. Med.* **186** (1997), 1407-1418.
28. Baars, P.A., Maurice, M.M., Rep, M.H.G., Hooibrink, B. and Van Lier, R.A.W.: Heterogeneity of the circulating human CD4$^+$ T-cell population: Further evidence that the CD4$^+$CD45RA$^-$ CD27$^-$ T-cell subset contains specialized primed cells, *J. Immunol.* **154** (1994), 17-25.
29. Fauci, A.S.: Multifactorial nature of human immunodeficiency virus disease: Implications for therapy, *Science* **262** (1993), 1011-1018.
30. Mocroft, A., Bofill, M., Lipman, M., Medina, E., Borthwick, N., Timms, A., Batista, L., Winter, M., Sabin, C.A., Johnson, M., Lee, C.A., Phillips, A.N. and Janossy, G.: CD8$^+$,CD38$^+$ lymphocyte percent: a useful immunological marker for monitoring HIV-1-infected patients, *J. AIDS and Human Retrovirol.* **14** (1997), 158-162.
31. Giorgi, J.V., Liu, Z., Hultin, L.E., Cumberland, W.G., Hennessey, K. and Detels, R.: Elevated levels of CD38$^+$ CD8$^+$ T cells in HIV infection add to the prognostic value of low CD4$^+$ T-cell levels: results of 6 years of follow-up. The Los Angeles Center, Multicenter AIDS Cohort Study, *J. Acq. Immune Def. Syndr. Human Retrovir.* **6** (1993), 904-912.
32. Liu, Z., Hultin, L.E., Cumberland, W.G., Hultin, P., Schmid, I., Matud, J.L., Detels, R. and Giorgi, J.V.: Elevated relative fluorescence intensity of CD38 antigen expression on CD8$^+$ T cells is a marker of poor prognosis in HIV infection: results of 6 years of follow-up, *Cytometry* **26** (1996), 1-7.
33. Gruters, R.A., Terpstra, F.G., De Jong, R., Van Noesel, C.J.M., Van Lier, R.A.W. and

Miedema, F.: Selective loss of T-cell functions in different stages of HIV infection, *Eur. J. Immunol.* **20** (1990), 1039-1044.

34. Borthwick, N.J., Bofill, M., Gombert, W.M., Akbar, A.N., Medina, E., Sagawa, K., Lipman, M.C., Johnson, M.A. and Janossy, G.: Lymphocyte activation in HIV-1 infection. II. Functional defects of CD28$^-$ T cells, *AIDS* **8** (1994), 431-441.
35. Brinchmann, J.E., Dobloug, J.H., Heger, B.H., Haaheim, L.L., Sannes, M. and Egeland, T.: Expression of costimulatory molecule CD28 on T cells in human immunodeficiency virus type 1 infection: functional and clinical correlations, *J. Infect. Dis.* **169** (1994), 730-738.
36. Vingerhoets, J.H., Vanham, G.L., Kestens, L.L., Penne, G.G., Colebunders, R.L., Vandenbruaene, M.J., Goeman, J., Gigase, P.L., De Boer, M. and Ceuppens, J.L.: Increased cytolytic T lymphocyte activity and decreased B7 responsiveness are associated with CD28 down-regulation on CD8$^+$ T cells from HIV-infected subjects, *Clin. Exp. Immunol.* **100** (1995), 425-433.
37. Fuchs, D., Shearer, G.M., Boswell, R.N., Clerici, M., Reibnegger, G., Werner, E.R., Zajac, R.A. and Wachter, H.: Increased serum neopterin in patients with HIV-1 infection is correlated with reduced *in-vitro* interleukin-2 production, *Clin. Exp. Immunol.* **80** (1990), 44-48.
38. Fahey, J.L., Taylor, J.M.G., Detels, R., Hofmann, B., Melmed, R., Nishanian, P. and Giorgi, J.V.: The prognostic value of cellular and serologic markers in infection with human immunodeficiency virus type 1, *New Engl. J. Med.* **322** (1990), 166-172.
39. Osmond, D.H., Shiboski, S., Bacchetti, P., Winger, E.E. and Moss, A.R.: Immune activation markers and AIDS prognosis, *AIDS* **5** (1991), 505-511.
40. Lifson, A R., Hessol, N.A., Buchbinder, S.P., O'Malley, P.M., Barnhart, L., Segal, M., Katz, M.H. and Holmberg, S.D.: Serum b_2-microglobulin and prediction of progression to AIDS in HIV infection, *Lancet* **339** (1992), 1436-1440.
41. Allen, J.B., McCartney-Francis, N., Smith, P.D., Simon, G., Gartner, S., Wahl, L.M., Popovic, M. and Wahl, S.M.: Expression of interleukin 2 receptors by monocytes from patients with acquired immunodeficiency syndrome and induction of monocyte interleukin 2 receptors by human immunodeficiency virus *in vitro*, *J. Clin. Invest.* **85** (1990), 192-199.
42. Walker, C.M. and Levy, J.A.: A diffusible lymphokine produced by CD8$^+$ T lymphocytes suppresses HIV replication, *Immunology* **66** (1989), 628-630.
43. Godfried, M.H., Van der Pol, T., Weverling, G.J., Mulder, J.W., Jansen, J., Van Deventer, S.J.H. and Sauerwein, H.P.: Soluble receptors for tumour necrosis factor as predictors of progression to AIDS in asymptomatic human immunodeficiency virus type 1 infection, *J. Infect. Dis.* **169** (1994), 739-745.
44. Moss, A.R., Bachetti, P., Osmond, D., Krampf, W., Chaisson, R.E., Stites, D., Wilber, J., Allain, J.P. and Carlson, J.: Seropositivity of HIV and the development of AIDS or AIDS-related conditions: Three year follow-up of the San Francisco General Hospital cohort, *Brit. Med. J.* **296** (1988), 745-750.
45. Melmed, R.N., Taylor, J M.G., Detels, R., Bozorgmehri, M. and Fahey, J.L.: Serum Neopterin Changes in HIV-infected Subjects: Indicator of Significant Pathology, CD4 T cell changes, and the development of AIDS, *J. Acq. Immune Def. Syndr. Human Retrovir.* **2** (1989), 70-76.
46. Saksela, K., Stevens, C.E., Rubinstein, P., Taylor, P.E. and Baltimore, D.: HIV-1 messenger RNA in peripheral blood mononuclear cells as an early marker of risk for progression to AIDS, *Ann. Intern. Med.* **123** (1995), 641-648.
47. Hofman, B., Bygbjerg, I. and Dickmeiss, E.: Prognostic value of immunologic abnormalities and HIV antigenemia in asymptomatic HIV-infected individuals: proposal of immunologic staging, *Scan. J. Infect. Dis.* **21** (1991), 633-643.
48. Shearer, G.M., Payne, S.M., Joseph, L.J. and Biddison, W.E.: Functional T lymphocyte immune deficiency in a population of homosexual men who do not exhibit symptoms of acquired immune deficiency syndrome, *J. Clin. Invest.* **74** (1984), 496-506.
49. Shearer, G.M., Salahuddn, S.Z., Markham, P.D., Joseph, L.J., Payne, S.M., Kriebel, P.W., Bernstein, D.C., Biddison, W.E., Sarngadharan, M.G. and Gallo, R.C.: Prospective study of cytotoxic T lymphocyte responses to influenza and antibodies to human T lymphotropic virus-III in homosexual men, *J. Clin. Invest.* **76** (1985), 1699-1704.
50. Shearer, G.M., Bernstein, D.C., Tung, K.S.K., Via, C.S., Redfield, R., Salahuddin, S.Z. and

Gallo, R.C.: A model for the selective loss of major histocompatibility complex self-restricted T cell immune responses during the development of acquired immune deficiency syndrome (AIDS), *J. Immunol.* **137** (1986), 2514-2521.

51. Clerici, M., Stocks, N., Zajac, R.A., Boswell, R.N., Lucey, D.R., Via, C.S. and Shearer, G.M.: Detection of three different patterns of T helper cell dysfunction in asymptomatic, human immundeficiency virus-seropositive patients, *J. Clin. Invest.* **84** (1989), 1892-1899.
52. Mueller, D.L., Jenkins, M.K. and Schwartz, R.H.: Clonal expansion *versus* functional inactivation: A costimulatory signalling pathway determines the outcome of T-cell antigen receptor occupancy, *Ann. Rev. Immunol.* **7** (1989), 445-480.
53. Schellekens, P.Th.A., Roos, M.Th.L., De Wolf, F., Lange, J.M.A. and Miedema, F.: Low T-cell responsiveness to activation via CD3/TCR is a prognostic marker for AIDS in HIV-1-infected men, *J. Clin. Immunol.* **10** (1990), 121-127.
54. Roos, M.Th.L., Lange, J.M.A., De Goede, R.E.Y., Coutinho, R.A., Schellekens, P.Th.A., Miedema, F. and Tersmette, M.: Viral phenotype and immune response in primary human immunodeficiency virus type 1 (HIV-1) infection, *J. Infect. Dis.* **165** (1992), 427-432.
55. Roos, M.Th.L., Miedema, F., Meinesz, A.A.P., De Leeuw, N.A.S.M., Pakker, N.G., Lange, J.M.A., Coutinho, R.A. and Schellekens, P.Th.A.: Low T cell reactivity to combined CD3 plus CD28 stimulation is predictive for progression to AIDS: correlation with decreased CD28 expression, *Clin. Exp. Immunol.* **105** (1996), 409-415.
56. Roos, M.Th.L., Prins, M., Koot, M., De Wolf, F., Bakker, M., Coutinho, R.A., Miedema, F. and Schellekens, P.Th.A.: Low T-cell responses to CD3 plus CD28 monoclonal antibodies are predictive of development of AIDS, *AIDS* **12** (1998), 1745-1751.
57. Farzadegan, H., Chmiel, J., Odaka, N., Ward, L., Poggensee, L., Saah, A. and Phair, J.P.: Association of antibody to human immunodeficiency virus type 1 core protein (p24), $CD4^+$ lymphocyte number and AIDS-free time, *J. Infect. Dis.* **166** (1992), 1217-1222.
58. Spijkerman, I.J.B., Prins, M., Goudsmit, J., Veugelers, P.J., Coutinho, R.A., Miedema, F. and De Wolf, F.: Early and late HIV-1 RNA level and its association with other markers and disease progression in long-term AIDS-free homosexual men, *AIDS* **11** (1997), 1383-1388.
59. Graham, N.M.H., Hoover, D.R., Park, L.P., Stein, D.S., Phair, J.P., Mellors, J.W., Detels, R. and Saah, A.J.: Survival in HIV-infected patients who have received zidovudine: comparison of combination therapy with sequential monotherapy and continued zidovudine monotherapy, *Ann. Intern. Med.* **124** (1996), 1031-1038.
60. Carr, A., Vella, S., De Jong, M.D., Sorice, F., Imrie, A., Boucher, C.A.B., Cooper, D.A. and for the Dutch-Italian-Australian Nevirapine Study Group: A controlled trial of nevirapine plus zido-vudine versus zidovudine alone in p24 antigenaemic HIV-infected patients, *AIDS* **10** (1996), 635-641.
61. Hammer, S.M., Katzenstein, D.A., Hughes, M.D., Gundacker, H., Schooley, R.T., Haubrich, R.H., Henry, W K., Lederman, M.M., Phair, J.P., Niu, M., Hirsch, M.S., Merigan, T.C. and for the AIDS Clinical Trials Group Study 175 Study Team: A trial comparing nucleoside monotherapy with combination therapy in HIV-infected adults with CD4 cell counts from 200 to 500 per cubic millimeter, *New Engl. J. Med.* **335** (1996), 1081-1090.
62. Schooley, R.T., Ramirez-Ronda, C., Lange, J M.A., Cooper, D.A., Lavelle, J., Lefkovits, L., Moore, M., Larder, B.A., St. Clair, M.H., Mulder, J.W., McKinnis, R., Pennington, K.N., Harrigan, P.R., Kinghorn, I., Steel, H. and Rooney, J.F.: Virologic and immunologic benefits of initial therapy with zidovudine and zalcitabine or didanosine compared with zidovudine monotherapy, *J. Infect. Dis.* **173** (1996), 1354-1366.
63. Delta Coordinating Committee Delta: A randomised double-blind controlled trial comparing combinations of zidovudine plus didanosine or zalcitabine with zidovudine alone in HIV-infected individuals, *Lancet* **348** (1996), 283-291.
64. Reijers, M.H.E., Weverling, G.J., Jurriaans, S., Wit, F.W.N.M., Weigel, H., Ten Kate, R W., Mulder, J.W., Frissen, P.H.J., Van Leeuwen, R., Reiss, P., Schuitemaker, H., De Wolf, F. and Lange, J.M.A.: Maintenance therapy after quadruple induction therapy in HIV-1-infected individuals: Amsterdam Duration of Antiretroviral Medication (ADAM) study, *Lancet* **352** (1998), 185-190.
65. Hall, D., Montaner, J.G., Reiss, P., Cooper, D.A., Vella, S., Dohnanyi, C., Myers, M., Lange,

J.M.A. and Conway, B.: Induction-maintenance antiretroviral therapy: proof of concept, *AIDS* **12** (1998), F41-F44.

66. Katzenstein, D.A., Hammer, S.M., Hughes, M.D., Gundacker, H., Jackson, J.B., Fiscus, S., Rasheed, S., Elbeik, T., Reichman, R., Japour, A.J., Merigan, T.C., Hirsch, M.S. and AIDS Clinical Trials Group 175 Virology Study Team: The relation of virologic and immunologic markers to clinical outcomes after nucleoside therapy in HIV-infected adults with 200 to 500 CD4 cells per cubic millimeter, *New Engl. J. Med.* **335** (1996), 1091-1098.

67. D'Aquila, R.T., Hughes, M.D., Johnson, V.A., Fischl, M.A., Sommadossi, J.P., Liou, S., Timpone, J., Myers, M., Basgoz, N., Niu, M., Hirsch, M.S. and the National Institutes of Allergy and Infectious Disease AIDS Clinical Trials Group Protocol 241 Investigators: Nevirapine, zidovudine, and didanosine compared with zidovudine and didanosine in patients with HIV infection, *Ann. Intern. Med.* **124** (1996), 1019-1030.

68. Markowitz, M., Saag, M.S., Powderly, W.G., Hurley, A.M., Hsu, A., Valdes, J.M., Henry, D., Sattler, F., La Marca, A., Leonard, J.M. and Ho, D.D.: A preliminary study of Ritonavir, an inhibitor of HIV-1 protease, to treat HIV-1 infection, *New Engl. J. Med.* **333** (1995), 1534-1540.

69. Danner, S.A., Carr, A., Leonard, J.M., Lehman, L.M., Gudiol, F., Gonzales, J., Raventos, A., Rubio, R., Bouza, E., Pintado, V., Gil Aguado, A., de Lomas, J.G., Delgado, R., Borleffs, J.C.C., Hsu, A., Valdes, J.M., Boucher, C.A.B. and Cooper, D.A.: A short-term study of the safety, pharmacokinetics, and efficacy of Ritonavir, an inhibitor of HIV-1 protease, *New Engl. J. Med.* **333** (1995), 1528-1533.

70. Collier, A.C., Coombs, R.W., Schoenfeld, D.A., Basset, R.L., Timpone, J., Baruch, A., Jones, M., Facey, K., Whitacre, C., McAuliffe, V.J., Friedman, H.M., Merigan, T.C., Reichman, R.C., Hooper, C., Corey, L. and for the AIDS Clinical Trials Group: Treatment of human immunodeficiency infection with saquinavir, zidovudine, and zalcitabine, *New Engl. J. Med.* **334** (1996), 1011-1017.

71. Notermans, D.W., Jurriaans, S., Wolf, F. de., Foudraine, N.A., Jong, J.J. de., Cavert, W., Schuwirth, C.M., Kauffmann, R.H., Meenhorst, P.L., McDade, H., Goodwin, G., Leonard, J.M., Goudsmit, J. and Danner, S.A.: Decrease of HIV-1 RNA levels in lymphoid tissue and peri-pheral blood during treatment with ritonavir, lamivudine and zidovudine, *AIDS* **12** (1998), 167-173.

72. Pakker, N.G., Kroon, E.D.M.B., Roos, M.Th.L., Otto, S.A., Hall, D., Wit, F.W.N.M., Hamann, D., Van der Ende, M.E., Claessen, F.A.P., Kauffmann, R.H., Koopmans, P.P., Kroon, F.P., Ten Napel, C.H.H., Sprenger, H.G., Weigel, H M., Montaner, J.S.G., Lange, J.M A., Reiss, P., Schellekens, P.Th.A. and Miedema, F.: Immune restoration does not invariably occur following long-term HIV-1 suppression during antiretroviral therapy, *AIDS* **13** (1999), 203-212.

73. Cavert, W., Notermans, D.W., Staskus, K., Wietgrefe, S.W., Zupancic, M., Gebhard, K., Henry, K., Zhang, Z., Mills, R., McDade, H., Goudsmit, J., Danner, S.A. and Haase, A.T.: Kinetics of response in lymphoid tissues to antiretroviral therapy of HIV-1 infection, *Science* **276** (1997), 960-964.

74. Wong, J.K., Günthard, H.F., Havlir, D., Zhang, Z., Haase, A.T., Ignacio, C.C., Kwok, S., Emini, E.A. and Richman, D.D.: Reduction of HIV-1 in blood and lymph nodes following potent anti-retroviral therapy and the virologic correlates of treatment failure, *Proc. Natl. Acad. Sci. USA* **94** (1997), 12574-12579.

75. Foudraine, N.A., Hoetelmans, R.M.W., Lange, J.M.A., De Wolf, F., van Benthem, B.H.B., Maas, J.J., Keet, I.P.M. and Portegies, P.: Cerebrospinal-fluid HIV-1 RNA and drug concentrations after treatment with lamivudine plus zidovudine or stavudine, *Lancet* **351** (1998), 1547-1551.

76. Lafeuillade, A., Chollet, L., Hittinger, G., Profizi, N., Costes, O. and Poggi, C.: Residual human immunodeficiency virus type 1 RNA in lymphoid tissue of patients with sustained plasma RNA of <200 copies/ml, *J. Infect. Dis.* **177** (1998), 238.

77. Chun, T.-W., Stuyver, L., Mizzel, S.B., Ehler, L.A., Mican, J.M., Baseler, M., Lloyd, A., Nowak, M. and Fauci, A.S.: Presence of an inducible HIV-1 latent reservoir during highly active antiretroviral therapy, *Proc. Natl. Acad. Sci. USA* **94** (1997), 13193-13197.

78. Wong, J.K., Hezareh, M., Günthard, H.F., Havlir, D., Ignacio, C.C., Spina, C.A. and Richman,

D.D.: Recovery of replication-competent HIV despite prolonged suppression of plasma viremia, *Science* **278** (1997), 1291-1295.

79. Finzi, D., Hermankova, M., Pierson, T., Carruth, L., Buck, C., Chaisson, R.E., Quinn, T.C., Chadwick, K., Margolick, J., Brookmeyer, R., Gallant, J. E., Markowitz, M., Ho, D.D., Richman, D.D. and Siliciano, R.F.: Identification of a reservoir for HIV-1 in patients on highly active antiretroviral therapy, *Science* **278** (1997), 1295-1300.

80. Bruisten, S.M., Reiss, P., van Swieten, P., Loeliger, A.E., Schuurman, R., Boucher, C.A.B., Weverling, G.J. and Huisman, J.G.: Cellular proviral HIV-1 DNA load persists after long-term RT-inhibitor therapy in HIV type-1 infected persons, *AIDS Res. Human Retrovir.* **14** (1998), 1053-1058.

81. Marschner, I.C., Collier, A.C., Coombs, R., D'Aquila, R.T., DeGruttola, V., Fischl, M.A., Hammer, S.M., Hughes, M.D., Johnson, V.A., Katzenstein, D.A., Richman, D., Smeaton, L., Spector, S.A. and Saag, M.S.: Use of changes in plasma levels of human immunodeficiency virus type 1 RNA to assess the clinical benefit of antiretroviral therapy, *J. Infect. Dis.* **177** (1998), 47-53.

82. Choi, S., Lagakos, S.W., Schooley, R.T. and Volberding, P.A.: CD4$^+$ lymphocytes are an incom-plete surrogate marker for clinical progression in persons with asymptomatic HIV infection taking zidovudine, *Ann. Intern. Med.* **118** (1993), 674-680.

83. Concorde coordinating committee: Concorde:MRC/ANRS randomised double-blind controlled trial of immediate and deferred zidovudine in symptom-free HIV infection, *Lancet* **343** (1994), 871-867.

84. DeGruttola, V., Beckett, L.A., Coombs, R.W., Arduino, J.M., Balfour, H.H., Rasheed, S., Hollinger, F.B., Fischl, M.A., Volberding, P. and the AIDS Clinical Trials Group Virology Laboratories: Serum p24 antigen level as an intermediate endpoint in clinical trials of zido-vudine in people infected with human immunodeficiency virus type 1, *J. Infect. Dis.* **169** (1994), 713-721.

85. Wei, X., Ghosh, S.K., Taylor, M.E., Johnson, V.A., Emini, E.A., Deutsch, P., Lifson, J.D., Bonhoeffer, S., Nowak, M.A., Hahn, B.H., Saag, M.S. and Shaw, G.M.: Viral dynamics in human immunodeficiency virus type-1 infection, *Nature* **373** (1995), 117-122.

86. Ho, D.D., Neumann, A.U., Perelson, A S., Chen, W., Leonard, J.M. and Markowitz, M.: Rapid turnover of plasma virions and CD4 lymphocytes in HIV-1 infection, *Nature* **373** (1995), 123-126.

87. Mosier, D.E., Sprent, J., Tough, D., Dimitrov, D.S. and Martin, M.A.: HIV results in the frame: CD4$^+$ cell turnover (scientific correspondence), *Nature* **375** (1995), 193-195.

88. Phillips, A.N., Sabin, C.A., Mocroft, A. and Janossy, G.: HIV results in the frame: Antiviral therapy (scientific correspondence), *Nature* **375** (1995), 195.

89. Pakker, N.G., Notermans, D.W., De Boer, R.J., Roos, M.Th.L., de Wolf, F., Hill, A., Leonard, J.M., Danner, S.A., Miedema, F. and Schellekens, P.Th.A.: Biphasic kinetics of peripheral blood T cells after triple combination therapy in HIV-1 infection: A composite of redistribution and proliferation, *Nature Med.* **4** (1998), 208-214.

90. Kelleher, A.D., Carr, A., Zaunders, J. and Cooper, D.A.: Alterations in the immune response of human immunodeficiency virus (HIV) infected subjects treated with an HIV-specific protease inhibitor, Ritonavir, *J. Infect. Dis.* **173** (1996), 321-329.

91. Pakker, N.G., Roos, M.Th.L., Van Leeuwen, R., De Jong, M.D., Koot, M., Reiss, P., Lange, J.M.A., Miedema, F., Danner, S.A. and Schellekens, P.Th.A.: Patterns of T-cell repopulation, viral load reduction and restoration of T-cell function in human immunodefiency virus-infected persons during therapy with different antiretrovirals, *J. AIDS and Human Retrovirol.* **16** (1997), 318-326.

92. Autran, B., Carcelain, G. and Li, T.S.: Positive effects of combined antiretroviral therapy on CD4$^+$ T-cell homeostasis and function in advanced HIV disease, *Science* **277** (1999), 112-116.

93. Lederman, M.M., Connick, E., Landay, A., Kuritzkes, D.R., Spritzler, J., St. Clair, M.H., Kotzin, B., Fox, L., Heath-Chiozzi, M., Leonard, J., Rousseau, F., Wade, M., D'Arc Roe, J., Martinez, A. and Kessler, H.: Immunologic responses associated with 12 weeks of combination antiretroviral therapy consisting of zidovudine, lamivudine and ritonavir: Results of AIDS Clinical Trials Group protocol 315, *J. Infect. Dis.* **178** (1998), 70-79.

94. Ottinger, H.D., Beelen, D.W., Scheulen, B., Schaefer, U.W. and Gross-Wilde, H.: Improved immune reconstitution after allotransplantation of peripheral blood stem cells instead of bone marrow, *Blood* **88** (1996), 2775-2779.
95. Heitger, A., Neu, N., Kern, H., Panzer-Grümayer, E.-R., Greinix, H., Nachbaur, D., Niederwieser, D. and Fink, F.M.: Essential role of the thymus to reconstitute naive (CD45RA+) T-helper cells after human allogeneic bone marrow transplantation, *Blood* **90** (1997), 850-857.
96. Mackall, C.L., Fleisher, T.A., Brown, M., Andrich, M.P., Chen, C., Feuerstein, I.M., Horowitz, M.E., Magrath, I.T., Shad, A.T., Steinberg, S.M., Wexler, L.H. and Gress, R.E.: Age, thymopoiesis, and CD4+ T-lymphocyte regeneration after intensive chemotherapy, *New Engl. J. Med.* **332** (1995), 143-149.
97. Rep, M., Van Oosten, B.W., Roos, M.Th.L., Adèr, H.J., Polman, C.H. and Van Lier, R.A.W.: Treatment with depleting CD4 monoclonal antibody results in a preferential loss of circulating naive T cells but does not affect IFN-γ-secreting TH1 cells in humans, *J. Clin. Invest.* **99** (1997), 2225-2231.
98. Bouscarat, F., Levacher, M., Landman, R., Muffat-Joly, M., Girard, P., Saimot, A.-G., Brun-Vezinet, F. and Sinet, M.: Changes in blood CD8+ lymphocyte activation status and plasma HIV RNA levels during antiretroviral therapy, *AIDS* **12** (1998), 1267-1273.
99. Gray, C.M., Schapiro, J.M., Winters, M.A. and Merigan, T.C.: Changes in CD4+ and CD8+ T-cell subsets in response to highly active antiretroviral therapy in HIV type-1-infected patients with prior protease inhibitor experience, *AIDS Res. Human Retrovir.* **14** (1998), 561-569.
100. Steinberg, H.N., Crumpacker, C.S. and Chatis, P.A.: *In-vitro* suppression of normal human bone marrow progenitor cells by human immunodeficiency virus, *J. Virol.* **65** (1991), 1765-1769.
101. Stanley, S.K., Kessler, S.W., Justement, J.S., Schnittman, S.M., Greenhouse, J.J., Brown, C.C., Musongela, L., Musey, K., Kapita, B. and Fauci, A.S.: CD34+ bone marrow cells are infected with HIV in a subset of seropositive individuals, *J. Immunol.* **149** (1992), 689-697.
102. Zauli, G., Re, M.C., Visani, G., Furlini, G., Mazza, P., Vignoli, M. and La Placa, M.: Evidence for a human immunodeficiency virus type 1-mediated suppression of uninfected hematopoietic (CD34+) cells in AIDS patients, *J. Infect. Dis.* **166** (1992), 710-716.
103. Zauli, G., Vitale, M., Gibellini, M. and Capitani, S.: Inhibition of purified CD34+ hematopoietic progenitor cells by human immunodeficiency virus 1 or gp120 mediated by endogenous transforming growth factor b1, *J. Exp. Med.* **183** (1996), 99-108.
104. Clark, D.R., Repping, S., Pakker, N.G., Prins, J.M., Notermans, D.W., Wit, F.W.N.M., Reiss, P., Danner, S.A., Coutinho, R.A., Lange, J.M.A. and Miedema, F.: Diminished T-cell renewal in HIV-1 infection contributes to CD4+ T-cell depletion and is reversed by antiretroviral therapy,
Unpublished Work, 1998.
105. Hakim, F.T., Cepeda, R., Kaimei, S., Mackall, C.L., McAtee, N., Zujewski, J., Cowan, K. and Gress, R.E.: Constraints on CD4 recovery postchemotherapy in adults: thymic insufficiency and apoptotic decline of expanded peripheral CD4 cells, *Blood* **90** (1997), 3789-3798.
106. McCune, J.M., Loftus, R., Schmidt, D.K., Carroll, P., Webster, D., Swor-Yim, L.B., Francis, I.R., Gross, B.H. and Grant, R.M.: High prevalence of thymic tissue in adults with human immunodeficiency virus-1 infection, *J. Clin. Invest.* **101** (1998), 2301-2308.
107. Walker, R.E., Carter, C., Muul, L., Natarajan, V., Herpin, B., Leitman, S.F., Klein, H.G., Mullen, C.A., Metcalf, J.A., Baseler, M., Falloon, J., Davey Jr., R.T., Kovacs, J., Polis, M., Masur, H., Blaese, R.M. and Lane, C.M.: Peripheral expansion of pre-existing mature T cells is an important means of CD4+ T-cell regeneration in HIV-infected adults, *Nature Med.* **4** (1998), 852-856.
108. Imberti, L., Sottini, A., Bettinardi, A., Puoti, M. and Primi, D.: Selective depletion in HIV infection of T cells that bear specific T-cell receptor Vb sequences, *Science* **254** (1991), 860-862.
109. Rebai, N., Pantaleo, G., Demarest, J.F., Ciurli, C., Soudeyns, H., Adelsberger, J.W., Vaccarezza, M., Walker, R.E., Sekaly, R.P. and Fauci, A.S.: Analysis of the T-cell receptor β-chain variable (Vb) repertoire in monozygotic twins discordant for human immunodeficiency virus: Evidence for perturbations of specific Vb segments in CD4+ T cells of the virus-positive

twins, *Proc. Natl.Acad. Sci. USA* **91** (1994), 1529-1533.

110. Gea-Banacloche, J.C., Weiskopf, E.E., Hallahan, C., Lopez Bernaldo de Quiros, J.C., Flanigan, M., Mican, J.M., Falloon, J., Baseler, M., Stevens, R., Lane, H.C. and Connors, M.: Progression of human immunodeficiency virus disease is associated with increasing disruptions within the CD4⁺ T-cell receptor repertoire, *J. Infect. Dis.* **177** (1998), 579-585.

111. Connors, M., Kovacs, J.A., Krevat, S., Gea-Banacloche, J.C., Sneller, M.C., Flanigan, M., Metcalf, J.A., Walker, R.E., Falloon, J., Baseler, M., Stevens, R., Feuerstein, I.M., Masur, H. and Lane, H.C.: HIV infection induces changes in $CD4^+$ T-cell phenotype and depletions within the $CD4^+$ T-cell repertoire that are not immediately restored by antiviral or immune-based therapies, *Nature Med.* **3** (1997), 533-540.

112. Gorochov, G., Neumann, A.U., Kereveur, A., Parizot, C., Li, T., Katlama, C., Karmochkine, M., Raguin, G., Autran, B. and Debré, P.: Perturbation of $CD4^+$ and $CD8^+$ T-cell repertoires during progression to AIDS and regulation of the $CD4^+$ repertoire during antiviral therapy, *Nature Med.* **4** (1998), 215-221.

113. Kostense, S., Raaphorst, F.M., Notermans, D.W., Joling, J., Hooibrink, B., Pakker, N.G., Danner, S.A., Teale, J.M. and Miedema, F.: Diversity of the TCRBV repertoire in HIV-1-infected patients reflects the biphasic $CD4^+$ T-cell repopulation kinetics during HAART, *AIDS* **12** (1998), F235-F240.

114. Li, T.S., Tubiana, R., Katlama, C., Calvez, V., Ait Mohand, H. and Autran, B.: Long-lasting recovery in CD4 T-cell function and viral-load reduction after highly active antiretroviral therapy in advanced HIV-1 disease, *Lancet* **351** (1998), 1682-1686.

115. Angel, J.B., Kumar, A. and Parato, K.: Improvement in cell-mediated immune function during potent anti-human immunodeficiency virus therapy with ritonavir and saquinavir, *J. Infect. Dis.* **177** (1998), 898-904.

116. Komanduri, K.V., Viswanathan, M.N., Wieder, E.D., Schmidt, D.K., Bredt, B.M., Jacobson, M.A. and McCune, J.M.: Restoration of cytomegalovirus-specific $CD4^+$ T-lymphocyte responses after ganciclovir and highly active antiretroviral therapy in individuals infected with HIV-1, *Nature Med.* **4** (1998), 953-956.

117. Cameron, D.W., Heath-Chiozzi, M., Danner, S.A., Cohen, C., Kravcik, S., Maurath, C., Sun, E., Henry, D., Rode, R., Potthoff, A. and Leonard, J.: Randomised placebo-controlled trial of ritonavir in advanced HIV-1 disease. The advanced HIV disease ritonavir study group, *Lancet* **351** (1998), 536-537.

118. Hammer, S.M., Squires, K.E., Hughes, M.D., Grimes, J.M., Demeter, L.M., Currier, J.S., Eron Jr., J.J., Feinberg, J.E., Balfour, H.H., Deyton, L.R., Chodakewitz, J.A. and Fischl, M.A.: A controlled trial of two nucleoside analogues plus indinavir in persons with human immunodeficiency virus infection and CD4 cell counts of 200 per cubic millimeter or less, *New Engl. J. Med.* **337** (1997), 725-733.

119. Bräu, N., Leaf, H.L., Wieczorek, L. and Margolis, D.M.: Severe hepatitis in three AIDS patients treated with indinavir, *Lancet* **349** (1997), 924-925.

120. Jacobson, M.A., Zegans, M., Pavan, P.R., O'Donnell, J.J., Sattler, F., Rao, N., Owens, S. and Pollard, R.: Cytomegalovirus retinitis after initiation of highly active antiretroviral therapy, *Lancet* **349** (1997), 1443-1445.

121. Carr, A. and Cooper, D.A.: Restoration of immunity to chronic hepatitis B infection in HIV-infected patient on protease inhibitor, *Lancet* **349** (1997), 995-996.

122. Gill, J., Moyle, G. and Nelson, M.: Discontinuation of *Mycobacterium avium* complex profylaxis in patients with a rise in CD4 cell count following highly active antiretroviral therapy, *AIDS* **12** (1998), 680-688.

CHAPTER 14

HIV VIRAL LOAD

FRANK DE WOLF[1] AND INGRID SPIJKERMAN[2]

[1]*Department of Human Retrovirology, University of Amsterdam, Academic Medical Centre, Amsterdam, The Netherlands, and* [2]*Department of Public Health, Municipal Health Service, Amsterdam, The Netherlands*

1. Introduction

Identification and weighting of factors that correlate with and possibly contribute to the outcome of infection with human immunodeficiency virus (HIV) is important for our understanding of the pathogenesis and natural history of HIV infection and in designing strategies for the treatment of infection. HIV infection runs a variable course in the population of infected individuals. Some develop AIDS within a few years post-infection; others do not develop AIDS in a decade.

Several virological and immunological markers, like p24 antigenaemia [1-3] and the syncytium-inducing (SI) phenotype [4,5] predict poor clinical outcome, as do poor T-cell function *in vitro* [6-12] and low $CD4^+$ T-cell counts [2,3,13-18]. Studies of the dynamics of HIV infection show a continuously high rate of viral replication, believed to be the main factor driving $CD4^+$ T-cell depletion [19,20]. At seroconversion, HIV-RNA levels in plasma or serum are relatively high and comparable for groups progressing to AIDS or remaining symptom-free [21]. Only after seroconversion do levels of serum HIV-RNA become stable [22] and correlate with disease outcome [23,24]. Cross-sectional and longitudinal studies show serum RNA levels to be strongly associated with progression to AIDS, $CD4^+$ T-cell decline, or death. Longitudinal studies showed the predictive value of HIV-RNA to be independent of the predictive value of $CD4^+$ T-cell counts [25].

The efficacy of anti-HIV treatment is strongly associated with the RNA plasma or serum concentration, as is the viral phenotype [26,27]. Higher base-line concentrations of HIV-RNA, less suppression of RNA levels by anti-HIV drugs and presence of SI viruses are predictive for the clinical outcome of treatment.

H. Schuitemaker and F. Miedema (ed.), Aids Pathogenesis, 261–272.

2. The Amount of Virus Particles and Prediction of Disease Outcome

2.1 THE RELATIONSHIP BETWEEN CULTURE, AMOUNT OF VIRAL DNA AND AMOUNT OF VIRAL RNA

Previous studies demonstrated a clear association between the titre of virus cultured from plasma and the clinical stage of disease [28,29]. By using end-point-dilution cultures it was shown that plasma tissue-culture infectious dose (TCID) ranged from 5-100 in asymptomatic HIV-infected individuals and from 25-50000 TCID/ml plasma in symptomatic patients [28]. The PBMC virus titre appeared to be associated to disease stage as well, ranging from 5-50 TCID/10^6 cells in asymptomatic and 50-10000 TCID/10^6 cells in symptomatic patients. During acute infection high titres are found, ranging from 1000-10000 TCID/ml plasma and 100-10000 TCID/10^6 cells [30,31], and the changes in the amount of HIV during primary infection was confirmed by quantitative PCR methods measuring HIV-DNA copies in PBMC. The amount of HIV-RNA in peripheral blood does reflect the amount of circulating particles [21]. Quantitative competitive PCR methods to quantify HIV-RNA showed that plasma RNA levels correlated with, but exceeded by an average of 60000 fold, the virus titres measured by end-point-dilution culture [32] and it was shown that determination of plasma RNA levels represents a marker of virus replication [33].

2.2 DETECTION OF VIRAL RNA IN PERIPHERAL BLOOD

Since the beginning of 1995 standardised tests for the detection and quantification of particle-associated HIV-RNA in peripheral blood are commercially available [34, 35]. Most widely used for clinical purposes are Quantiplex (Chiron Corporation), NucliSens (Organon Teknika) and Amplicor (Roche Molecular Systems). Both NucliSens and Amplicor are based on enzymatic methods for the amplification of a conserved region in the gag genome of HIV in combination with co-amplification of known amounts of one or more calibrators. These calibrators (*i.e.* sequences only differing from the HIV gag sequence at specific nucleotide positions) are necessary to compute the amount of wild-type RNA. In essence, the difference between NucliSens and Amplicor is the amplification method: Amplicor is based on the amplification of cDNA obtained from HIV-RNA through reverse transcription, whereas NucliSens is based on amplification of HIV-RNA. Quantiplex is based on the branched DNA technique (bDNA). The bDNA technique amplifies the signal from a captured viral RNA target by sequential oligonucleotide hybridisation steps in stead of HIV-RNA or cDNA amplification. All three tests have their pro's and contra's.

The sensitivity of the tests is the same with a lower detection limit, ranging between 200 to 400 HIV-RNA copies/ml serum or plasma. Also the dynamic range is the same. There is a difference with respect to the input volume related to the technique used. NucliSens, as well as Amplicor need an input of at least 200 ml plasma or serum, whereas in the Quantiplex test an amount of 500-1000 ml has to be used in order to reach the same sensitivity. Results of the assays may differ, when used for the same sample, but is limited in most cases to values <0,5log [22,36,37] and results of different assays correlate strongly [38,39]. Notwithstanding this small difference, it is recommended to use the same test when monitoring a patient for HIV-RNA. Protocols to enhance the sensitivity of the assays are available, resulting in lower detection limits of 20, 5-50 and 50 HIV-RNA copies/ml for the Amplicor, the NucliSens and the Quantiplex assay, respectively.

2.3 THE RELATIONSHIP BETWEEN VIRAL RNA AND DISEASE OUTCOME AND WITH OTHER MARKERS OF DISEASE PROGRESSION

Previously, it was assumed that HIV infection was latent during the asymptomatic phase of the infection and before HIV-related disease symptoms became apparent. Recent studies showed that there is no latent infection. Instead, both the asymptomatic and the symptomatic phase of the infection are virologically dynamic processes [19,20,32,40,41]. Every day billions of virus particles are produced with a virus half-life in plasma of about 6 hours. This high virus production drives the destruction of the immune system.

The amount of HIV-RNA in peripheral blood is the result of virus production and clearance, where the clearance is relatively constant [19]. Consequently, individual differences of plasma or serum HIV-RNA levels are the result of differences in virus production. Most of the HIV particles are produced by infected activated $CD4^+$ T cells with a short half-life [40,41]. In addition, cells chronically producing virus over an extended period of time may add to the total amount of HIV produced. Next to these virus-producing cells, non-activated or resting infected $CD4^+$ T cells and possibly immature monocytes may serve as a virus reservoir capable of producing virus upon activation [40-45].

During primary HIV infection, a large number of infectious virus particles, as well as high levels of viral RNA, are found [30,31,46]. After the initial HIV-RNA peak, the concentration of HIV-RNA falls. It is assumed that this decrease of HIV-RNA is the result of the host's early immune response [47-49]. However, mathematical modelling of the events early in infection indicates that reduction of HIV concentration during acute infection is independent of the ability of the HIV-specific immune response to control virus replication [50].

After primary infection, variable HIV-RNA can be found in peripheral blood, indicating ongoing viral replication throughout the course of infection by a limited number of productively infected cells [28]. The number of infected resting $CD4^+$ T cells with replication-competent integrated HIV is small [42]. In a study among 20 individuals seroconverting for HIV antibodies (with a median seroconversion interval of 3.2 months) RNA levels ranged from 10^4 to 10^6 copies/ml at 3 months after seroconversion [21]. Subsequently, different patterns of changing HIV-RNA levels in peripheral blood can be found. Rapid progressors to AIDS do have high HIV-RNA levels in serum or plasma in contrast to long-term asymptomatic individuals showing stable low or slowly rising levels [23]. When measured longitudinally, different patterns of HIV-RNA are related to the time to AIDS. In the Amsterdam cohort of homosexual men seroconverting for HIV antibodies two patterns appeared (Fig. 1): *i*) a high steady-state level (10^5 copies/ml) from seroconversion on in a group developing AIDS within 52 months after seroconversion, and

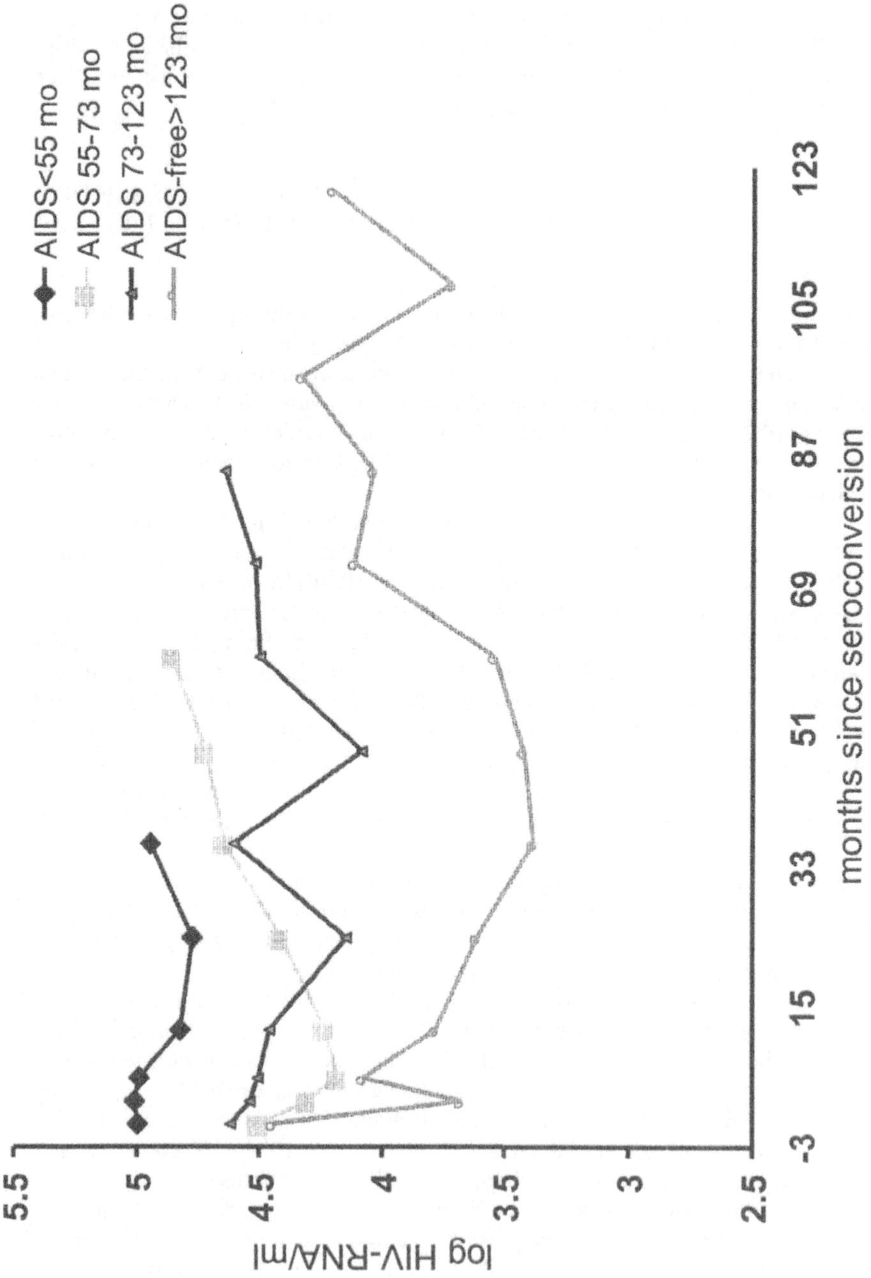
AIDS<55 mo
AIDS 55-73 mo
AIDS 73-123 mo
AIDS-free>123 mo
log HIV-RNA/ml
5.5
5
4.5
4
3.5
3
2.5
-3
15
33
51
69
87
105
123
months since seroconversion

ii) an U-shaped curve [51]. An U-shaped curve showing a limited decline of RNA levels (from $10^{4.5}$ to $10^{4.2}$ copies/ml) in the first 6 months after and a rise thereafter was associated with AIDS within 4-6 years after seroconversion. The U-shaped curve characterised by a steeper decline (from $10^{4.6}$ to $10^{4.1}$ copies/ml) in the first 2 to 4 years after seroconversion and a rise to $10^{4.6}$ copies/ml thereafter appeared to be associated with progression to AIDS 6 to 10 years after seroconversion. Apparently, in these groups the production and clearance of virus never reach equilibrium. Thus viral load accumulates, with disease progression as the final result. In contrast, the low U-shaped curve, characterised by a decline from $10^{4.5}$ to $10^{3.5}$ copies/ml 3 to 6 years after seroconversion and a subsequent increase to levels $<10^4$ copies/ml, was associated with an AIDS-free period of more than 10 years. In this particular group, equilibrium between virus production and clearance is set almost 2 years after infection and maintained for at least years thereafter. For as yet unknown reasons, this equilibrium is then lost as viral load slowly increases, probably indicating the ultimate development towards disease.

The risk of AIDS and death appeared to be directly related to plasma viral RNA levels. In our study among 123 homosexual men with a known seroconversion date, the risk of development of AIDS could be distinguished between high and low HIV-RNA concentrations already at one year after infection. Those with HIV-RNA levels $>10^{4.0}$ copies/ml showed a significant difference in progression to AIDS compared to those with HIV-RNA levels $<10^{4.0}$ copies/ml (Fig. 2). Moreover, stable low and, especially, declining RNA to levels $<10^{4.5}$ copies/ml within one year after seroconversion were associated with a prolonged AIDS-free period.

The results from our study confirmed those of Mellors [24,25] and Henrard [22]. In addition, we found that at one year after seroconversion no other known marker was predictive for AIDS and for the five years following seroconversion, RNA levels predicted AIDS irrespective the $CD4^+$ T-cell number. Later in infection, $CD4^+$ T-cell counts and low T-cell function became predictive [52] and HIV-RNA levels no longer added predictive power to these two markers with respect to disease outcome.

Development of AIDS is also associated with the existence or emergence of the syncytium-forming (SI) phenotype of HIV [4,5]. Thus, where AIDS is associated with RNA levels as well, SI viruses are related with high plasma RNA levels, as do non-syncytium-inducing (NSI) variants when found in those progressing to AIDS but not in those remaining symptom-free [21]. The shift from NSI to SI viruses during infection, however, is accompanied by a substantial rise in plasma HIV-RNA, when compared to matched controls without a switch [53]. The reciprocal prognostic value in time of HIV-RNA *versus* $CD4^+$ T-cell count reflects the progressive development of immune deficiency which, in later stages of HIV infection, is most predictive for disease progression. Since patients generally present to the clinician several years after HIV seroconversion, patient management and treatment decisions should not be based on viral RNA measurements alone.

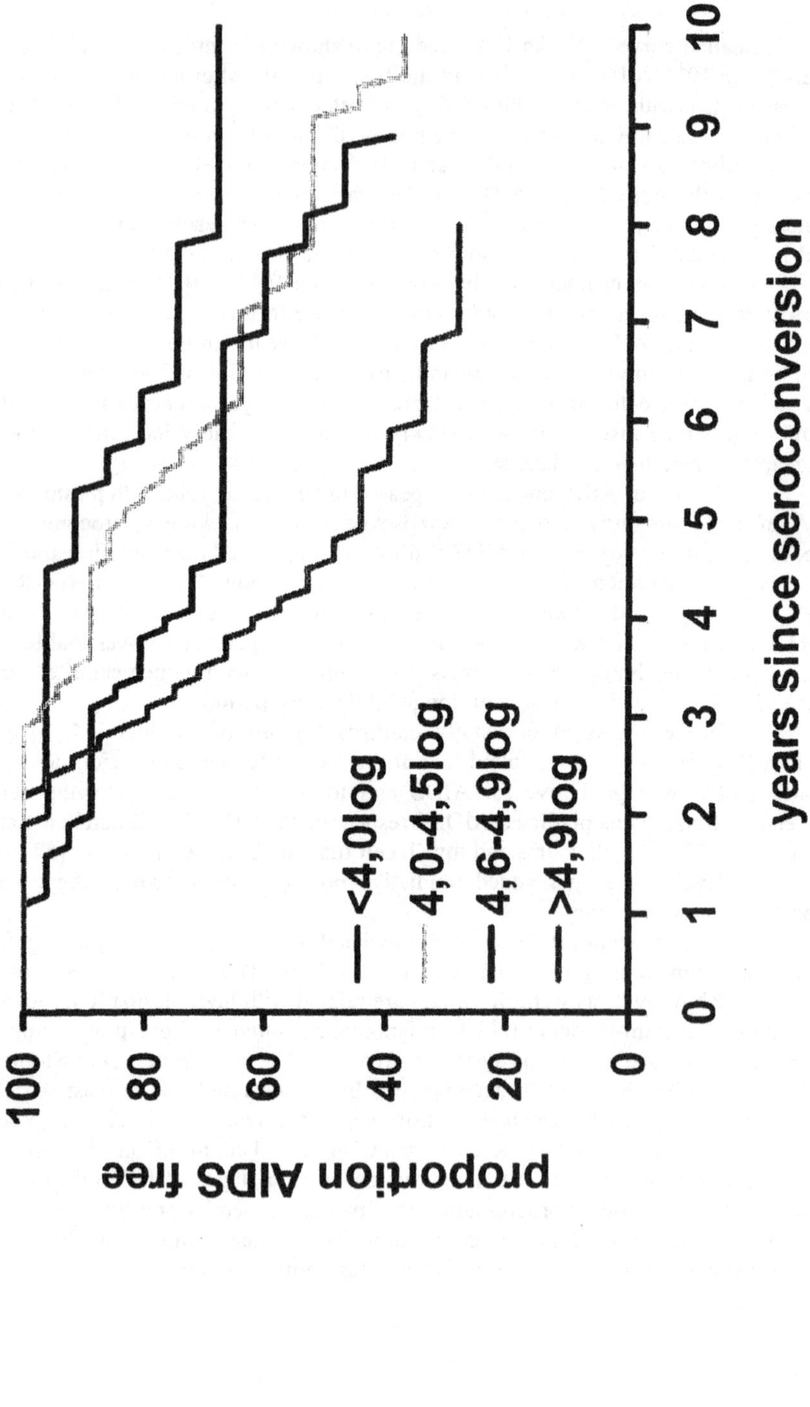

Fig. 2 - Kaplan-Meier analysis for AIDS per HIV-RNA quartile at one year after seroconversion [51].

3. HIV-RNA as a Tool in Monitoring the Effect of Treatment with Anti-retroviral Drugs

3.1 CHANGES OF PLASMA RNA LEVELS IN RESPONSE TO ANTI-HIV TREATMENT

The concept of continuous virus production has changed the strategies for the treatment of HIV-infected individuals with anti-HIV drugs. Already early in infection, that is before symptoms of deterioration of the immune system occur, treatment should be started in order to lower the amount of HIV and to keep it as low as possible [54]. Plasma HIV-RNA measurements are useful for the evaluation of the anti-retroviral effect of drugs [20,26,27]. Effective anti-HIV treatment significantly decreases HIV-RNA levels in plasma within one week of the start of treatment. No significant decrease of the plasma levels within this time period suggests that the regimen used have no anti-HIV activity. Mono-therapy with AZT results in a 0,5 to 1,0 log reduction in the plasma HIV-RNA level within two weeks, which returns towards baseline 24 to 48 weeks after start of treatment [27,55]. Combination of two HIV-RT inhibitors usually gives better results [56-58]. HIV-RNA levels decline with median 1,5 log and $CD4^+$ T-cell peaks are observed and the responses are more durable, often persisting for more than a year. HIV-protease inhibitor mono-therapy results in plasma RNA declines of 1,0 to 2,0 log [59-61] and combination of protease inhibitors with RT inhibitors, in most cases one protease inhibitor and two nucleoside RT inhibitors, results in plasma RNA reductions of 2,0 to 3,0 log [62-65].

In general, after start of effective anti-retroviral treatment, HIV-RNA plasma levels decrease rapidly during the first phase of about a week, followed by a slower decline towards values below the lower detection limit of quantitative HIV-RNA assays [19,20,40,41]. The first-phase decline is determined by the short half-life of productively infected activated $CD4^+$ T cells. The slow second-phase decline of plasma HIV-RNA levels represents the suppression of HIV production of low-level-producing cells and from the small fraction of latently infected resting $CD4^+$ T cells [66-68]. Decreases in plasma HIV-RNA levels generally correlate with increases in $CD4^+$ T-cell count in peripheral blood, the initial rapid rise most probably being the result of redistribution of cells from lymphoid tissue, followed by a continuous slow repopulation with newly produced naive $CD4^+$ T cells [69].

3.2 MONITORING OF PLASMA HIV-RNA: VIROLOGIC FAILURES OF ANTI-HIV TREATMENT

Triple-drug combination treatment has rapidly become the standard of care for the treatment of HIV infection [70]. However, one has to take into account that reported reductions in HIV-RNA plasma levels of 2,0 to 3,0 log are found in well-controlled trials among anti-HIV treatment naive individuals. Starting with baseline HIV-RNA plasma levels of at least 30000 copies/ml, the percentage of individuals on triple combination treatment having HIV-RNA plasma levels below 200-400 copies/ml ranged from 80 to 90% in the first 24 weeks of treatment [63]. Longer-term virologic effects of triple-drug combination treatment revealed ongoing low-level virus replication, and long-term clinical outcomes of treatment are as yet unknown. Other, more extensive and complicated combinations of anti-retroviral drugs did result in an improvement of the virologic effect [71]. These drug regimens did not result in a significantly steeper initial plasma HIV-RNA decline. However, they did result in a shorter time period to reach HIV-RNA levels <50 copies/ml.

When evaluating the virologic effect of anti-HIV treatment in a non-trial setting, again plasma HIV-RNA does not decline below the lower detection limit of the assay in 15 to 20% of treatment-naive individuals on triple-drug combination treatment . However, among the treatment-experienced individuals results are worse: in 35 to 40% of these patients RNA levels can still be determined despite triple-drug combination treatment for almost 6 months [67]. In these substantial numbers of individuals having detectable amounts of HIV-RNA in plasma strains, resistant to one or more anti-HIV drugs, will eventually dominate unless alternative triple- or quadruple-drug regimes do suppress the viral replication sufficiently.

3.3 MONITORING HIV-RNA AND-DNA: VIROLOGIC SUCCESSES OF ANTI-HIV TREATMENT?

When defining virological success of anti-HIV treatment as the induction of a sustained decline of HIV-RNA to below the detection limit of the assay, then success per definition is dependent on the sensitivity of the assay used. The lower detection limit of the current assays in the standard format ranges between 200-400 HIV-RNA copies/ml plasma. When using this lower detection limit, 88% of the individuals having 24 weeks triple-drug combination treatment became "undetectable"; however, with a protocol resulting in a lower detection limit of 50 copies/ml in only 75% HIV-RNA levels below the lower detection limit were seen [63]. The lymphoid tissue viral burden, ranging from 9.2 to 8.5 log copies RNA per gram tissue at baseline, was markedly reduced with at least 2.1 log over the same 24 weeks in the 5 patients analysed and all were below the detection limit of the RT PCR used. Also, the frequency of the productive mononuclear cells decreased, as well as the amount of HIV-1 RNA in virus trapped on follicular dendritic cells. However, after 24 weeks still a very few copies of HIV-RNA could be detected in mononuclear cells, as was proviral DNA [73].

Recently, it was shown that even after triple- or quadruple-drug combination treatment, resulting in long-term reduction of HIV-RNA in plasma to levels <200 or <50 copies/ml, still replication-competent virus could be recovered routinely from resting $CD4^+$ T cells. The frequency of resting $CD4^+$ T cells harbouring latent HIV-1 was low, 0,2 to 16,4 per 10^6 cells and did not decrease with increasing time of treatment. The recovered virus did not show mutations associated with resistance to the relevant drugs [66,67]. Thus, it appeared that, with the development and usage of more sensitive protocols for the detection and quantification of HIV-RNA and/or DNA, highly active anti-retroviral therapy is not always as successful as was thought previously.

4. References

1. De Wolf, F., Goudsmit, J., Paul, D.A. *et al.*: Risk of AIDS-related complex and AIDS in homosexual men with persistent HIV antigenaemia, *Brit. Med. J.* **295** (1987), 569-572.
2. De Wolf, F., Lange, J.M.A., Houweling, J.T.M. *et al.*: Numbers of CD4$^+$ cells and the levels of core antigens of and antibodies to the human immunodeficiency virus as predictors of AIDS among seropositive homosexual men, *J. Infect. Dis.* **158** (1988), 615-622.
3. Moss, A.R., Bacchetti, P., Osmond, D. *et al.*: Seropositivity for HIV and the development of AIDS or AIDS-related conditions: three year follow up of the San Francisca General Hospital cohort, *Brit. Med. J.* **296** (1988), 745-750.
4. Koot, M., Keet, I.P.M., Vos, A.H.V. *et al.*: Prognostic value of HIV-1 syncytium-inducing phenotype for rate of CD4$^+$ cell depletion and progression to AIDS, *Ann. Intern. Med.* **118** (1993), 681-688.
5. Schuitemaker, H., Koot, M., Kootstra, N.A. *et al.*: Biological phenotype of human immunodeficiency virus type-1 clones at different stages of infection: progression to disease is associated with a shift from monocytotropic to T-cell-tropic virus populations, *J. Virol.* **66** (1992), 1354-1360.
6. Miedema, F., Petit, C.A.J., Terpstra, F.G. *et al.*: Immunological abnormalities in human immunodeficiency virus (HIV)-infected asymptomatic homosexual men, *J. Clin. Invest.* **82** (1988), 1908-1914.
7. Shearer, G.M., Bernstein, D.C., Tung, K.S.K. *et al.*: A model for the selective loss of major histocompatibility complex self-restricted T-cell immune responses during the development of acquired immune deficiency syndrome (AIDS), *J. Immunol.* **137** (1986), 2514-2521.
8. Giorgi, J.V., Fahey, J.L., Smith, D.C. *et al.*: Early effects of HIV on CD4$^+$ lymphocytes *in vivo*, *J. Immunol.* **138** (1987), 3725-3730.
9. Hofmann, B., Orskov Lindhardt, B., Gerstoft, J. *et al.*: Lymphocyte transformation response to pokeweed mitogen as a marker for development of AIDS and AIDS-related symptoms in homosexual men with HIV antibodies, *Brit. Med. J.* **295** (1987), 293-296.
10. Terpstra, F.G., Al, B.J.M., Roos, M.Th.L. *et al.*: Longitudinal study of leukocyte functions in homosexual men seroconverted for HIV: rapid and persistent loss of B-cell function after HIV infection, *Eur. J. Immunol.* **19** (1989), 667-673.
11. Schellekens, P.Th.A., Roos, M.Th.L., De Wolf, F. *et al.*: Low level responsiveness to activation via CD3/TCR is a prognostic marker for AIDS in HIV-1-infected men, *J. Clin. Immunol.* **10** (1990), 121-127.
12. Roos, M.Th.L., Miedema, F., Koot, M. *et al.*: T-cell function *in vitro* is an independent progression marker for AIDS in human immunodeficiency virus-infected asymptomatic subjects, *J. Infect. Dis.* **171** (1995), 531-536.
13. Schellekens, P.Th.A., Tersmette, M., Roos, M.Th.L. *et al.*: Biphasic rate of CD4$^+$ cell count decline during progression to AIDS correlates with HIV-1 phenotype, *AIDS* **6** (1992), 665-669.
14. Hofmann, B., Wang, Y., Cumberland, W.G. *et al.*: Serum beta$_2$-microglobulin level increases in HIV infection: relation to seroconversion, CD4 T-cell fall and prognosis, *AIDS* **4** (1990), 207-214.
15. Fahey, J.L., Taylor, J.M.G., Detels, R. *et al.*: The prognostic value of cellular and serologic markers in infection with human immunodeficiency virus type 1, *New Engl. J. Med.* **322** (1990), 166-172.
16. Polk, B.F., Fox, R., Brookmeyer, R. *et al.*: Predictors of the acquired immunodeficiency syndrome developing in a cohort of seropositive homosexual men, *New Engl. J. Med.* **316** (1987), 61-66.
17. Felsenstein, J.: Parsimony in systematics: biological and statistical issues, *Ann. Rev. Syst.* **14** (1983), 313-333.
18. Stein, D.S., Korvick, J.A. and Vermund, S.H.: CD4$^+$ lymphocyte cell enumeration for prediction of clinical course of human immunodeficiency virus disease: a review, *J. Infect. Dis.* **165** (1992), 352-356.
19. Ho, D.D., Neumann, A.U., Perelson, A.S. *et al.*: Rapid turnover of plasma virions and CD4$^+$ lymphocytes in HIV-1 infection, *Nature* **373** (1995), 123-126.
20. Wei, X., Ghosh, S.K., Taylor, M.E. *et al.*: Viral dynamics in human immunodeficiency virus type-1 infection, *Nature* **373** (1995), 117-122.

21. Jurriaans, S., Van Gemen, B., Weverling, G.J. *et al.*: The natural history of HIV-1 infection: virus load and virus phenotype independent determinants of clinical course?, *Virology* **204** (1994), 223-233.
22. Henrard, D.R., Phillips, J.F., Muenz, L.R. *et al.*: Natural history of HIV-1 cell-free viremia, *JAMA* **274** (1995), 554-558.
23. Hogervorst, E., Jurriaans, S., De Wolf, F. *et al.*: Predictors for non- and slow progression in human immunodeficiency virus (HIV) type-1 infection: low viral RNA copy numbers in serum and maintenance of high HIV-1 p24-specific but not V3-specific antibody levels, *J. Infect. Dis.* **171** (1995), 811-821.
24. Mellors, J.W., Kingsley, L.A., Rinaldo Jr., C.R. *et al.*: Quantitation of HIV-1 RNA in plasma predicts outcome after seroconversion, *Ann. Intern. Med.* **122** (1995), 573-579.
25. Mellors, J.W., Rinaldo Jr., C.R., Gupta, P. *et al.*: Prognosis for HIV-1 infection predicted by the quantity of virus in plasma, *Science* **272** (1996), 1167-1170.
26. Katzenstein, D.A., Hammer, S.M., Hughes, M.D. *et al.*: The relation of virologic and immunologic markers to clinical outcomes after nucleoside therapy in HIV-infected adults with 200 to 500 CD4 cells per cubic millimeter, *New Engl. J. Med.* **335** (1996), 1091-1098.
27. O'Brien, W.A., Hartigan, P.M., Martin, D. *et al.*: Changes in plasma HIV-1 RNA and $CD4^+$ lymphocyte counts and the risk of progression to AIDS, *New Engl. J. Med.* **334** (1996), 426-431.
28. Ho, D.D., Moudgil, T. and Alam, M.: Quantitation of human immunodeficiency virus type 1 in the blood of infected persons, *New Engl. J. Med.* **321** (1989), 1621-1625.
29. Coombs, R.W., Collier, A.C., Allain, J.P. *et al.*: Plasma viremia in human immunodeficiency virus infection, *New Engl. J. Med.* **321** (1989), 1626-1631.
30. Clark, S.J., Saag, M.S., Decker, W.D. *et al.*: High titers of cytopathic virus in plasma of patients with symptomatic primary HIV-1 infection, *New Engl. J. Med.* **324** (1991), 954-960.
31. Daar, E.S., Moudgil, T., Meyer, R.D. *et al.*: Transient high levels of viremia in patients with primary human immunodeficiency virus type-1 infection, *New Engl. J. Med.* **324** (1991), 961-965.
32. Piatak Jr., M., Saag, M.S., Yang, L.C. *et al.*: High levels of HIV-1 in plasma during all stages of infection determined by competitive PCR, *Science* **259** (1993), 1749-1754.
33. Blaak, H., De Wolf, F., Van 't Wout, A.B. *et al.*: Temporal relationship between human immunodeficiency virus type-1 RNA levels in serum and cellular infectious load in peripheral blood, *J. Infect. Dis.* **176** (1997), 1383-1387.
34. Huisman, J.G., Bruisten, S.M. and Cuypers, H.T.M.: Kwantitatieve bepaling van HIV-1-RNA/DNA, *CLB Bull.* **March** (1995), 4-6.
35. Volberding, P.A.: HIV quantification: clinical applications, *Lancet* **347** (1996), 71-73.
36. Weverling, G.J., Lange, J.M.A., De Jong, M.D. *et al.*: A comparison of serum HIV-1 RNA levels are measured by two quantitative assays in Zidovudine-treated asymptomatic individuals, *Antiviral Treatment* **1** (1996), 255-263.
37. De Wolf, F. and Goudsmit, J.: AIDS; nieuwe ontwikkelingen. III. Voorspellende waarde van de hoeveelheid HIV-RNA voor het beloop van de HIV infectie en het effect van de behandeling, *Ned. Tijdschr. Geneeskd.* **141** (1997), 1043-1050.
38. Cao, Y., Ho, D.D., Todd, J. *et al.*: Clinical evaluation of branched DNA signal amplification for quantifying HIV type 1 in human plasma, *AIDS Res. Human Retrovir.* **11** (1995), 353-361.
39. Revets, H., Marissens, D., De Wit, S. *et al.*: Comparative evaluation of NASBA HIV-1 RNA QT, Amplicor-HIV monitor, and Quantiplex HIV RNA assay, three methods for quantification of human immunodeficiency virus type-1 RNA in plasma, *J. Clin. Microbiol.* **34** (1996), 1058-1064.
40. Coffin, J.M.: HIV population dynamics *in vivo*: implications for genetic variation, pathogenesis, and therapy, *Science* **267** (1995), 483-489.
41. Perelson, A.S., Neumann, A.U., Markowitz, M. *et al.*: HIV-1 dynamics *in vivo*: virion clearance rate, infected cell life-span, and viral generation time, *Science* **271** (1996), 1582-1586.
42. Chun, T.-W., Carruth, L.M., Finzi, D. *et al.*: Quantification of latent tissue reservoirs and total body viral load in HIV-1 infection, *Nature* **387** (1997), 183-188.
43. Stevenson, M., Stanwick, T.L., Dempsey, M.P. and Lamonica, C.A.: HIV-1 replication is controlled at the level of T-cell activation and proviral integration, *EMBO J.* **9** (1990), 1551-1560.
44. Bukrinsky, M.I., Stanwick, T.L., Dempsey, M.P. and Stevenson, M.: Quiescent T lymphocytes as an inducible virus reservoir in HIV-1 infection, *Science* **254** (1991), 423-427.

45. Stevenson, M.: Molecular mechanisms for the regulation of HIV replication, persistence and latency, *AIDS* **11 (suppl. A)** (1997), S25-S33.
46. Van Gemen, B., Kievits, T., Schukkink, R. *et al.*: Quantification of HIV-1 RNA in plasma using NASBA during HIV-1 primary infection, *J. Virol. Methods* **43** (1993), 177-188.
47. Koup, R.A., Safrit, J.T., Cao, Y. *et al*,: Temporal association of cellular immune responses with the initial control of viremia in primary human immunodeficiency virus type-1 syndrome, *J. Virol.* **68** (1994), 4650-4655.
48. Lange, J.M.A., Coutinho, R.A., Krone, W.J.A. *et al.*: Distinct IgG recognition patterns during progression of subclinical and clinical infection with lymphadenopathy-associated virus/human T-lymphotropic virus, *Brit. Med. J.* **292** (1986), 228-230.
49. Goudsmit, J., De Wolf, F., Paul, D.A. *et al.*: Expression of human immunodeficiency virus antigen (HIV-Ag) in serum and cerebrospinal fluid during acute and chronic infection, *Lancet* **ii** (1986), 177-180.
50. Phillips, A.N.: Reduction of HIV concentration during acute infection: independence of a specific immune response, *Science* **271** (1996), 497-499.
51. De Wolf, F., Spijkerman, I., Schellekens, P.Th.A. *et al.*: AIDS prognosis based on HIV-1 RNA, $CD4^+$ T-cell count and function: markers with reciprocal predictive value over time after seroconversion, *AIDS* **11** (1997), 1799-1806.
52. Spijkerman, I.J.B., Prins, M., Goudsmit, J. *et al.*: Early and late HIV-1 RNA level and its association with T-cell reactivity and disease progression in long-term AIDS-free homosexual men, *AIDS* **11** (1997), 1383-1388.
53. Spijkerman, I., De Wolf, F., Langendam, M. *et al.*: Emergence of syncytium-inducing HIV-1 variants coincides with a transient increase in viral RNA level and is an independent predictor for progression to AIDS, *J. Infect. Dis.* **178** (1998), 397-403.
54. Ho, D.D.: Time to hit HIV, early and hard, *New Engl. J. Med.* **333** (1995), 450-451.
55. Loveday, C., Kaye, S., Tenant-Flowers, M. *et al.*: HIV-1 RNA serum load and resistant viral genotypes during early zidovudine therapy, *Lancet* **345** (1995), 820-824.
56. Eron, J.J., Benoit, S.L. and Jemsek, J.: Treatment with lamivudine, zidovudine, or both in HIV-positive patients with 200 to 500 $CD4^+$ cells per cubic millimeter, *New Engl. J. Med.* **333** (1995), 1662-1669.
57. Delta Coordinating Committee. Delta: A randomized double-blind controlled trial comparing combinations of zidovudine plus didanosine or zalcitabine with zidovudine alone in HIV-infected individuals, *Lancet* **348** (1996), 283-291.
58. Hammer, S.M., Katzenstein, D.A., Hughes, M.D. *et al.*: A trial comparing nucleoside monotherapy with combination therapy in HIV-infected adults with CD4 counts from 200 to 500 per cubic millimeter. AIDS Clinical Trial Group Study 175 Study Team, *New Engl. J. Med.* **335** (1996), 1081-1090.
59. Danner, S.A., Carr, A., Leonard, J.M. *et al.*: A short-term study of safety, pharmacokinetics, and efficacy of Ritonavir, an inhibitor of HIV-1 protease, *New Engl. J. Med.* **333** (1995), 1528-1533.
60. Markowitz, M., Saag, M., Powderly, W.G. *et al.*: A preliminary study of ritonavir, an inhibitor of HIV-1 protease, to treat HIV-1 infection, *New Engl. J. Med.* **333** (1995), 1534-1539.
61. Kitchen, V.S., Skinner, C., Ariyoshi, K. *et al.*: Safety and activity of saquinavir in HIV infection, *Lancet* **345** (1995), 952-955.
62. Notermans, D.W., De Wolf, F., Foudraine, N.A. *et al.*: The effects of an antiretroviral combination with ritonavir, AZT and 3TC (Abstract), *AIDS* **10 (suppl. 2)** (1996), S17.
63. Notermans, D.W., Jurriaans, S., De Wolf, F. *et al.*: Decrease of HIV-1 RNA levels in lymphoid tissue and peripheral blood during treatment with ritonavir, lamivudine and zidovudine, *AIDS* **12** (1998), 167-173.
64. Hammer, S.M., Squires, K.E., Hughes, M.D. *et al.*: A controlled trial of two nucleoside analogues plus indinavir in persons with human immunodeficiency virus infection and CD4 cell counts of 200 per cubic millimeter or less, *New Engl. J. Med.* **337** (1997), 725-733.
65. Gulick, R.M., Mellors, J.W., Havlir, D. *et al.*: Treatment with indinavir, zidovudine and lamivudine in adults with human immunodeficiency virus infection and prior antiretroviral therapy, *New Engl. J. Med.* **337** (1997), 734-739.

66. Wong, J.K., Hezareh, M., Gunthard, H.F. *et al.*: Recovery of replication-competent HIV despite prolonged suppression of plasma viremia, *Science* **278** (1997), 1291-1295.
67. Finzi, D., Hermankova, M., Pierson, T. *et al.*: Identification of a reservoir for HIV-1 in patients on highly active antiretroviral therapy, *Science* **278** (1997), 1295-1300.
68. Finzi, D. and Siliciano, R.F.: Viral dynamics in HIV-1 infection, *Cell* **93** (1998), 666-671.
69. Pakker, N.G., Notermans, D.W., De Boer, R.J. *et al.*: Biphasic kinetics of peripheral blood T cells after triple combination therapy in HIV infection: a composite of redistribution and proliferation, *Nature Med.* **4** (1998), 208-214.
70. Carpenter, C.J., Fischl, M.A., Hammer, S.A. *et al.*: Antiviral therapy for HIV infection in 1997: update recommendations of the International AIDS Society USA panel, *JAMA* **277** (1997), 19629.
71. Weverling, G.J., Lange, J.M.A., Jurriaans, S. *et al.*: Alternative multidrug regimen provides improved suppression of HIV-1 replication over triple therapy, *AIDS* **12** (1998), F117-F122.
72. De Wolf, F., De Jong, J., Hertogs, K. *et al.*: Virologische evaluatie van HIV-geïnfecteerden behandeld met (combinaties van) anti-retrovirale middelen in het AMC 1996/1997: preliminaire waarnemingen, *Ned. Tijdschr. Geneeskd.* **142** (1998), 573-578.
73. Cavert, W., Notermans, D.W., Staskus, K. *et al.*: Kinetics of response in lymphoid tissues to antiretroviral therapy of HIV-1 infection, *Science* **276** (1997), 960-964.

CHAPTER 15

A - THE SCID-hu MOUSE: AN IN-VIVO MODEL FOR HIV-1 INFECTION IN HUMANS

HIDETO KANESHIMA, M.D., Ph.D.

HIV Gene Therapy, SyStemix, Inc.
Palo Alto, USA

1. Introduction

The decline of $CD4^+$ T-cell number in the peripheral circulation has been considered an indicative barometer of HIV-1 disease progression. A variety of possible mechanisms for the loss of $CD4^+$ T cells have been proposed, yet the main mechanism for immuno-pathogenesis remained unknown.

Theoretically, this phenomenon can be attributed, at least in part, to three major mechanisms: (*i*) failure to generate new T cells from a progenitor pool; (*ii*) failure of the normal expansion of existing periphe-ral T cells; and/or (*iii*) altered representation in the blood owing to redistribution to peripheral lymphoid tissues. According to Roederer *et al.* [1], the numbers of naive T cells defined by CD45R and CD62L antigen expression in $CD4^+$ as well as $CD8^+$ T cells were found to decrease con-stantly before and after AIDS manifestation. This observation highly suggests that a lack of new T-cell development, rather than T-cell destruction in the peripheral pool, could be a major factor for the systematic T-cell depletion toward AIDS.

The thymus is a major site for both $CD4^+$ and $CD8^+$ naive T lymphopoiesis from $CD34^+$ haematopoietic progenitor cells and is clearly a permissive organ for HIV-1 replication *in vivo* [2]. Especially in the report using a quantitative flow cyto-metric analysis [3], the thymus was found to have a severe depletion of $CD4^+$ thymocytes in a HIV-positive newborn. A similar pathology has been demon-strated in SIV-infected rhesus macaques [4], in FIV-infected cats [5], and in the SCID-hu animal model [6,7]. If so, then abrogation of thymopoiesis may represent a central lesion in HIV-1 disease. In this review, the recent studies of the mechanism for HIV-1-induced thymocyte depletion in the SCID-hu mice will be summarised.

H. Schuitemaker and F. Miedema (ed.), Aids Pathogenesis, 273–278.

2. The Validations of SCID-hu Mouse as a Model for Human Active Thymuses in vivo

In the case of animal models used to investigate human pathophysiology, the physiology of the model must first be shown to approximate that in humans. Thereafter, perturbation of the normal state might permit observation of pathophysiological processes. If these observations are found to be relevant to disease, then the model could serve as a model. With regard to the SCID-hu mice transplanted with human thymus and liver fragments (Thy/Liv), the fused organ has been shown to provide the stromal micro-environment required for the normal differentiation of human haematopoietic stem/progenitor cells into T-cell lineages [8, 9]. The T cells generated in the graft appear to be physiological in terms of distribution in phenotypes and T-cell receptor V beta chain repertoires [10] as well as the expected alteration of those by bacterial toxins act as super antigens in human [11]. Further, human T cells developing in this system have been shown recently to be functional in allogeneic rejection models [12].

The modifications to the model, in which increased amounts of fetal tissue were implanted under each of the two kidney capsules of SCID mouse recipients, were reported to give mice with increased levels of human circulating lymphocytes outside the grafts [13]. Substantial numbers of circulating human $CD4^+$ and $CD8^+$ T cells were found in the peripheral blood of engrafted animals, as well as in the spleen and lymph nodes. The conventional SCID-hu Thy/Liv model can only be infected with HIV-1 if the virus is injected directly into the human grafts. In contrast, the modified model permits infection of HIV-1 to spread to the graft following intraperitoneal injection. In addition to human haematolymphoid tissues in SCID mice, the segment of human intestine was found to be transplanted and provided the permissive transmission of HIV-1 across the mucosa of the human intestinal implant [14].

3. HIV-1-induced Pathology in the SCID-hu Thy/Liv Model

Following direct injection of HIV-1 into Thy/Liv implants, a dramatic shrinkage of thymic cortical regions was observed in association with severe loss of graft cellularity by more than two-log order of magnitude [7]. Because the loss of $CD4^+$ thymocyte is faster than $CD8^+$ cells, CD4/CD8 cell ratios within the thymus are demonstrated to get inverted. It should be stressed, however, that the total $CD8^+$ thymocytes are also decreasing severely and gradually, thus HIV-1 replication results in the suppression of thymopoiesis for both $CD4^+$ and $CD8^+$ T cells. This pathology happen always depending on the HIV-1 replication because the degree of thymocyte depletion can be delayed as associated with the suppression of viral load by the use of anti-HIV-1 compound, such as AZT and ddI [15]. Using a variety of methods, such as DNA and RNA PCR, *in-situ* hybridisation techniques, and immunohistochemistry, all T-cell compartments, including $CD4^+CD8^+$, $CD4^+CD8^-$, $CD4^-CD8^+$, appear to be infected [16-18]. In addition, myeloid cells, such as dendritic-like cells as well as thymic keratin-positive epithelial cells, are also found to be infected with HIV-1 [16]. Thymic T-cell progenitors, which have no expression of the CD3 molecule on the cell surface but do express CD4 ($CD3^-$ $CD4^+CD8^-$) were found to be infected and decreased in number after HIV-1 infection [19]. The severe reduction in total cell number especially in double positive cortical thymocytes, as a major cell type of the next progeny should be attributed at least in part to the reduction of this progenitor population. Furthermore, thymic progenitor cells, which are similar to those in bone marrow, were found to be suppressed by number and in their function [20,21]. Thus, the mechanism of thymocyte depletion seems to be multiple, which is relating to those in the normal T-cell differentiation steps.

As has been shown in peripheral T cells, HIV-1-induced thymocyte death might be mediated, at least in part, by processes consistent with programmed cell death (PCD). After HIV-1 infection in SCID-hu Thy/Liv grafts, increasing number of dying thymocytes were observed by histology, by electron microscopy, and by flow cytometric visualisation of DNA-breaks using propidium iodide [7], and have been confirmed as PCD by positively labeling with biotinylated dUTP when terminal deoxynucleotidyl transferase was added [19,22] This work has provided evidence that some of the cells that are undergoing PCD might not actually be infected (as measured by DNA PCR for proviral genome). Thus, HIV-1-induced effects on thymocytes could, in part, be mediated by indirect facilitation of PCD. Jamieson *et al.*, however, has reported that the majority of thymocytes were infected with HIV-1, which suggested direct viral killing during high viral burden *in vivo* [23].

4. HIV-1 Virology in the SCID-hu Thy/Liv Model

In the SCID-hu mouse system, primary clinical isolates from HIV-1-positive patients are invariably infectious, but with variable kinetics (slow *vs.* rapid) [7]; in contrast, viral replication generally is hard to be observed when tissue culture-adapted isolates, such as IIIb and MN, are inoculated into the grafts [24]. Regarding viral replication kinetics, it has been known that clinical deterioration in HIV-1-infected individuals is associated with an increased viral burden in the peripheral blood as well as lymph nodes. HIV-1 isolates obtained before this stage of disease often have a slow, non-syncytium-inducing (NSI) phenotype, whereas those obtained afterwards are often characterized as rapid and syncytium-inducing (SI). Paired NSI and SI isolates from two different patients were inoculated into the SCID-hu mouse [22]. The slow NSI isolates replicated to minimal levels in the grafts and did not induce thymocyte depletion. In contrast, the two SI isolates from the same patients showed high levels of viral replication and induced a marked degree of thymocyte depletion, accompanied by evidence of PCD. These observations provide a correlation between the replicative and cytopathic patterns of HIV-1 isolates *in vitro* and in the SCID-hu mouse *in vivo*, and provide direct evidence that the biological phenotype of HIV-1 switch could be a causal and not a derivative correlate of HIV-1 disease progression.

To dissect biological phenotypes of HIV-1 further in molecular level, the presence or absence of subgenomic region(s) of HIV-1 has been investigated for replication kinetics and pathogenicity *in vivo* by making recombinant viruses between infectious and non-infectious molecular clones [25,26]. as well as by making the mutation of known open reading frames in infectious viruses. [27,28]. These kinds of studies revealed a critical role for several HIV-1-encoded proteins, specifically nef and env (gp120, V3 region), as major contributors to the spread of infection and ensuing pathology. Similarly, other HIV-1-encoded proteins, such as vpu and vif, have been identified in the SCID-hu model as being less critical for maintaining pathogenicity [29].

5. SCID-hu Mice as a Model for Haematopoietic Stem Cell-based Gene Therapy

The expression of antiviral genes in human haematopoietic stem or progenitor cells has been proposed as a strategy for gene therapy of AIDS. This is primarily because both T cells and macrophages appear to be the major targets for HIV-1 infection and sites of viral replication, and because both cell types are derived from a common multipotent stem/progenitor cell. Thus, introduction of a 'protective' gene into stem cells should allow the long-term generation of 'protected' progeny T cells and macrophages. Most assays, which assess gene transfer into stem cells and analyze gene expression in their subsequently derived progeny utilize *in-vitro* systems, which are not always reflective of the clinical setting. Recently, several groups including ours have exploited the SCID-hu mouse to evaluate the stem cell-based gene therapy concept for AIDS [30-32]. The transduced $CD34^{+}$ cells by retroviral vectors were injected into SCID-hu Thy/Liv grafts, where they eventually gave rise to thymocytes that expressed the transgene products, which, in one case, was a potentially therapeutic protein, RevM10. Furthermore, T cells derived from those grafts can express the antiviral gene product RevM10 at levels sufficient to inhibit HIV-1 replication after the expansion *in vitro*. In similar studies, retrovirus gene-modified $CD34^{+}$ cells from umbilical cord blood were injected into bone grafts in the SCID-hu bone model, where they were shown to differentiate into cells of the macrophage lineage expres-sing gene products from the vector [33]. All of these studies suggest that the SCID-hu Thy/Liv and bone models might eventually play important roles in assessing various gene therapy approaches as well as in dissecting pathogenic mechanisms in further.

6. Future Directions

The recent demonstration that murine and other non-human cells can be rendered permissive for HIV-1 infection, through expression of the human receptors for HIV-1, CD4 plus CXCR4 or CCR5 [34,35], makes possible for the first time the creation of transgenic mouse models which might support HIV-1 infection and, possibly, replication. Such a model certainly would offer new avenues for exploring a variety of features related to HIV-1 infection. However, because so many aspects of HIV-1 biology (*e.g.* activation, replication, and interaction with host cellular factors) are dependent upon its human host's unique genetic and biochemical configuration, it is unlikely that a transgenic mouse model could replace the SCID-hu model. Moreover, the results of the first feasibility studies for stem cell-based gene therapy performed in SCID-hu Thy/Liv and bone mice are laying the foundation for future studies that will certainly begin to address the *in-vivo* safety and efficacy of such approaches in treating HIV-1 infection.

7. References

1. Roederer, M., Dubs, J.G., Anderson, M.T. *et al.*: CD8 naive T cell counts decrease progressively in HIV-infected adults, *J. Clin. Invest.* **95** (1995), 2061-2066.
2. Schuurman, H.J., Krone, W.J.A., Broekhuizen, R. *et al.*: The thymus in acquired immune deficiency syndrome: comparison with other types of immunodeficiency disease, and presence of components of human immunodeficiency virus type 1, *Am. J. Pathol.* **131** (1989), 1329-1334.
3. Rosenzweig, M., Clark, D.P., and Gaulton, G.N.: Selective thymocyte depletion in neonatal HIV-1 thymic infection, *AIDS* **7** (1993), 1601-1605.
4. Baskin, G.B., Murphey, C.M., Martin, L.N. *et al.*: Thymus in simian immunodeficiency virus-infected rhesus monkeys, *Lab. Invest.* **65** (1991), 400-407.
5. Beebe, A.M., Dua, N., Faith, T.G. *et al.*: Primary stage of feline immunodeficiency virus infection: viral dissemination and cellular targets, J. *Virol.* **68** (1994), 3080-3091.
6. Aldrovandi, G.M., Feuer, G., Gao, L. *et al.*: The SCID-hu mouse as a model for HIV-1 infection, *Nature* **363** (1993), 732-736.
7. Bonyhadi, M.L., Rabin, L., Salimi, S. *et al.*: HIV induces thymus depletion *in vivo*, *Nature* **363** (1993), 728-732.
8. Mc Cune, J.M., Kaneshima, H., Krowka, J. *et al.*: The SCID-hu mouse: a small animal model for HIV infection and pathogenesis, *Ann. Rev. Immunol.* **9** (1991), 399-429.
9. Kaneshima, H., Namikawa, R. and McCune, J.M.: Human hematolymphoid cells in SCID mice, *Curr. Opin. Immunol.* **6** (1994), 327-333.
10. Vandekerckhove, B.A, Baccala, R., Jones, D. *et al.*: Thymic selection of the human T cell receptor V beta repertoire in SCID-hu mice, *J. Exp. Med.* **176** (1992), 1619-1624.
11. Baccala, R., Vandekerckhove, B.A., Jones, D. *et al.*: Bacterial superantigens mediate T cell deletions in the mouse severe combined immunodeficiency-human liver/thymus model, *J. Exp. Med.* **177** (1993), 1481-1485.
12. Rouleau, M., Namikawa, R., Antonenko, S. *et al.*: Antigen-specific cytotoxic T cells mediated human fetal pancreas allorejection in SCID-hu mice, *J. Immunol.* **157** (1996), 5710-5720.
13. Kollmann, T.R., Pettoello, M.M., Zhuang, X. *et al.*: Disseminated human immunodeficiency virus 1 (HIV-1) infection in SCID-hu mice after peripheral inoculation with HIV-1, *J. Exp. Med.* **179** (1994), 513-522.
14. Gibbons, C., Kollmann, T.R., Pettoello-Mantovani, M., *et al.*: Thy/Liv-SCID-Hu mice implanted with human intestine: An *in-vivo* model for investigation of mucosal transmission of HIV, *AIDS Res. Human Retrovir.* **13** (1997), 1453-1460.
15. Rabin, L., Hincenbergs, M., Moreno, M.B. *et al.*: Use of standardized SCID-hu Thy/Liv mouse model for preclinical efficacy testing of anti-human immunodeficiency virus type 1 compounds, *Antimicrob. Agents Chemother.* **40** (1996), 755-762.
16. Stanley, S.K., McCune, J.M., Kaneshima, H. *et al.*: Human immunodeficiency virus infection of the human thymus and disruption of the thymic microenvironment in the SCID-hu mouse, *J. Exp. Med.* **178** (1993), 1151-1163.
17. Lee, S., Goldstein, H., Baseler, M., *et al.*: Human immunodeficiency virus type 1 infection of mature CD3hiCD8$^+$ thymocytes, *J Virol.* **71** (1997), 6671-6676.
18. Jamieson, B. and Zack, J.: *In-vivo* pathogenesis of a human immunodeficiency virus type 1 reporter virus, *J Virol.* **72** (1998), 6520-6526.
19. Su, L., Kaneshima, H., Bonyhadi, M.L. *et al.*: HIV-1-induced thymocyte depletion is associated with indirect cytopathicity and infection of progenitor cells *in vivo*, *Immunity* **2** (1995), 25-36.
20. Koka, P., Fraser, J., Bryson, Y., *et al.*: Human immunodeficiency virus inhibits multilineage hematopoiesis *in vivo*, *J Virol.* **72** (1998), 5121-5127.
21. Jenkins, M., Hanley, M.B., Moreno, M.B., *et al.*: Human immunodeficiency virus-1 infection interrupts thymopoiesis and multilineage hematopoiesis *in vivo*, *Blood* **91** (1998), 2672-2678.

22. Kaneshima, H., Su, L., Bonyhadi, M.L. *et al.*: Rapid-high, syncytium-inducing isolates of human immunodeficiency virus type 1 induce cytopathicity in the human thymus of the SCID-hu mouse, *J. Virol.* **68** (1994), 8188-8192.
23. Jamieson, B., Uittenbogaart, C., Schmid, I., *et al.*: High viral burden and rapid CD4⁺ cell depletion in human immunodeficiency virus type 1-infected SCID-hu mice suggest direct viral killing of thymocytes *in vivo*, *J Virol.* **71** (1997), 8245-8253.
24. Kaneshima, H., Shih, C.C., Namikawa, R. *et al.*: Human immunodeficiency virus infection of human lymph nodes in the SCID-hu mouse, *Proc. Natl. Acad. Sci. USA* **88** (1991), 4523-4527.
25. Jamieson, B.D., Pang, S., Aldrovandi, G.M., *et al.*: *In-vivo* pathogenic properties of two clonal human immunodeficiency virus type 1 isolates, *J. Virol.* **69** (1995), 6259-6264.
26. Su, L., Kaneshima, H., Bonyhadi, M.L. *et al.*: Identification of HIV-1 determinants for replication *in vivo*, *Virology* **227** (1997), 45-52.
27. Jamieson, B.D., Aldrovandi, G.M., Planelles, V. *et al.*: Requirement of human immunodeficiency virus type 1 nef for *in-vivo* replication and pathogenicity, *J. Virol.* **68** (1994), 3478-3485.
28. Aldrovandi, G.M., Gao, L.Y., Bristol, G., *et al.*: Regions of human immunodeficiency virus type 1 nef required for function *in vivo*, *J Virol.* **72** (1998), 7032-7039.
29. Aldrovandi, G.M., and Zack, J.A.: Replication and pathogenicity of human immunodeficiency virus type 1 accessory gene mutants in SCID-hu mice, *J. Virol.* **70** (1996), 1505-1511.
30. An, D.S., Koyanagi, Y., Zhao, J.Q., *et al.*: High-efficiency transduction of human lymphoid progenitor cells and expression in differentiated T cells, *J. Virol.* **71** (1997), 1397-1404.
31. Champseix, C., Marechal, V., Khazaal, I. *et al.*: A cell surface marker gene transferred with a retroviral vector into CD34⁺ cord blood cells is expressed by their T-cell progeny in the SCID-hu thymus, *Blood* **88** (1996), 107-113.
32. Bonyhadi, M.L., Moss, K., Voytovich, A. *et al.*: RevM10-expressing T cells derived *in vivo* from transduced human hematopoietic stem-progenitor cells inhibit human immunodeficiency virus replication, *J. Virol.* **71** (1997), 4707-4716.
33. Su, L.S., Lee, R., Bonyhadi, M.L. *et al.*: Hematopoietic stem cell-based gene therapy for acquired immunodeficiency syndrome: Efficient transduction and expression of RevM10 in myeloid cells *in vivo* and *in vitro*, *Blood* **89** (1997), 2283-2290.
34. Feng, Y., Broder, C.C., Kennedy, P.E. *et al.*: HIV-1 entry cofactor: Functional cDNA cloning of a seven-transmembrane, G-protein-coupled receptor, *Science* **272** (1996), 872-877.
35. Dragic, T., Litwin, V., Allaway, G.P. *et al.*: HIV-1 entry into CD4⁺ cells is mediated by the chemokine receptor CC-CKR-5, *Nature* **381** (1996), 667-673.

CHAPTER 15

B - NON-HUMAN PRIMATE MODELS FOR HIV-1 INFECTION

NORMAN L. LETVIN, M.D.

Professor of Medicine, Harvard Medical School
Chief, Division of Viral Pathogenesis, Beth Israel Deaconess Medical Center, Boston, MA 02215, USA

1. HIV Infection in Chimpanzees

The only animal species readily infectable with HIV-1 is the chimpanzee. However, following infection with most HIV-1 isolates, chimpanzees do not develop disease [28]. Infection with some HIV-1 isolates results in seroconversion, but barely detectable plasma viral RNA. In these infected animals, virus isolations from peripheral blood are achievable only for a few weeks following HIV-1 inoculation [29]. Infection with other HIV-1 isolates results in a more intense viraemia with readily measurable viral RNA in the plasma and persistantly isolatable virus from the peripheral blood. However, the extent of viral replication *in vivo* with any of these usual HIV-1 isolates appears to decline over the first months following infection, resulting in an infectious process with a low level of viral replication and no disease pathogenesis.

Recently, an LAI isolate of HIV-1 passaged *in vivo* in chimpanzees has been shown to induce an AIDS-like syndrome in this species [30,31]. Blood from an infected animal has transferred an infection that causes a profound CD4 cell decline and eventual death due to opportunistic infections. Cell-free stocks of this virus, however, have been shown in a limited number of infected chimpanzees to cause a less marked decline in CD4 cell numbers and, accordingly, a more indolent disease process.

While it is now clear that HIV-1 infection can lead to AIDS in chimpanzees, this model has limitations for the study of AIDS pathogenesis. These animals are endangered and can, therefore, be used only for restricted studies. It is difficult to do invasive experimental procedures in these animals. Finally, chimpanzee studies are extremely costly.

H. Schuitemaker and F. Miedema (ed.), Aids Pathogenesis, 279–282.

2. Simian Immunodeficiency Viruses in African Non-human Primate Species

HIV-1 is a member of a large family of primate immunodeficiency viruses [32]. The other members of this family of lentiviruses, the simian immunodeficiency viruses (SIVs), infect a variety of African non-human primate species. Thus, for example, the sootey mangabey SIV isolate (SIVsm) infects greater than 90% of all sootey mangabeys in the wild; more than 50% of all wild-caught *Cercopithecus* species are infected with African green monkey SIV isolates. These various SIVs represent an extremely diverse genetic population of viruses that are quite distinct one from another.

Interestingly, these viruses do not induce disease in their natural host species [33]. The reason for the absence of pathogenicity of these viruses in African monkey species remains unclear [34]. In some instances, the virus replicates at extremely low levels in the natural host species; this low level of virus replication, perhaps, explains the absence of disease induction. In other instances, for example, in sootey mangabeys infected with SIVsm, virus persistently replicates to high levels yet does not induce disease. While elucidating how viruses of this type can infect but be non-pathogenic is of great interest, most studies related to AIDS pathogenesis cannot be addressed in these animal species.

3. SIV Infection of Macaque Species

Although isolates of SIVs derived from African non-human species do not cause disease in their natural hosts, a number of these SIV isolates do cause an AIDS-like syndrome in Asian macaque species [35]. The prototype of these pathogenic viruses is a sootey mangabey-derived isolate referred to as SIVmac. This virus was first isolated from a macaque that had developed an AIDS-like disease process [36]. SIVmac infection of macaques induces a syndrome characterised by lymphadenopathy, wasting and CD4 cell depletion [37]. The CD4 cell loss results in the eventual development of a variety of opportunistic infections, including *Pneumocystis pneumonia*, systemic spread of cytomegalovirus and adenovirus, and infection with *mycobacterium avium-intracellulare*. SIV-infected animals also frequently die with lymphomas.

SIVmac-induced disease in macaque monkeys has a clinical course that is temporally more compressed than HIV-induced disease in man. Monkeys develop an intense viraemia that is partially controlled within 2-3 weeks following infection, but eventually die by 1-2 years after infection. A number of other SIV isolates cause infections in Asian monkeys with different dynamics of viral replication and different tempos of disease progression. The SIVmne isolate replicates only to low levels and, accordingly, induces an AIDS-like disease with a relatively indolent clinical course [38]. The Pbj14 isolate of SIV can cause intense early viral replication, leading to death in pigtail macaques in under two weeks following infection [39]. This diversity of SIV isolates and clinical disease provides a number of powerful model systems for studying AIDS disease progression in non-human primate species.

4. Simian/Human Immunodeficiency Virus Infection of Macaques

While there are substantial nucleotide sequence homologies between the SIVs and HIV-1, the envelope genes of these viruses are substantially divergent. Since the envelope glycoproteins of these viruses play a critical role in their pathogenic potential, the SIVs have provided an imperfect model system for studying envelope-related issues in AIDS pathogenesis. To overcome these limitations, viruses have been constructed in laboratories that express HIV-1 envelopes on an SIVmac backbone [40-42]. Such viruses have been created employing envelope genes of laboratory-adapted and primary patient isolates of HIV-1 [43].

Interestingly, these constructed viruses persistently replicate in macaques but do not induce disease. However, rapid passage of such viruses from one infected animal to another has been shown to increase the *in-vivo* replication capacity of these viruses and their pathogenicity [44]. While these pathogenic SHIVs replicate *in vivo* with dynamics similar to those of SIVs, their pathogenicity has proven somewhat different. These SHIVs induce a precipitous CD4 cell decline during the first few weeks of infection with no rebound in circulating CD4 cell number. Nevertheless, infected animals die with opportunistic infections only a number of months following infection.

5. References

28. Fultz, P.N.: Animal models for human immunodeficiency virus infection and disease. In: Wormser GP, ed. *AIDS and Other Manifestations of HIV Infection*, 3rd ed. New York: Raven Press, 1997 pp. 201-215.

29. Girard, M., Meignier, B., Barre-Sinoussi, F. *et al.*: Vaccine-induced protection of chimpanzees against infection by a heterologous human immunodeficiency virus type 1, *J. Virol.* **69** (1995), 6239-6248.

30. Novembre, F.J., Saucier, M., Anderson, D.C. *et al.*: Development of AIDS in a chimpanzee infected with human immunodeficiency virus type 1, *J. Virol.* **71** (1997), 4086-4091.

31. Wei, Q. and Fultz, P.N.: Extensive diversification of human immunodeficiency virus type 1 subtype B strains during dual infection of a chimpanzee that died of AIDS, *J. Virol.* **72** (1998), in press.

32. Hayami, M., Ido, E. and Miura, T.: Survey of simian immunodeficiency virus among nonhuman primate populations. In: Letvin NL, Desrosiers RC, eds. *Simian Immunodeficiency Virus*, Berlin: Springer-Verlag, 1994, pp. 1-20.

33. Hirsch, V.M., Olmsted, R.A., Murphey-Corb, M. *et al.*: An African primate lentivirus (SIV_{sm}) closely related to HIV-2, *Nature* **339** (1989), 389-392.

34. Kurth, R. and Norley, S.: Animal immunodeficiency viruses. In: Levy JA, Weiss R, Neu HC, eds. *Focus on HIV*, Edinburgh: Churchill Livingstone, 1993.

35. Letvin, N.L. and King, N.W.: Immunologic and pathologic manifestations of the infection of rhesus monkeys with simian immunodeficiency virus of macaques, *J. AIDS* **3** (1990), 1023-1040.

36. Daniel, M.D., Letvin, N.L., King, N.W. *et al.*: Isolation of T-cell tropic HTLV-III-like retrovirus from macaques, *Science* **228** (1985), 1201-1204.

37. Letvin, N.L., Daniel, M.D., Sehgal, P.K. *et al.*: Induction of AIDS-like disease in macaque monkeys with T-cell tropic retrovirus STLV-III, *Science* **230** (1985), 71-73.

38. Benveniste, R.E., Arthur, L.O., Tsai, C. *et al.*: Isolation of a lentivirus from a macaque with lymphoma: comparison with HTLV-III/LAV and other lentiviruses, *J. Virol.* **60** (1986), 483-490.

39. Dewhurst, S., Embretson, J.E., Anderson, D.C. *et al.*: Sequence analysis and acute pathogenicity of molecularly cloned $SIV_{smm\text{-}PBj14}$, *Nature* **345** (1990), 636-640.

40. Li, J., Lord, C.I., Haseltine, W. *et al.*: Infection of cynomolgus monkeys with a chimeric HIV-1/SIVmac virus that expresses the HIV-1 envelope glycoproteins, *J. AIDS* **5** (1992), 639-646.
41. Shibata, R., Kawamura, M., Sakai, H. *et al.*: Generation of a chimeric human and simian immunodeficiency virus infectious to monkey peripheral blood mononuclear cells, *J. Virol.* **65** (1991), 3514-3520.
42. Luciw, P.A., Pratt-Lowe, E., Shaw, K.E. *et al.*: Persistent infection of rhesus macaques with T-cell-line-tropic and macrophage-tropic clones of simian/human immunodeficiency viruses (SHIV), *Proc. Natl. Acad. Sci. USA* **92** (1995), 7490-7494.
43. Reimann, K.A., Li, J.T., Voss, G. *et al.*: An env gene derived from a primary human immunodeficiency virus type 1 isolate confers high *in-vivo* replicative capacity to a chimeric simian/human immunodeficiency virus in rhesus monkesy, *J. Virol.* **70** (1996), 3198-3206.
44. Reimann, K.A., Li, J.T., Veazey, R. *et al.*: A chimeric simian/human immunodeficiency virus expressing a primary patient human immunodeficiency virus type 1 isolate env causes an AIDS-like disease after *in-vivo* passage in rhesus monkeys, *J. Virol.* **70** (1996), 6922-6928.

Immunology and Medicine Series

1. A.M. McGregor (ed.). *Immunology of Endocrine Diseases*. 1986 ISBN: 0-85200-963-1
2. L. Ivanyi (ed.). *Immunological Aspects of Oral Diseases*. 1986 ISBN: 0-85200-961-5
3. M.A.H. French (ed.). *Immunoglobulins in Health and Disease*. 1986 ISBN: 0-85200-962-3
4. K. Whaley (ed.). *Complement in Health and Disease*. 1987 ISBN: 0-85200-954-2
5. G.R.D. Catto (ed.). *Clinical Transplantation: Current Practice and Future Prospects*. 1987 ISBN: 0-85200-960-7
6. V.S. Byers and R.W. Baldwin (eds.). *Immunology of Malignant Diseases*. 1987 ISBN: 0-85200-964-X
7. S.T. Holgate (ed.). *Mast Cells, Mediators and Disease*. 1988 ISBN: 0-85200-968-2
8. D.J.M. Wright (ed.). *Immunology of Sexually Transmitted Diseases*. 1988 ISBN: 0-74620-087-0
9. A.D.B. Webster (ed.). *Immunodeficiency and Disease*. 1988 ISBN: 0-85200-688-8
10. C. Stern (ed.). *Immunology of Pregnancy and Its Disorders*. 1989 ISBN: 0-7462-0065-X
11. M.S. Klempner, B. Styrt and J. Ho (eds.). *Phagocytes and Disease*. 1989 ISBN: 0-85200-842-2
12. A.J. Zuckerman (ed.). *Recent Developments in Prophylactic Immunization*. 1989 ISBN: 0-7923-8910-7
13. S. Lightman (ed.). *Immunology of Eye Disease*. 1989 ISBN: 0-7923-8908-5
14. T.J. Hamblin (ed.). *Immunotherapy of Disease*. 1990 ISBN: 0-7462-0045-5
15. D.B. Jones and D.H. Wright (eds.). *Lymphoproliferative Diseases*. 1990 ISBN: 0-85200-965-8
16. C.D. Pusey (ed.). *Immunology of Renal Diseases*. 1991 ISBN: 0-7923-8964-6
17. A.G. Bird (ed.). *Immunology of HIV Infection*. 1991 ISBN: 0-7923-8962-X
18. J.T. Whicher and S.W. Evans (eds.). *Biochemistry of Inflammation*. 1992 ISBN: 0-7923-8985-9
19. T.T. MacDonald (ed.). *Immunology of Gastrointestinal Diseases*. 1992 ISBN: 0-7923-8961-1
20. K. Whaley, M. Loos and J.M. Weiler (eds.). *Complement in Health and Disease, 2nd Edn*. 1993 ISBN: 0-7923-8823-2
21. H.C. Thomas and J. Waters (eds.). *Immunology of Liver Disease*. 1994 ISBN: 0-7923-8975-1
22. G.S. Panayi (ed.). *Immunology of Connective Tissue Diseases*. 1994 ISBN: 0-7923-8988-3
23. G. Scadding (ed.). *Immunology of ENT Disorders*. 1994 ISBN: 0-7923-8914-X
24. R. Hohlfeld (ed.). *Immunology of Neuromuscular Disease*. 1994 ISBN: 0-7923-8844-5
25. J.G.P. Sissons, L.K. Borysiewicz and J. Cohen (eds.). *Immunology of Infection*. 1994 ISBN: 0-7923-8968-9
26. J.G.J. van de Winkel and P.M. Hogarth (eds.). *The Immunoglobulin Receptors and Their Physiological and Pathological Roles in Immunity*. 1998 ISBN: 0-7923-5021-9
27. A.P. Weetman (ed.). *Endocrine Autoimmunity and Associated Conditions*. 1998 ISBN: 0-7923-5042-1
28. H. Schuitemaker and F. Miedema (eds.). *AIDS Pathogenesis*. 2000 ISBN: 0-7923-6196-2

Kluwer Academic Publishers – Dordrecht / Boston / London

Made in the USA
Middletown, DE
05 May 2023

29918857R10170